To Alina and to Our Mothers

Titu Andreescu
Răzvan Gelca

Mathematical Olympiad Challenges

SECOND EDITION

Foreword by Mark Saul

Birkhäuser
Boston • Basel • Berlin

Titu Andreescu
University of Texas at Dallas
School of Natural Sciences
   and Mathematics
Richardson, TX 75080
USA
titu.andreescu@utdallas.edu

Răzvan Gelca
Texas Tech University
Department of Mathematics
   and Statistics
Lubbock, TX 79409
USA
rgelca@gmail.com

ISBN: 978-0-8176-4528-1   e-ISBN: 978-0-8176-4611-0
DOI: 10.1007/978-0-8176-4611-0

Mathematics Subject Classification (2000): 00A05, 00A07, 05-XX, 11-XX, 51XX

Printed on acid-free paper

springer.com

Contents

Matematică, matematică, matematică, atâta matematică?
Nu, mai multă.[1]
Grigore Moisil

---

[1]Mathematics, mathematics, mathematics, that much mathematics? No, even more.

Foreword

Why Olympiads?

Working mathematicians often tell us that results in the field are achieved after long experience and a deep familiarity with mathematical objects, that progress is made slowly and collectively, and that flashes of inspiration are mere punctuation in periods of sustained effort.

The Olympiad environment, in contrast, demands a relatively brief period of intense concentration, asks for quick insights on specific occasions, and requires a concentrated but isolated effort. Yet we have found that participants in mathematics Olympiads have often gone on to become first-class mathematicians or scientists and have attached great significance to their early Olympiad experiences.

For many of these people, the Olympiad problem is an introduction, a glimpse into the world of mathematics not afforded by the usual classroom situation. A good Olympiad problem will capture in miniature the process of creating mathematics. It's all there: the period of immersion in the situation, the quiet examination of possible approaches, the pursuit of various paths to solution. There is the fruitless dead end, as well as the path that ends abruptly but offers new perspectives, leading eventually to the discovery of a better route. Perhaps most obviously, grappling with a good problem provides practice in dealing with the frustration of working at material that refuses to yield. If the solver is lucky, there will be the moment of insight that heralds the start of a successful solution. Like a well-crafted work of fiction, a good Olympiad problem tells a story of mathematical creativity that captures a good part of the real experience and leaves the participant wanting still more.

And this book gives us more. It weaves together Olympiad problems with a common theme, so that insights become techniques, tricks become methods, and methods build to mastery. Although each individual problem may be a mere appetizer, the table is set here for more satisfying fare, which will take the reader deeper into mathematics than might any single problem or contest.

The book is organized for learning. Each section treats a particular technique or topic. Introductory results or problems are provided with solutions, then related problems are presented, with solutions in another section.

The craft of a skilled Olympiad coach or teacher consists largely in recognizing similarities among problems. Indeed, this is the single most important skill that the coach can impart to the student. In this book, two master Olympiad coaches have offered the results of their experience to a wider audience. Teachers will find examples and topics for advanced students or for their own exercise. Olympiad stars will find practice material that will leave them stronger and more ready to take on the next challenge, from whatever corner of mathematics it may originate. Newcomers to Olympiads will find an organized introduction to the experience.

There is also something here for the more general reader who is interested in mathematics. Simply perusing the problems, letting their beauty catch the eye, and working through the authors' solutions will add to the reader's understanding. The multiple solutions link together areas of mathematics that are not apparently related. They often illustrate how a simple mathematical tool—a geometric transformation, or an algebraic identity—can be used in a novel way, stretched or reshaped to provide an unexpected solution to a daunting problem.

These problems are daunting on any level. True to its title, the book is a challenging one. There are no elementary problems—although there are elementary solutions. The content of the book begins just at the edge of the usual high school curriculum. The calculus is sometimes referred to, but rarely leaned on, either for solution or for motivation. Properties of vectors and matrices, standard in European curricula, are drawn upon freely. Any reader should be prepared to be stymied, then stretched. Much is demanded of the reader by way of effort and patience, but the reader's investment is greatly repaid.

In this, it is not unlike mathematics as a whole.

Mark Saul
Bronxville School

Preface to the Second Edition

The second edition is a significantly revised and expanded version. The introductions to many sections were rewritten, adopting a more user-friendly style with more accessible and inviting examples. The material has been updated with more than 70 recent problems and examples. Figures were added in some of the solutions to geometry problems. Reader suggestions have been incorporated.

We would like to thank Dorin Andrica and Iurie Boreico for their suggestions and contributions. Also, we would like to express our deep gratitude to Richard Stong for reading the entire manuscript and considerably improving its content.

Titu Andreescu
University of Texas at Dallas
Răzvan Gelca
Texas Tech University
April 2008

Preface to the First Edition

At the beginning of the twenty-first century, elementary mathematics is undergoing two major changes. The first is in teaching, where one moves away from routine exercises and memorized algorithms toward creative solutions to unconventional problems. The second consists in spreading problem-solving culture throughout the world. *Mathematical Olympiad Challenges* reflects both trends. It gathers essay-type, nonroutine, open-ended problems in undergraduate mathematics from around the world. As Paul Halmos said, "problems are the heart of mathematics," so we should "emphasize them more and more in the classroom, in seminars, and in the books and articles we write, to train our students to be better problem-posers and problem-solvers than we are."

The problems we selected are definitely not exercises. Our definition of an exercise is that you look at it and you know immediately how to complete it. It is just a question of doing the work. Whereas by a problem, we mean a more intricate question for which at first one has probably no clue to how to approach it, but by perseverance and inspired effort, one can transform it into a sequence of exercises. We have chosen mainly Olympiad problems, because they are beautiful, interesting, fun to solve, and they best reflect mathematical ingenuity and elegant arguments.

Mathematics competitions have a long-standing tradition. More than 100 years ago, Hungary and Romania instituted their first national competitions in mathematics. The Eőtvős Contest in elementary mathematics has been open to Hungarian students in their last years of high school since 1894. The Gazeta Matematică contest, named after the major Romanian mathematics journal for high school students, was founded in 1895. Other countries soon followed, and by 1938 as many as 12 countries were regularly organizing national mathematics contests. In 1959, Romania had the initiative to host the first International Mathematical Olympiad (IMO). Only seven European countries participated. Since then, the number has grown to more than 80 countries, from all continents. The United States joined the IMO in 1974. Its greatest success came in 1994, when all six USA team members won a gold medal with perfect scores, a unique performance in the 48-year history of the IMO.

Within the United States, there are several national mathematical competitions, such as the AMC 8 (formerly the American Junior High School Mathematics Examination), AMC 10 (the American Mathematics Contest for students in grades 10 or below), and AMC 12 (formerly the American High School Mathematics Examination), the American Invitational Mathematics Examination (AIME), the United States Mathematical Olympiad (USAMO), the W. L. Putnam Mathematical Competition, and a number of regional contests such as the American Regions Mathematics

League (ARML). Every year, more than 600,000 students take part in these competitions. The problems from this book are of the type that usually appear in the AIME, USAMO, IMO, and the W. L. Putnam competitions, and in similar contests from other countries, such as Austria, Bulgaria, Canada, China, France, Germany, Hungary, India, Ireland, Israel, Poland, Romania, Russia, South Korea, Ukraine, United Kingdom, and Vietnam. Also included are problems from international competitions such as the IMO, Balkan Mathematical Olympiad, Ibero-American Mathematical Olympiad, Asian-Pacific Mathematical Olympiad, Austrian-Polish Mathematical Competition, Tournament of the Towns, and selective questions from problem books and from the following journals: *Kvant (Quantum), Revista Matematică din Timişoara (Timişoara's Mathematics Gazette), Gazeta Matematică (Mathematics Gazette), Matematika v Škole (Mathematics in School), American Mathematical Monthly*, and *Matematika Sofia*. More than 60 problems were created by the authors and have yet to be circulated.

Mathematical Olympiad Challenges is written as a textbook to be used in advanced problem-solving courses or as a reference source for people interested in tackling challenging mathematical problems. The problems are clustered in 30 sections, grouped in 3 chapters: Geometry and Trigonometry, Algebra and Analysis, and Number Theory and Combinatorics. The placement of geometry at the beginning of the book is unusual but not accidental. The reason behind this choice is well reflected in the words of V. I. Arnol'd: "Our brain has two halves: one is responsible for the multiplication of polynomials and languages, and the other half is responsible for orientating figures in space and all things important in real life. Mathematics is geometry when you have to use both halves." (*Notices* of the AMS, April 1997).

Each section is self-contained, independent of the others, and focuses on one main idea. All sections start with a short essay discussing basic facts and with one or more representative examples. This sets the tone for the whole unit. Next, a number of carefully chosen problems are listed, to be solved by the reader. The solutions to all problems are given in detail in the second part of the book. After each solution, we provide the source of the problem, if known. Even if successful in approaching a problem, the reader is advised to study the solution given at the end of the book. As problems are generally listed in increasing order of difficulty, solutions to initial problems might suggest illuminating ideas for completing later ones. At the very end we include a glossary of definitions and fundamental properties used in the book.

Mathematical Olympiad Challenges has been successfully tested in classes taught by the authors at the Illinois Mathematics and Science Academy, the University of Michigan, the University of Iowa, and in the training of the USA International Mathematical Olympiad Team. In the end, we would like to express our thanks to Gheorghe

Eckstein, Chetan Balwe, Mircea Grecu, Zuming Feng, Zvezdelina Stankova-Frenkel, and Florin Pop for their suggestions and especially to Svetoslav Savchev for carefully reading the manuscript and for contributions that improved many solutions in the book.

Titu Andreescu
American Mathematics Competitions
Răzvan Gelca
University of Michigan
April 2000

Part I

Problems

Chapter 1

Geometry and Trigonometry

T. Andreescu and R. Gelca, *Mathematical Olympiad Challenges*, DOI: 10.1007/978-0-8176-4611-0_1,
© Birkhäuser Boston, a part of Springer Science+Business Media, LLC 2009

1.1   A Property of Equilateral Triangles

Given two points A and B, if one rotates B around A through 60° to a point B', then the triangle ABB' is equilateral. A consequence of this result is the following property of the equilateral triangles, which was noticed by the Romanian mathematician D. Pompeiu in 1936. Pompeiu's theorem is a simple fact, part of classical plane geometry. Surprisingly, it was discovered neither by Euler in the eighteenth century nor by Steinitz in the nineteenth.

Given an equilateral triangle ABC and a point P that does not lie on the circumcircle of ABC, one can construct a triangle of side lengths equal to PA, PB, and PC. If P lies on the circumcircle, then one of these three lengths is equal to the sum of the other two.

To understand why this property holds, let us rotate the triangle by an angle of 60° clockwise around C (see Figure 1.1.1).

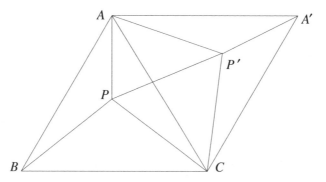

Figure 1.1.1

Let A' and P' be the images of A and P through this rotation. Note that B rotates to A. Looking at the triangle $PP'A$, we see that the side $P'A$ is the image of PB through the rotation, so $P'A = PB$. Also, the triangle $PP'C$ is equilateral; hence $PP' = PC$. It follows that the sides of the triangle $PP'A$ are equal to PA, PB, and PC.

Let us determine when the triangle $PP'A$ is degenerate, namely when the points P, P', and A are collinear (see Figure 1.1.2). If this is the case, then P is not interior to the triangle. Because A is on the line PP' and the triangle $PP'C$ is equilateral, the angle $\angle APC$ is 120° if P is between A and P', and 60° otherwise. It follows that A, P, and P' are collinear if and only if P is on the circumcircle. In this situation, $PA = PB + PC$ if P is on the arc $\overset{\frown}{BC}$, $PB = PA + PC$ if P is on the arc $\overset{\frown}{AC}$, and $PC = PA + PB$ if P is on the arc $\overset{\frown}{AB}$.

This property can be extended to all regular polygons. The proof, however, uses a different idea. We leave as exercises the following related problems.

1. Prove the converse of Pompeiu's theorem, namely that if for every point P in the interior of a triangle ABC one can construct a triangle having sides equal to PA, PB, and PC, then ABC is equilateral.

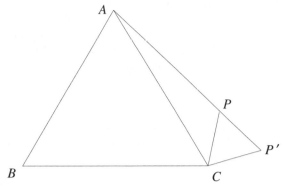

Figure 1.1.2

2. In triangle ABC, AB is the longest side. Prove that for any point P in the interior of the triangle, $PA + PB > PC$.

3. Find the locus of the points P in the plane of an equilateral triangle ABC that satisfy

$$\max\{PA, PB, PC\} = \frac{1}{2}(PA + PB + PC).$$

4. Let $ABCD$ be a rhombus with $\angle A = 120°$ and P a point in its plane. Prove that

$$PA + PC > \frac{BD}{2}.$$

5. There exists a point P inside an equilateral triangle ABC such that $PA = 3$, $PB = 4$, and $PC = 5$. Find the side length of the equilateral triangle.

6. Let ABC be an equilateral triangle. Find the locus of the points P in the plane with the property that PA, PB, and PC are the side lengths of a right triangle.

7. Given a triangle XYZ with side lengths x, y, and z, construct an equilateral triangle ABC and a point P such that $PA = x$, $PB = y$, and $PC = z$.

8. Using a straightedge and a compass, construct an equilateral triangle with each vertex on one of three given concentric circles. Determine when the construction is possible and when not.

9. Let ABC be an equilateral triangle and P a point in its interior. Consider XYZ, the triangle with $XY = PC$, $YZ = PA$, and $ZX = PB$, and M a point in its interior such that $\angle XMY = \angle YMZ = \angle ZMX = 120°$. Prove that $XM + YM + ZM = AB$.

10. Find the locus of the points P in the plane of an equilateral triangle ABC for which the triangle formed with PA, PB, and PC has constant area.

1.2   Cyclic Quadrilaterals

Solving competition problems in plane geometry often reduces to proving the equality of some angles. A good idea in such situations is to hunt for cyclic quadrilaterals because of two important facts (see Figure 1.2.1):

Theorem 1. A quadrilateral is cyclic if and only if one angle of the quadrilateral is equal to the supplement of its opposite angle.

Theorem 2. A quadrilateral is cyclic if and only if the angle formed by a side and a diagonal is equal to the angle formed by the opposite side and the other diagonal.

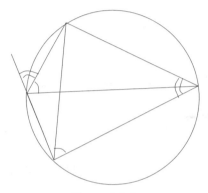

Figure 1.2.1

We illustrate with several examples how these properties can be used in solving an Olympiad problem.

Let AB be a chord in a circle and P a point on the circle. Let Q be the projection of P on AB and R and S the projections of P onto the tangents to the circle at A and B. Prove that PQ is the geometric mean of PR and PS.

We will prove that the triangles PRQ and PQS are similar. This will imply $PR/PQ = PQ/PS$; hence $PQ^2 = PR \cdot PS$.

The quadrilaterals $PRAQ$ and $PQBS$ are cyclic, since each of them has two opposite right angles (see Figure 1.2.2). In the first quadrilateral $\angle PRQ = \angle PAQ$ and in the second $\angle PQS = \angle PBS$. By inscribed angles, $\angle PAQ$ and $\angle PBS$ are equal. It follows that $\angle PRQ = \angle PQS$. A similar argument shows that $\angle PQR = \angle PSQ$. This implies that the triangles PRQ and PQS are similar, and the conclusion follows.

The second problem is from Gheorghe Țițeica's book *Probleme de Geometrie (Problems in Geometry)*.

Let A and B be the common points of two circles. A line passing through A intersects the circles at C and D. Let P and Q be the projections of B onto the tangents to the two circles at C and D. Prove that PQ is tangent to the circle of diameter AB.

After a figure has been drawn, for example Figure 1.2.3, a good guess is that the tangency point lies on *CD*. Thus let us denote by M the intersection of the circle of diameter AB with the line CD, and let us prove that PQ is tangent to the circle at M.

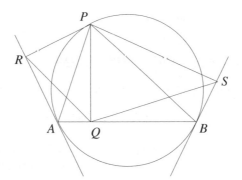

Figure 1.2.2

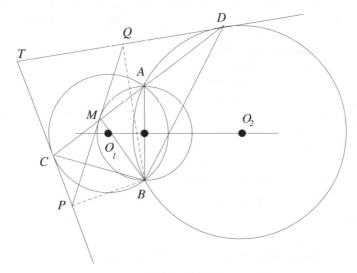

Figure 1.2.3

We will do the proof in the case where the configuration is like that in Figure 1.2.3; the other cases are completely analogous. Let T be the intersections of the tangents at C and D. The angles $\angle ABD$ and $\angle ADT$ are equal, since both are measured by half of the arc $\overset{\frown}{AD}$. Similarly, the angles $\angle ABC$ and $\angle ACT$ are equal, since they are measured by half of the arc $\overset{\frown}{AC}$. This implies that

$$\angle CBD = \angle ABD + \angle ABC = \angle ADT + \angle ACT = 180° - \angle CTD,$$

where the last equality follows from the sum of the angles in triangle TCD. Hence the quadrilateral $TCBD$ is cyclic.

The quadrilateral $TPBQ$ is also cyclic, since it has two opposite right angles. This implies that $\angle PBQ = 180° - \angle CTD$; thus $\angle PBQ = \angle DBC$ as they both have $\angle CTD$ as their supplement. Therefore, by subtracting $\angle CBQ$, we obtain $\angle CBP = \angle QBD$.

The quadrilaterals $BMCP$ and $BMQD$ are cyclic, since $\angle CMB = \angle CPB = \angle BQD = \angle DMB = 90°$. Hence

$$\angle CMP = \angle CBP = \angle QBD = \angle QMD,$$

which shows that M lies on PQ. Moreover, in the cyclic quadrilateral $QMBD$,

$$\angle MBD = 180° - \angle MQD = \angle QMD + \angle QDM = \angle QMD + \angle ABD,$$

because $\angle QDM$ and $\angle ABD$ are both measured by half of the arc $\overset{\frown}{AD}$. Since $\angle MBD = \angle MBA + \angle ABD$, the above equality implies that $\angle QMD = \angle MBA$; hence MQ is tangent to the circle, and the problem is solved.

Angle-chasing based on cyclic quadrilaterals is a powerful tool. However, angle-chasing has a major drawback: it may be case dependent. And if the argument is convincing when there are few cases and they appear very similar, what is to be done when several cases are possible and they don't look quite so similar? The answer is to use directed angles and work modulo 180°.

We make the standard convention that the angles in which the initial side is rotated counter-clockwise toward the terminal side are positive and the others are negative. Thus $\angle ABC = -\angle CBA$. We also work modulo 180°, which means that angles that differ by a multiple of 180° are identified. The condition that four points A, B, C, D lie on a circle translates to $\angle ABC \equiv \angle ADC \pmod{180°}$, regardless of the order of the points. This method is somewhat counter-intuitive, so we only recommend it for problems where many configurations are possible and these configurations look different from each other. Such is the case with the following example.

Four circles $\mathscr{C}_1, \mathscr{C}_2, \mathscr{C}_3, \mathscr{C}_4$ are given such that $\mathscr{C}_i$ intersects $\mathscr{C}_{i+1}$ at A_i and B_i, $i = 1, 2, 3, 4$ ($\mathscr{C}_5 = \mathscr{C}_1$). Prove that if $A_1A_2A_3A_4$ is a cyclic quadrilateral, then so is $B_1B_2B_3B_4$.

It is easy to convince ourselves that there are several possible configurations, two of which are illustrated in Figure 1.2.4.

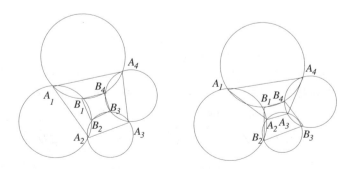

Figure 1.2.4

Thus we will work with oriented angles modulo 180°. We want to prove that $\angle B_1B_2B_3 = \angle B_1B_4B_3$, and in order to do this we examine the angles around the points

B_2 and B_4. In the cyclic quadrilateral $A_1B_1A_2B_2$, $\angle B_1B_2A_2 = \angle B_1A_1A_2$, and in the cyclic quadrilateral $B_2B_3A_3A_2$, $\angle A_2B_2B_3 = \angle A_2A_3B_3$. Looking at the vertex B_4, we obtain from a similar argument that $\angle B_1B_4A_4 = \angle B_1A_1A_4$ and $\angle A_4B_4B_3 = \angle A_4A_3B_3$.

Therefore

$$\angle B_1B_2B_3 = \angle B_1B_2A_2 + \angle A_2B_2B_3 = \angle B_1A_1A_2 + \angle A_2A_3B_3$$

and

$$\angle B_1B_4B_3 = \angle B_1B_4A_4 + \angle A_4B_4B_3 = \angle B_1A_1A_4 + \angle A_4A_3B_3,$$

where equalities are to be understood modulo $180°$. Consequently

$$\angle B_1B_2B_3 - \angle B_1B_4B_3 = \angle B_1A_1A_2 + \angle A_4A_1B_1 + \angle A_2A_3B_3 + \angle B_3A_3A_4$$
$$= \angle A_4A_1A_2 + \angle A_2A_3A_4 = 0°,$$

where the last equality follows from the fact that the quadrilateral $A_1A_2A_3A_4$ is cyclic.

Here are more problems that can be solved using the above-mentioned properties of cyclic quadrilaterals.

1. Let $\angle AOB$ be a right angle, M and N points on the half lines (rays) OA, respectively OB, and let $MNPQ$ be a square such that MN separates the points O and P. Find the locus of the center of the square when M and N vary.

2. An interior point P is chosen in the rectangle $ABCD$ such that $\angle APD + \angle BPC = 180°$. Find the sum of the angles $\angle DAP$ and $\angle BCP$.

3. Let $ABCD$ be a rectangle and let P be a point on its circumcircle, different from any vertex. Let X, Y, Z, and W be the projections of P onto the lines AB, BC, CD, and DA, respectively. Prove that one of the points X, Y, Z, and W is the orthocenter of the triangle formed by the other three.

4. Prove that the four projections of vertex A of the triangle ABC onto the exterior and interior angle bisectors of $\angle B$ and $\angle C$ are collinear.

5. Let $ABCD$ be a convex quadrilateral such that the diagonals AC and BD are perpendicular, and let P be their intersection. Prove that the reflections of P with respect to AB, BC, CD, and DA are concyclic (i.e., lie on a circle).

6. Let B and C be the endpoints and A the midpoint of a semicircle. Let M be a point on the line segment AC, and P, Q the feet of the perpendiculars from A and C to the line BM, respectively. Prove that $BP = PQ + QC$.

7. Points E and F are given on the side BC of a convex quadrilateral $ABCD$ (with E closer than F to B). It is known that $\angle BAE = \angle CDF$ and $\angle EAF = \angle FDE$. Prove that $\angle FAC = \angle EDB$.

8. In the triangle ABC, $\angle A = 60°$ and the bisectors BB' and CC' intersect at I. Prove that $IB' = IC'$.

9. In the triangle ABC, let I be the incenter. Prove that the circumcenter of AIB lies on CI.

10. Let ABC be a triangle and D the foot of the altitude from A. Let E and F be on a line passing through D such that AE is perpendicular to BE, AF is perpendicular to CF, and E and F are different from D. Let M and N be the midpoints of line segments BC and EF, respectively. Prove that AN is perpendicular to NM.

11. Let ABC be an isosceles triangle with $AC = BC$, whose incenter is I. Let P be a point on the circumcircle of the triangle AIB lying inside the triangle ABC. The lines through P parallel to CA and CB meet AB at D and E, respectively. The line through P parallel to AB meets CA and CB at F and G, respectively. Prove that the lines DF and EG intersect on the circumcircle of the triangle ABC.

12. Let ABC be an acute triangle, and let T be a point in its interior such that $\angle ATB = \angle BTC = \angle CTA$. Let M, N, and P be the projections of T onto BC, CA, and AB, respectively. The circumcircle of the triangle MNP intersects the lines BC, CA, and AB for the second time at M', N', and P', respectively. Prove that the triangle $M'N'P'$ is equilateral.

13. Let A be a fixed point on the side Ox of the angle xOy. A variable circle $\mathscr{C}$ is tangent to Ox and Oy, with D the point of tangency with Oy. The second tangent from A to $\mathscr{C}$ intersects $\mathscr{C}$ at E. Prove that when $\mathscr{C}$ varies, the line DE passes through a fixed point.

14. Let $A_0A_1A_2A_3A_4A_5$ be a cyclic hexagon and let P_0 be the intersection of A_0A_1 and A_3A_4, P_1 the intersection of A_1A_2 and A_4A_5, and P_2 the intersection of A_2A_3 and A_5A_0. Prove that P_0, P_1, and P_2 are collinear.

1.3   Power of a Point

In the plane, fix a point P and a circle, then consider the intersections A and B of an arbitrary line passing through P with the circle. The product $PA \cdot PB$ is called the power of P with respect to the circle. It is independent of the choice of the line AB, since if $A'B'$ were another line passing through P, with A' and B' on the circle, then the triangles PAA' and $PB'B$ would be similar. Because of that $PA/PB' = PA'/PB$, and hence $PA \cdot PB = PA' \cdot PB'$ (see Figure 1.3.1).

Considering a diameter through P, we observe that the power of P is really a measure of how far P is from the circle. Indeed, by letting O be the center of the circle and R the radius, we see that if P is outside the circle, its power is $(PO - R)(PO + R) = PO^2 - R^2$; if P is on the circle, its power is zero; and if P is inside the circle, its power is $(R - PO)(R + PO) = R^2 - PO^2$. It is sometimes more elegant to work with directed segments, in which case the power of P with respect to the circle is $PO^2 - R^2$ regardless of whether P is inside or outside. Here the convention is that two segments have the same sign if they point in the same direction and opposite signs if they point in opposite directions. In the former case, their product is positive, and in the latter case, it is negative.

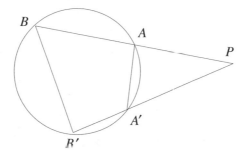

Figure 1.3.1

The locus of the points having equal powers with respect to two circles is a line perpendicular to the one determined by the centers of the circles. This line is called the *radical axis*. In this case, we need to work with directed segments, so the points on the locus are either simultaneously inside or simultaneously outside the circles.

Let us prove that indeed the locus is a line. Denote by O_1 and O_2 the centers and by R_1 and R_2 the radii of the circles. For a point P on the locus, $PO_1^2 - R_1^2 = PO_2^2 - R_2^2$; that is,

$$PO_1^2 - PO_2^2 = R_1^2 - R_2^2.$$

If we choose Q to be the projection of P onto O_1O_2 (see Figure 1.3.2), then the Pythagorean theorem applied to triangles QPO_1 and PQO_2 implies $QO_1^2 - QO_2^2 = R_1^2 - R_2^2$. Hence the locus is the line orthogonal to O_1O_2 passing through the point Q on O_1O_2 whose distances to the centers satisfy the above relation. If the two circles intersect, the radical axis obviously contains their intersection points. The radical axis cannot be defined if the two circles are concentric.

The power of a point that lies outside of the circle equals the square of the tangent from the point to the circle. For that reason, the radical axis of two circles that do not lie one inside the other passes through the midpoints of the two common tangents.

Given three circles with noncollinear centers, the radical axis of the first pair of circles and that of the second pair intersect. Their intersection point has equal powers with respect to the three circles and is called the *radical center*. Consequently, the radical axes of the three pairs of circles are concurrent. If the centers of the three circles are collinear, then the radical axes are parallel, or they might coincide. In the latter case, the circles are called *coaxial*.

The notion of *power of a point* can be useful in solving problems, as the next example shows.

Let C be a point on a semicircle of diameter AB and let D be the midpoint of the arc $\overset{\frown}{AC}$. Denote by E the projection of the point D onto the line BC and by F the intersection of the line AE with the semicircle. Prove that BF bisects the line segment DE.

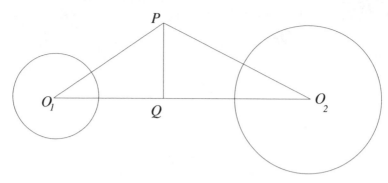

Figure 1.3.2

Here is a solution found by the student G.H. Baek during the 2006 Mathematical Olympiad Summer Program. First note that DE is tangent to the circle (see Figure 1.3.3). To see why this is true, let O be the center of the circle. Since D is the midpoint of the arc AC, $OD \perp AC$. The angle $\angle BCA$ is right; hence DE is parallel to AC. This implies that $DE \perp OD$, so DE is tangent to the circle.

Note also that DE is tangent to the circumcircle of BEF because it is perpendicular to the diameter BE. The radical axis BF of the circumcircles of ABD and BEF passes through the midpoint of the common tangent DE, and we are done.

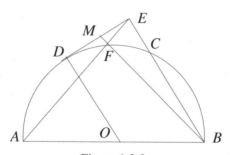

Figure 1.3.3

The second example is a proof of the famous Euler's relation in a triangle.

In a triangle with circumcenter O and incenter I,

$$OI^2 = R(R - 2r),$$

where R is the circumradius and r is the inradius.

In the usual notation, let the triangle be ABC. Let also A' be the second intersection of the line IA with the circumcircle. The power of the point I with respect to the circumcircle is $IA \cdot IA' = R^2 - OI^2$, where R is the circumradius. Now, AI is the bisector of $\angle BAC$, and the distance from I to AB is r, the inradius. We obtain $AI = r/\sin(A/2)$.

On the other hand, in triangle $A'IB$, $\angle IA'B = \angle AA'B = \angle ACB$ and $\angle IBA' = \angle IBC + \angle CBA' = \frac{1}{2}\angle ABC + \angle CAA' = \angle ABC/2 + \angle BAC/2$. It follows that $A'IB$ is isosceles; hence $IA' = BA'$. The law of sines in the triangle ABA' gives $BA' = 2R\sin(A/2)$; hence

$$OI^2 = R^2 - IA \cdot BA' = R^2 - (r/\sin(A/2)) \cdot 2R\sin(A/2) = R(R - 2r).$$

Here is a list of problems whose solutions use either the power of a point or the properties of the radical axis.

1. Let $\mathscr{C}_1, \mathscr{C}_2, \mathscr{C}_3$ be three circles whose centers are not collinear, and such that $\mathscr{C}_1$ and $\mathscr{C}_2$ intersect at M and N, $\mathscr{C}_2$ and $\mathscr{C}_3$ intersect at P and Q, and $\mathscr{C}_3$ and $\mathscr{C}_1$ intersect at R and S. Prove that MN, PQ, and RS intersect at one point.

2. Let P be a point inside a circle such that there exist three chords through P of equal length. Prove that P is the center of the circle.

3. For a point P inside the angle xOy, find $A \in Ox$ and $B \in Oy$ such that $P \in AB$ and $AP \cdot BP$ is minimal. (Here Ox and Oy are two given rays.)

4. Given a plane $\mathscr{P}$ and two points A and B on different sides of it, construct a sphere containing A and B and meeting $\mathscr{P}$ along a circle of the smallest possible radius.

5. Given an acute triangle ABC, let O be its circumcenter and H its orthocenter. Prove that

$$OH^2 = R^2 - 8R^2 \cos A \cos B \cos C,$$

where R is the circumradius. What if the triangle has an obtuse angle?

6. Let ABC be a triangle and let A', B', C' be points on sides BC, CA, AB, respectively. Denote by M the point of intersection of circles ABA' and $A'B'C'$ other than A', and by N the point of intersection of circles ABB' and $A'B'C'$ other than B'. Similarly, one defines points P, Q and R, S, respectively. Prove that:

   (a) At least one of the following situations occurs:

      (i) The triples of lines $(AB, A'M, B'N)$, $(BC, B'P, C'Q)$, $(CA, C'R, A'S)$ are concurrent at C'', A'', and B'', respectively;

      (ii) $A'M$ and $B'N$ are parallel to AB, or $B'P$ and $C'Q$ are parallel to BC, or $C'R$ and $A'S$ are parallel to CA.

   (b) In the case where (i) occurs, the points A'', B'', C'' are collinear.

7. Among the points A, B, C, D, no three are collinear. The lines AB and CD intersect at E, and BC and DA intersect at F. Prove that either the circles with diameters AC, BD, and EF pass through a common point or no two of them have any common point.

8. Let $\mathscr{C}_1$ and $\mathscr{C}_2$ be concentric circles, with $\mathscr{C}_2$ in the interior of $\mathscr{C}_1$. From a point A on $\mathscr{C}_1$, draw the tangent AB to $\mathscr{C}_2$ ($B \in \mathscr{C}_2$). Let C be the second point of intersection of AB and $\mathscr{C}_1$, and let D be the midpoint of AB. A line passing through A intersects $\mathscr{C}_2$ at E and F in such a way that the perpendicular bisectors of DE and CF intersect at a point M on AB. Find, with proof, the ratio AM/MC.

9. Let ABC be an acute triangle. The points M and N are taken on the sides AB and AC, respectively. The circles with diameters BN and CM intersect at points P and Q. Prove that P, Q, and the orthocenter H are collinear.

10. Let $ABCD$ be a convex quadrilateral inscribed in a semicircle s of diameter AB. The lines AC and BD intersect at E and the lines AD and BC at F. The line EF intersects semicircle s at G and the line AB at H. Prove that E is the midpoint of the line segment GH if and only if G is the midpoint of the line segment FH.

11. Let ABC be a triangle and let D and E be points on the sides AB and AC, respectively, such that DE is parallel to BC. Let P be any point interior to triangle ADE, and let F and G be the intersections of DE with the lines BP and CP, respectively. Let Q be the second intersection point of the circumcircles of triangles PDG and PFE. Prove that the points A, P, and Q lie on a straight line.

12. Let A be a point exterior to a circle $\mathscr{C}$. Two lines through A meet the circle $\mathscr{C}$ at points B and C, respectively at D and E (with D between A and E). The parallel through D to BC meets the circle $\mathscr{C}$ for the second time at F. The line AF meets $\mathscr{C}$ again at G, and the lines BC and EG meet at M. Prove that

$$\frac{1}{AM} = \frac{1}{AB} + \frac{1}{AC}.$$

13. Let A, B, C, and D be four distinct points on a line, in that order. The circles with diameters AC and BD intersect at X and Y. The line XY meets BC at Z. Let P be a point on the line XY other than Z. The line CP intersects the circle with diameter AC at C and M, and the line BP intersects the circle with diameter BD at B and N. Prove that the lines AM, DN, and XY are concurrent.

14. Consider a semicircle of center O and diameter AB. A line intersects AB at M and the semicircle at C and D in such a way that $MB < MA$ and $MD < MC$. The circumcircles of triangles AOC and DOB intersect a second time at K. Show that MK and KO are perpendicular.

15. The quadrilateral $ABCD$ is inscribed in a circle. The lines AB and CD meet at E, and the diagonals AC and BD meet at F. The circumcircles of the triangles AFD and BFC meet again at H. Prove that $\angle EHF = 90°$.

16. Given two circles that intersect at X and Y, prove that there exist four points with the following property. For any circle $\mathscr{C}$ tangent to the two given circles, we let A and B be the points of tangency and C and D the intersections of $\mathscr{C}$ with the

line *XY*. Then each of the lines *AC*, *AD*, *BC*, and *BD* passes through one of these four points.

1.4 Dissections of Polygonal Surfaces

The following graphical proof (Figure 1.4.1) of the Pythagorean theorem shows that one can cut any two squares into finitely many pieces and reassemble these pieces to get a square. In fact, much more is true.

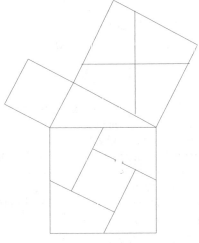

Figure 1.4.1

Any two polygonal surfaces with the same area can be transformed one into the other by cutting the first into finitely many pieces and then assembling these pieces into the second polygonal surface.

This property was proved independently by F. Bolyai (1833) and Gerwin (1835). Its three-dimensional version was included by Hilbert in the list of 23 problems that he presented to the International Congress of Mathematicians in 1900. Hilbert stated that this property does not hold for polyhedra and asked for a complete invariant that gives the obstruction to transforming one polyhedron into another. The problem was solved by M. Dehn, who constructed the required invariant.

Figure 1.4.2

Let us prove the Bolyai–Gerwin theorem. First, note that using diagonals, one can cut any polygon into finitely many triangles. A triangle can be transformed into a rectangle as shown in Figure 1.4.2. We showed that two squares can be cut and reassembled into a single square; thus it suffices to show that from a rectangle one can produce a square.

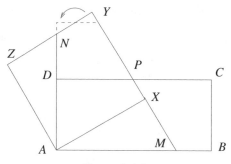

Figure 1.4.3

Let $ABCD$ be the rectangle. By eventually cutting the rectangle $ABCD$ into smaller rectangles and performing the construction below for each of them, we can assume that $AB/4 < BC < AB/2$. Choose the square $AXYZ$ with the same area as the rectangle such that XY intersects CD at its midpoint P (Figure 1.4.3). Let M be the intersection of AB and XY, and N that of AD and YZ. The triangles AZN and AXM are congruent, so the quadrilaterals $MBCP$ and $DNYP$ have the same area. A cut and a flip allows us to transform the second quadrilateral into a trapezoid congruent to the first (the two are congruent, since $PC = PD$, $\angle DPY = \angle CPM$, and they have the same area).

We have proved that any polygon can be transformed into a square. But we can also go backwards from the square to the polygon, and hence we can transform any polygon into any other polygon of the same area, with a square as an intermediate step.

Show that for $n \geq 6$, an equilateral triangle can be dissected into n equilateral triangles.

An equilateral triangle can be dissected into six, seven, and eight equilateral triangles as shown in Figure 1.4.4. The conclusion follows from an inductive argument, by noting that if the triangle can be decomposed into n equilateral triangles, then it can be decomposed into $n+3$ triangles by cutting one of the triangles of the decomposition in four.

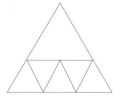 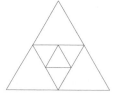 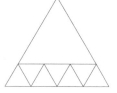

Figure 1.4.4

Here is a problem from the 2006 Mathematical Olympiad Summer Program.

From a 9 × 9 chess board, 46 unit squares are chosen randomly and are colored red. Show that there exists a 2 × 2 block of squares, at least three of which are colored red.

The solution we present was given by A. Kaseorg. Assume the property does not hold, and dissect the board into 25 polygons as shown in Figure 1.4.5. Of these, 5 are unit squares and they could be colored red. Each of the 20 remaining polygons can contain at most 2 colored squares. Thus there are at most $20 \times 2 + 5 = 45$ colored squares, a contradiction. The conclusion follows.

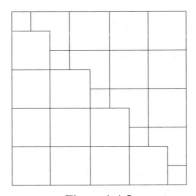

Figure 1.4.5

We conclude the introduction with a problem from the 2007 USAMO, proposed by Reid Barton.

An animal with n cells is a connected figure consisting of n equal-sized square cells. A dinosaur is an animal with at least 2007 cells. It is said to be primitive if its cells cannot be partitioned into two or more dinosaurs. Find with proof the maximum number of cells in a primitive dinosaur.

For the following solution, Andrew Geng received the prestigious Clay prize. Start with a primitive dinosaur and consider the graph whose vertices are the centers of cells. For each pair of neighboring cells, connect the corresponding vertices by an edge. Now cut open the cycles to obtain a tree. The procedure is outlined in Figure 1.4.6. Note that if, by removing an edge, this tree were cut into two connected components each of which having at least 2007 vertices, then the original dinosaur would not be primitive.

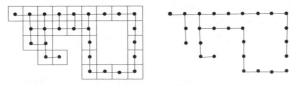

Figure 1.4.6

A connected component obtained by deleting a vertex and its adjacent edges will be called *limb* (of that vertex). A limb that has at least 2007 vertices (meaning that it represents a subdinosaur) is called a *big limb*. Because the dinosaur is primitive, a vertex has at most one big limb.

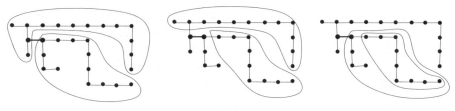

Figure 1.4.7

We claim that the tree of a primitive dinosaur has a vertex with no big limbs. If this is not the case, let us look at a pair of adjacent vertices. There are three cases, outlined in Figure 1.4.7:

- Each of the two vertices has a big limb that contains the other vertex. Then cut along the edge determined by the vertices and obtain two dinosaurs. However, this is impossible because the dinosaur is primitive.

- Each of the two vertices has a big limb that does not contain the other vertex. Cut again along the edge determined by the vertices and obtain two dinosaurs. Again, this is impossible because the dinosaur is primitive.

- One of the vertices is included in the other's big limb.

As we have seen, only the third case can happen. If v and v' are the vertices and v' lies in the big limb of v, then consider next the pair (v', v'') where v'' is adjacent to v' and lies inside its big limb. Then repeat. Since there are no cycles, this procedure must terminate. It can only terminate at a vertex with no big limbs, and the claim is proved.

The vertex with no big limbs has at most 4 limbs, each of which has therefore at most 2006 vertices. We conclude that a primitive dinosaur has at most $4 \cdot 2006 + 1 = 8025$ cells. A configuration where equality is attained is shown in Figure 1.4.8.

And now some problems for the reader.

1. Cut the region that lies between the two rectangles in Figure 1.4.9 by a straight line into two regions of equal areas.

2. Dissect a regular hexagon into 8 congruent polygons.

3. Given three squares with sides equal to 2, 3, and 6, perform only two cuts and reassemble the resulting 5 pieces into a square whose side is equal to 7 (by a cut we understand a polygonal line that decomposes a polygon into two connected pieces).

4. Prove that every square can be dissected into isosceles trapezoids that are not rectangles.

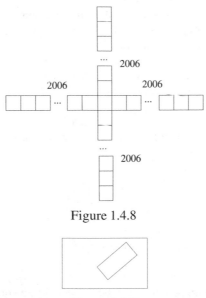

Figure 1.4.8

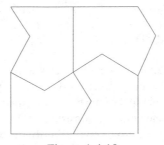

Figure 1.4.9

5. (a) Give an example of a triangle that can be dissected into 5 congruent triangles.

   (b) Give an example of a triangle that can be dissected into 12 congruent triangles.

6. Given the octagon from Figure 1.4.10, one can see how to divide it into 4 congruent polygons. Can it be divided into 5 congruent polygons?

Figure 1.4.10

7. Show that any cyclic quadrilateral can be dissected into n cyclic quadrilaterals for $n \geq 4$.

8. Show that a square can be dissected into n squares for all $n \geq 6$. Prove that this cannot be done for $n = 5$.

9. Show that a cube can be dissected into n cubes for $n \geq 55$.

10. Determine all convex polygons that can be decomposed into parallelograms.

11. Prove that given any $2n$ distinct points inside a convex polygon, there exists a dissection of the polygon into $n + 1$ convex polygons such that the $2n$ points lie on the boundaries of these polygons.

12. Let ABC be an acute triangle with $\angle A = n\angle B$ for some positive integer n. Show that the triangle can be decomposed into isosceles triangles whose equal sides are all equal.

13. Prove that a 10×6 rectangle cannot be dissected into L-shaped 3×2 tiles such as the one in Figure 1.4.11.

Figure 1.4.11

14. Prove that if a certain rectangle can be dissected into equal rectangles similar to it, then the rectangles of the dissection can be rearranged to have all equal sides parallel.

15. A regular $4n$-gon of side-length 1 is dissected into parallelograms. Prove that there exists at least one rectangle in the dissection. Find the sum of the areas of all rectangles from the dissection.

16. Find with proof all possible values of the largest angle of a triangle that can be dissected into five disjoint triangles similar to it.

1.5   Regular Polygons

This section discusses two methods for solving problems about regular polygons. The first method consists in the use of symmetries of these polygons. We illustrate it with the following fascinating fact about the construction of the regular pentagon. Of course, there exists a classical ruler and compass construction, but there is an easier way to do it. Make the simplest knot, the trefoil knot, on a ribbon of paper, then flatten it as is shown in Figure 1.5.1. After cutting off the two ends of the ribbon, you obtain a regular pentagon.

To convince yourself that this pentagon is indeed regular, note that it is obtained by folding the chain of equal isosceles trapezoids from Figure 1.5.2. The property that explains this phenomenon is that the diagonals of the pentagon can be transformed into one another by rotating the pentagon, and thus the trapezoids determined by three sides and a diagonal can be obtained one from the other by a rotation.

More in the spirit of mathematical Olympiads is the following problem of Z. Feng that appeared at a training test during the 2006 Mathematical Olympiad Summer Program.

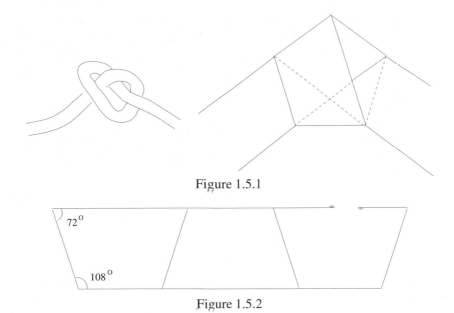

Figure 1.5.1

Figure 1.5.2

Given a triangle ABC with $\angle ABC = 30°$, let H be its orthocenter and G the centroid of the triangle AHB. Find the angle between the lines AH and CG.

The solution is based on an equilateral triangle that is hidden somewhere in the picture. We first replace the line CG by another that is easier to work with. To this end, we choose P on HC such that C is between H and P and $CP/HC = 1/2$ (Figure 1.5.3). If M denotes the midpoint of AB, then $M \in HG$ and $MG/GH = 1/2$, so by Thales' theorem MP is parallel to GC. As the figure suggests, MP passes through D, the foot of the altitude AH. This is proved as follows.

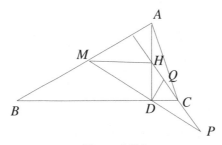

Figure 1.5.3

The angle $\angle DCH$ is the complement of $\angle ABC$, so $\angle DCH = 60°$. Hence in the right triangle DCH, the midpoint Q of HC forms with D and C an equilateral triangle. This is the equilateral triangle we were looking for, which we use to deduce that $DC = QC = CP$. We thus found out that the triangle CDP is isosceles, and, as $\angle CDP = 120°$, it follows that $\angle CDP = 30°$.

Note on the other hand that in the right triangle DAB, M is the circumcenter, so MBD is isosceles, and therefore $\angle MDB = 30°$. This proves that M, D, and P are collinear, as the angles $\angle MDB$ and CDP are equal. It follows that the angle between MP and AD is $\angle ADM = 60°$, and consequently the angle between AD and CG is $60°$.

The second method discussed in this section consists in the use of trigonometry. It refers either to reducing metric relations to trigonometric identities or to using complex numbers written in trigonometric form. We exemplify the use of trigonometry with the following problem, which describes a relation holding in the regular polygon with 14 sides.

Let $A_1A_2A_3\ldots A_{14}$ be a regular polygon with 14 sides inscribed in a circle of radius R. Prove that

$$A_1A_3^2 + A_1A_7^2 + A_3A_7^2 = 7R^2.$$

Let us express the lengths of the three segments in terms of angles and the circumradius R. Since the chords A_1A_3, A_3A_7, and A_1A_7 are inscribed in arcs of measures $\pi/7$, $2\pi/7$, and $3\pi/7$, respectively (see Figure 1.5.4), their lengths are equal to $2R\sin\pi/7$, $2R\sin 2\pi/7$, and $2R\sin 3\pi/7$. Hence the identity to be proved is equivalent to

$$4R^2\left(\sin^2\frac{\pi}{7} + \sin^2\frac{2\pi}{7} + \sin^2\frac{3\pi}{7}\right) = 7R^2.$$

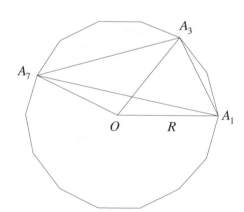

Figure 1.5.4

Using double-angle formulas, we obtain

$$4R^2\left(\sin^2\frac{\pi}{7} + \sin^2\frac{2\pi}{7} + \sin^2\frac{3\pi}{7}\right)$$
$$= 2R^2\left(1 - \cos\frac{2\pi}{7} + 1 - \cos\frac{4\pi}{7} + 1 - \cos\frac{6\pi}{7}\right).$$

To compute the sum

$$\cos\frac{2\pi}{7} + \cos\frac{4\pi}{7} + \cos\frac{6\pi}{7}$$

we multiply it by $\sin 2\pi/7$ and use product-to-sum formulas. We obtain

$$\frac{1}{2}\left(\sin\frac{4\pi}{7} + \sin\frac{6\pi}{7} - \sin\frac{2\pi}{7} + \sin\frac{8\pi}{7} - \sin\frac{4\pi}{7}\right) = -\frac{1}{2}\sin\frac{2\pi}{7}.$$

Here we used the fact that $\sin 8\pi/7 = \sin(2\pi - 6\pi/7) = -\sin 6\pi/7$. Hence the above sum is equal to $-\frac{1}{2}$, and the identity follows.

Here is a list of problems left to the reader.

1. Let ABC and BCD be two equilateral triangles sharing one side. A line passing through D intersects AC at M and AB at N. Prove that the angle between the lines BM and CN is $60°$.

2. On the sides AB, BC, CD, and DA of the convex quadrilateral $ABCD$, construct in the exterior squares whose centers are $M,N,P,$ and Q, respectively. Prove that MP and NQ are perpendicular and have equal lengths.

3. Let $ABCDE$ be a regular pentagon and M a point in its interior such that $\angle MBA = \angle MEA = 42°$. Prove that $\angle CMD = 60°$.

4. On the sides of a hexagon that has a center of symmetry, construct equilateral triangles in the exterior. The vertices of these triangles that are not vertices of the initial hexagon form a new hexagon. Prove that the midpoints of the sides of this hexagon are vertices of a regular hexagon.

5. Let $A_1A_2A_3A_4A_5A_6A_7$ be a regular heptagon. Prove that

$$\frac{1}{A_1A_2} = \frac{1}{A_1A_3} + \frac{1}{A_1A_4}.$$

6. Let $A_1A_2A_3A_4A_5A_6A_7$, $B_1B_2B_3B_4B_5B_6B_7$, $C_1C_2C_3C_4C_5C_6C_7$ be regular heptagons with areas S_A, S_B, and S_C, respectively. Suppose that $A_1A_2 = B_1B_3 = C_1C_4$. Prove that

$$\frac{1}{2} < \frac{S_B + S_C}{S_A} < 2 - \sqrt{2}.$$

7. Let $P_1P_2P_3\ldots P_{12}$ be a regular dodecagon. Prove that P_1P_5, P_4P_8, and P_3P_6 are concurrent.

8. Inside a square $ABCD$, construct the equilateral triangles ABK, BCL, CDM, and DAN. Prove that the midpoints of the segments KL, LM, MN, NK and those of AN, AK, BK, BL, CL, CM, DM, and DN are the vertices of a regular dodecagon.

9. On a circle with diameter AB, choose the points C, D, E on one side of AB, and F on the other side, such that $\overset{\frown}{AC}=\overset{\frown}{CD}=\overset{\frown}{BE}= 20°$ and $\overset{\frown}{BF}= 60°$. Let M be the intersection of BD and CE. Prove that $FM = FE$.

10. Let $A_1A_2A_3\ldots A_{26}$ be a regular polygon with 26 sides, inscribed in a circle of radius R. Denote by A_1', A_7', and A_9' the projections of the orthocenter H of the triangle $A_1A_7A_9$ onto the sides A_7A_9, A_1A_9, and A_1A_7, respectively. Prove that

$$HA_1' - HA_7' + HA_9' = \frac{R}{2}.$$

11. Let $A_1A_2A_3\ldots A_n$ be a regular polygon inscribed in the circle of center O and radius R. On the half-line OA_1 choose P such that A_1 is between O and P. Prove that

$$\prod_{i=1}^{n}PA_i = PO^n - R^n.$$

12. Let $A_1A_2A_3\ldots A_{2n+1}$ be a regular polygon with an odd number of sides, and let A' be the point diametrically opposed to A_{2n+1}. Denote $A'A_1 = a_1$, $A'A_2 = a_2$, $\ldots$, $A'A_n = a_n$. Prove that

$$a_1 - a_2 + a_3 - a_4 + \cdots \pm a_n = R,$$

where R is the circumradius.

13. Given a regular n-gon and M a point in its interior, let x_1, x_2, $\ldots$, x_n be the distances from M to the sides. Prove that

$$\frac{1}{x_1} + \frac{1}{x_2} + \cdots + \frac{1}{x_n} > \frac{2\pi}{a},$$

where a is the side length of the polygon.

14. Let $n > 2$ be an integer. Under the identification of the plane with the set of complex numbers, let $f : \mathbf{C} \to \mathbf{R}$ be a function such that for any regular n-gon $A_1A_2\ldots A_n$,

$$f(A_1)+f(A_2)+\cdots+f(A_n) = 0.$$

Prove that f is the zero function.

15. A number of points are given on a unit circle so that the product of the distances from any point on the circle to the given points does not exceed 2. Prove that the points are the vertices of a regular polygon.

16. For which integers $n \geq 3$ does there exist a regular n-gon in the plane such that all of its vertices have integer coordinates?

1.6 Geometric Constructions and Transformations

In this section, we look at some geometric constructions from the perspective of geometric transformations. Let us illustrate what we have in mind with an easy example.

Let A and B be two points on the same side of line l. Construct a point M on l such that AM + MB is the shortest possible.

The solution is based on a reflection across l. Denote by C the reflection of B across l, and let M be the intersection of AC and l (see Figure 1.6.1). Then $AM + MB = AC$, and for any other point N on l, $AN + NB = AN + NC > AC$ by the triangle inequality. The point C can be effectively constructed by choosing two points P and Q on l and then taking the second intersection of the circle centered at P and of radius PB with the circle centered at Q and radius QB.

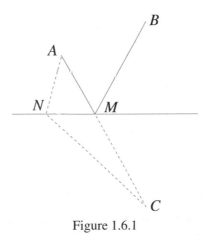

Figure 1.6.1

And now we present an example from the German Mathematical Olympiad (Bundeswettbewerb Mathematik) in 1977.

Given three points A, B, and C in the plane and a compass with fixed opening such that the circumradius of the triangle ABC is smaller than the opening of the compass, construct a fourth point D such that ABCD is a parallelogram.

The problem requires us to translate the point A by the vector $\overrightarrow{BC}$. Let a be the length of the opening of the compass. The construction is very simple in the particular case where $AB = BC = a$. Indeed, if we construct the two circles of radius a centered at A and C, one of their intersections is B and the other is the desired point D. Our intention is to reduce the general case to this particular one.

We have already solved the problem when both the segment and the vector have length a. Let us show how to translate A by the arbitrary vector $\overrightarrow{BC}$ when $AB = a$. The restriction on the size of the triangle implies that there is a point P at distance a from B and C. Construct P as one of the intersections of the circles centered at B and C with radii a. The point D is obtained from A by a translation of vector $\overrightarrow{BP}$ followed by a translation of vector $\overrightarrow{PC}$, both vectors having length a (Figure 1.6.2).

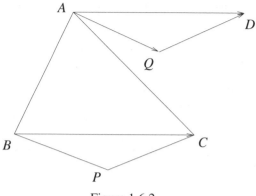

Figure 1.6.2

If AB has arbitrary length, construct Q such that $QA = QB = a$. Then translate Q to R by the vector $\overrightarrow{BC}$ and finally translate A to D by the vector $\overrightarrow{QR}$.

Here are more problems of this kind.

1. Given a polygon in the plane and M a point in its interior, construct two points A and B on the sides of the polygon such that M is the midpoint of the segment AB.

2. Given a polygon in the plane, construct three points A, B, C on the sides of this polygon such that the triangle ABC is equilateral.

3. Using a straightedge and a template in the shape of an equilateral triangle, divide a given segment into (a) 2 equal parts, (b) 3 equal parts.

4. With a straightedge and a compass, construct a trapezoid given the lengths of its sides.

5. Construct a trapezoid given the lengths of its diagonals, the length of the line segment connecting the midpoints of the nonparallel sides, and one of the angles adjacent to the base.

6. Let $\mathscr{C}_1$ and $\mathscr{C}_2$ be two concentric circles. Construct, whenever possible, a line d that intersects these circles consecutively in A, B, C, and D, such that $AB = BC = CD$.

7. The lines m and n pass through a point M situated equidistantly from two parallel lines. Construct a line d that intersects the four lines at A, B, C, D such that $AB = BC = CD$.

8. Given two arbitrary points A and B, with a compass with fixed opening construct a third point C such that the triangle ABC is equilateral.

9. Given two arbitrary points A and B, with a compass with fixed opening construct a point C such that the triangle ABC is right.

10. Given a circle in the plane, construct its center using only a compass.

1.7 Problems with Physical Flavor

In this section, we discuss some elementary problems that can be easily solved at the physical level of rigor, and we explain how physical intuition helps us find a mathematical solution. We exemplify this with the problem of the Toricelli point and leave the rest of the problems as exercises.

Find the point in the plane of an acute triangle that has the smallest sum of the distances to the vertices of the triangle.

The idea of Leibniz was to place the triangle on a table, drill holes at each vertex, and suspend through each hole a ball of weight 1 hanging on a thread, then tie the three threads together (Figure 1.7.1). The system reaches its equilibrium when the

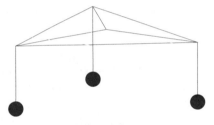

Figure 1.7.1

gravitational potential is minimal, hence when the sum of the lengths of the parts of the threads that are on the table is minimal. The point P where the three threads are tied together is the one we are looking for. On the other hand, the three equal forces that act at P, representing the weights of the balls, add up to zero, because there is equilibrium; hence $\angle APB = \angle BPC = \angle APC = 120°$. This way physical intuition helped us locate the point P.

Let us now prove rigorously that if $\angle APB = \angle BPC = \angle APC = 120°$, then $AP + BP + CP$ is minimal. Let D be such that BCD is equilateral and such that BC separates A and D (Figure 1.7.2). By Pompeiu's theorem (Section 1.1), we have $BQ + CQ \geq QD$ for any point Q in the plane, with equality if and only if Q is on the circumcircle $\mathscr{C}$ of the triangle BCD. By the triangle inequality, $AQ + QD \geq AD$, with equality if and only if Q is on the line segment AD. Hence $AQ + BQ + CQ \geq AD$, and the equality is attained only if Q coincides with the intersection of $\mathscr{C}$ and AD, that is, with point P. It follows that P is a minimum and the only minimum for the sum of the distances to the vertices, and we are done.

We invite the reader to consider the three-dimensional version of the Toricelli point (problem 4), along with some other problems of the same type.

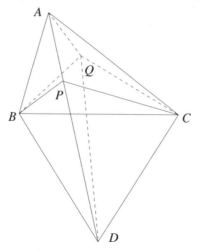

Figure 1.7.2

1. Consider on the sides of a polygon orthogonal vectors of lengths proportional to the lengths of the sides, pointing outwards. Show that the sum of these vectors is equal to zero.

2. Orthogonal to each face of a polyhedron, consider a vector of length numerically equal to the area of that face, pointing outwards. Prove that the sum of these vectors is equal to zero.

3. Prove that the sum of the cosines of the dihedral angles of a tetrahedron does not exceed 2; moreover, it equals 2 if and only if the faces of the tetrahedron have the same area.

4. Let *ABCD* be a tetrahedron and assume there is a point *P* in its interior such that the sum of the distances from *P* to the vertices is minimal. Prove that the bisectors of the angles $\angle APB$ and $\angle CPD$ are supported by the same line. Moreover, this line is orthogonal to the line determined by the bisectors of $\angle BPC$ and $\angle APD$.

5. Given a point inside a convex polyhedron, show that there exists a face of the polyhedron such that the projection of the point onto the plane of that face lies inside the face.

6. The towns *A* and *B* are separated by a straight river. In what place should we construct the bridge *MN* to minimize the length of the road *AMNB*? (The bridge is supposed to be orthogonal to the shore of the river, and the river is supposed to have nonnegligible width.)

7. Five points are given on a circle. A perpendicular is drawn through the centroid of the triangle formed by any three of them to the chord connecting the remaining

two. Such a perpendicular is drawn for each triplet of points. Prove that the 10 lines obtained in this way have a common point. Generalize this statement to n points.

8. Find all finite sets S of at least three points in the plane such that for all distinct points A, B in S, the perpendicular bisector of AB is an axis of symmetry for S.

9. The vertices of the n-dimensional cube are n-tuples $(x_1, x_2, \ldots, x_n)$ where each x_i is either 0 or 1. Two vertices are joined by an edge if they differ by exactly one coordinate. Suppose an electrical network is constructed from the n-dimensional cube by making each edge a 1 ohm resistance. Show that the resistance between $(0, 0, \ldots, 0)$ and $(1, 1, \ldots, 1)$ is

$$R_n = \sum_{j=1}^{n} \frac{1}{j 2^{n-j}}.$$

1.8   Tetrahedra Inscribed in Parallelepipeds

There is no equilateral triangle in the plane with vertices of integer coordinates. In space, however, the tetrahedron with vertices $(1, 0, 0)$, $(0, 1, 0)$, $(0, 0, 1)$, and $(1, 1, 1)$ is regular. Thus there is a regular tetrahedron inscribed in the unit cube as shown in Figure 1.8.1.

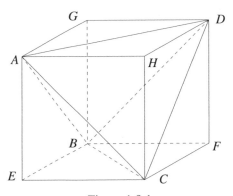

Figure 1.8.1

This allows a quick computation of the volume of a regular tetrahedron of edge a. All we have to do is subtract from the volume of the cube with edge equal to $a\sqrt{2}/2$ the volumes of the four tetrahedra we cut out. Thus the volume of the regular tetrahedron is $a^3\sqrt{2}/4 - 4 \cdot a^3\sqrt{2}/24 = a^3\sqrt{2}/12$.

By deforming Figure 1.8.1, we see that any tetrahedron can be inscribed in a parallelepiped in this way. It is worth remarking that, as we just saw, the tetrahedron $ABCD$ has volume $\frac{1}{3}$ the volume of the cube, and each of the four deleted tetrahedra has volume $\frac{1}{6}$ the volume of the cube. Since volumes are preserved by affine maps, these ratios hold for all tetrahedra inscribed in parallelepipeds.

We are interested in the relationship between the properties of the tetrahedron and those of the associated parallelepiped. Two cases will be considered: that of the orthogonal tetrahedron and that of the isosceles tetrahedron.

A tetrahedron is called *orthogonal* if the opposite edges are orthogonal. Consequently, a tetrahedron is orthogonal if and only if the associated parallelepiped is rhomboidal, i.e., all of its faces are rhombi. Indeed, the tetrahedron is orthogonal if and only if the two diagonals of each face of the parallelepiped are orthogonal. On the other hand, a parallelogram whose diagonals are orthogonal is a rhombus, hence the conclusion.

An *isosceles* tetrahedron is one in which the opposite edges are equal. It is sometimes called equifacial, since its faces are congruent triangles. A tetrahedron is isosceles if and only if the diagonals of each face of the associated parallelepiped are equal, hence if and only if the associated parallelepiped is right.

Let us show how these considerations can be applied to solve the following problem, given in 1984 at a Romanian selection test for the Balkan Mathematical Olympiad.

Let ABCD be a tetrahedron and let d_1, d_2, and d_3 be the common perpendiculars of AB and CD, AC and BD, AD and BC, respectively. Prove that the tetrahedron is isosceles if and only if d_1, d_2, and d_3 are pairwise orthogonal.

For the solution, note that if we inscribe the tetrahedron in a parallelepiped as in Figure 1.8.2, then d_1, d_2, and d_3 are orthogonal to the diagonals of the three pairs of faces, respectively, and hence they are orthogonal to the faces. The parallelepiped is right if and only if every two faces sharing a common edge are orthogonal. On the other hand, two planes are orthogonal if and only if the perpendiculars to the planes are orthogonal. Hence the parallelepiped is right if and only if d_1, d_2, and d_3 are pairwise orthogonal, from which the claim follows.

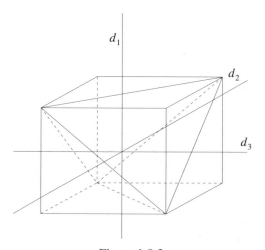

Figure 1.8.2

The problems below describe properties of orthogonal and isosceles tetrahedra that can be proved using the same technique.

1. Express the volume of an isosceles tetrahedron in terms of its edges.

2. Express the circumradius of the isosceles tetrahedron in terms of its edges.

3. Let $ABCD$ be a tetrahedron and M, N, P, Q, R, and S the midpoints of AB, CD, AC, BD, AD, and BC, respectively. Prove that the segments MN, PQ, and RS intersect.

4. Prove that if two pairs of opposite edges in a tetrahedron are orthogonal, then the tetrahedron is orthogonal.

5. Prove that in an orthogonal tetrahedron, the altitudes intersect.

6. Prove that in a rhomboidal parallelepiped $A_1B_1C_1D_1A_2B_2C_2D_2$, the common perpendiculars of the pairs of lines A_1C_1 and B_2D_2, A_1B_2 and C_1D_2, A_1D_2 and B_2C_1 intersect.

7 Let $ABCD$ be an orthogonal tetrahedron. Prove that $AB^2 + CD^2 = AC^2 + BD^2 = AD^2 + BC^2$.

8. Express the volume of an orthogonal tetrahedron in terms of its edges.

9. Prove that if all four faces of a tetrahedron have the same area, then the tetrahedron is isosceles.

10. In a tetrahedron, all altitudes are equal, and one vertex projects orthogonally in the orthocenter of the opposite face. Prove that the tetrahedron is regular.

1.9   Telescopic Sums and Products in Trigonometry

This section is about telescopic sums and products in trigonometry. Problems about the telescopic principle in algebra are the object of a later section.

For problems involving sums, the idea is to use trigonometric identities to write the sum in the form

$$\sum_{k=2}^{n} [F(k) - F(k-1)]$$

then cancel out terms to obtain $F(n) - F(1)$.

Here is an easy example.

Compute the sum $\sum_{k=1}^{n} \cos kx$.

Assuming that $x \neq 2m\pi$, m an integer, we multiply by $2 \sin \frac{x}{2}$. From the product-to-sum formula for the product of a sine and a cosine, we get

$$2 \sum_{k=1}^{n} \sin \frac{x}{2} \cos kx = \sum_{k=1}^{n} \left(\sin \left(k + \frac{1}{2}\right) x - \sin \left(k - \frac{1}{2}\right) x \right)$$

$$= \sin \left(n + \frac{1}{2}\right) x - \sin \frac{1}{2}x.$$

It follows that the original sum is equal to $(\sin(n+\frac{1}{2})x)/(2\sin(x/2)) - \frac{1}{2}$. Of course, when $x = 2m\pi$, m an integer, the answer is n.

In the second example, we apply a formula for the tangent function.

Evaluate the sum

$$\sum_{k=0}^{n} \tan^{-1} \frac{1}{k^2+k+1},$$

where $\tan^{-1}$ *stands for the arctangent function.*

In the solution, we will use the subtraction formula for the tangent

$$\tan(a-b) = \frac{\tan a - \tan b}{1 + \tan a \tan b},$$

which gives the formula for the arctangent

$$\tan^{-1} u - \tan^{-1} v = \tan^{-1} \frac{u-v}{1+uv}.$$

For simplicity, set $a_k = \tan^{-1} k$. Then

$$\begin{aligned}
\tan(a_{k+1} - a_k) &= \frac{\tan a_{k+1} - \tan a_k}{1 + \tan a_{k+1} \tan a_k} \\
&= \frac{k+1-k}{1+k(k+1)} = \frac{1}{k^2+k+1}.
\end{aligned}$$

Hence the sum we are evaluating is equal to

$$\begin{aligned}
\sum_{k=0}^{n} \tan^{-1}(\tan(a_{k+1} - a_k)) &= \sum_{k=0}^{n}(a_{k+1} - a_k) = a_{n+1} - a_0 \\
&= \tan^{-1}(n+1).
\end{aligned}$$

Similar ideas can be used to solve the following problems. In the expressions containing denominators that can vanish, consider only the cases where this does not happen.

1. Prove that

$$\frac{\sin x}{\cos x} + \frac{\sin 2x}{\cos^2 x} + \cdots + \frac{\sin nx}{\cos^n x} = \cot x - \frac{\cos(n+1)x}{\sin x \cos^n x},$$

   for all $x \neq k\frac{\pi}{2}$, k an integer.

2. Prove

$$\frac{1}{\cos 0° \cos 1°} + \frac{1}{\cos 1° \cos 2°} + \cdots + \frac{1}{\cos 88° \cos 89°} = \frac{\cos 1°}{\sin^2 1°}.$$

3. Let n be a positive integer and a a real number, such that a/π is an irrational number. Compute the sum

$$\frac{1}{\cos a - \cos 3a} + \frac{1}{\cos a - \cos 5a} + \cdots + \frac{1}{\cos a - \cos(2n+1)a}.$$

4. Prove the identity

$$\sum_{k=1}^{n} \tan^{-1} \frac{1}{2k^2} = \tan^{-1} \frac{n}{n+1}.$$

5. Evaluate the sum

$$\sum_{n=1}^{\infty} \frac{1}{2^n} \tan \frac{a}{2^n},$$

where $a \neq k\pi$, k an integer.

6. Prove that

$$\sum_{n=1}^{\infty} 3^{n-1} \sin^3 \frac{a}{3^n} = \frac{1}{4}(a - \sin a).$$

7. Prove that the average of the numbers $n \sin n°$, $n = 2, 4, 6, \ldots, 180$, is $\cot 1°$.

8. Prove that for every positive integer n and for every real number $x \neq \frac{k\pi}{2^m}$ $(m = 0, 1, \ldots, n$, k an integer),

$$\frac{1}{\sin 2x} + \frac{1}{\sin 4x} + \cdots + \frac{1}{\sin 2^n x} = \cot x - \cot 2^n x.$$

9. Compute

$$\frac{\tan 1}{\cos 2} + \frac{\tan 2}{\cos 4} + \cdots + \frac{\tan 2^n}{\cos 2^{n+1}}.$$

10. Prove that for any nonzero real number x,

$$\prod_{n=1}^{\infty} \cos \frac{x}{2^n} = \frac{\sin x}{x}.$$

11. Prove that for any integer $n > 1$,

$$\cos \frac{2\pi}{2^n - 1} \cos \frac{4\pi}{2^n - 1} \cdots \cos \frac{2^n \pi}{2^n - 1} = \frac{1}{2^n}.$$

12. Evaluate the product

$$\prod_{k=1}^{n} \left(1 - \tan^2 \frac{2^k \pi}{2^n + 1} \right).$$

13. Evaluate the product

$$(1 - \cot 1°)(1 - \cot 2°) \cdots (1 - \cot 44°).$$

14. Prove the identity

$$\left(\frac{1}{2} + \cos \frac{\pi}{20} \right) \left(\frac{1}{2} + \cos \frac{3\pi}{20} \right) \left(\frac{1}{2} + \cos \frac{9\pi}{20} \right) \left(\frac{1}{2} + \cos \frac{27\pi}{20} \right) = \frac{1}{16}.$$

15. Let n be a positive integer, and let x be a real number different from $2^{k+1}(\frac{\pi}{3} + l\pi)$, $k = 1, 2, \ldots, n$, l an integer. Evaluate the product

$$\prod_{k=1}^{n} \left(1 - 2\cos\frac{x}{2^k}\right).$$

16. Prove that

$$\prod_{k=1}^{n} \left(1 + 2\cos\frac{2\pi \cdot 3^k}{3^n + 1}\right) = 1.$$

1.10   Trigonometric Substitutions

Because of the large number of trigonometric identities, the choice of a clever trigonometric substitution often leads to a simple solution. This is the case with all the problems presented below. The substitution is usually suggested by the form of an algebraic expression, as in the case of the following problem.

Find all real solutions to the system of equations

$$x^3 - 3x = y,$$
$$y^3 - 3y = z,$$
$$z^3 - 3z = x.$$

Here the presence of $x^3 - 3x$ recalls the triple-angle formula for the cosine. Of course, the coefficient in front of x^3 is missing, but we take care of that by working with the double of the cosine instead of the cosine itself. We start by finding the solutions between -2 and 2. Writing $x = 2\cos u$, $y = 2\cos v$, $z = 2\cos w$, with $u, v, w \in [0, \pi]$, the system becomes

$$2\cos 3u = 2\cos v,$$
$$2\cos 3v = 2\cos w,$$
$$2\cos 3w = 2\cos u.$$

By use of the triple-angle formula for both $\cos 3u$ and $\cos v$, the first equation becomes $\cos 9u = \cos 3v$. Combining this with the second equation, we obtain $\cos 9u = \cos w$. As before, $\cos 27u = \cos 3w$, and the third equation yields $\cos 27u = \cos u$. This equality holds if and only if $27u = 2k\pi \pm u$ for some integer k. The solutions in the interval $[0, \pi]$ are $u = k\pi/14$, $k = 0, 1, \ldots, 14$ and $u = k\pi/13$, $k = 1, 2, \ldots, 12$.

Consequently,

$$x = 2\cos k\pi/14, \; y = 2\cos 3k\pi/14, \; z = 2\cos 9k\pi/14, \; k = 0, 1, \ldots, 14,$$

and

$$x = 2\cos k\pi/13, \; y = 2\cos 3k\pi/13, \; z = 2\cos 9k\pi/13, \; k = 1, 2, \ldots, 12$$

are solutions to the given system of equations. Since there are at most $3 \times 3 \times 3 = 27$ solutions (note the degree of the system), and we have already found 27, these are all the solutions.

We proceed with an example where the tangent function is used.

Let $\{x_n\}_n$ be a sequence satisfying the recurrence relation

$$x_{n+1} = \frac{\sqrt{3}x_n - 1}{x_n + \sqrt{3}}, \, n \geq 1.$$

Prove that the sequence is periodic.

Recall the formula for the tangent of a difference:

$$\tan(a - b) = \frac{\tan a - \tan b}{1 - \tan a \tan b}.$$

Note also that $\tan \frac{\pi}{6} = \frac{1}{\sqrt{3}}$.

If we rewrite the recurrence relation as

$$x_{n+1} = \frac{x_n - \frac{1}{\sqrt{3}}}{1 + x_n \frac{1}{\sqrt{3}}},$$

it is natural to substitute $x_1 = \tan t$, for some real number t. Then $x_2 = \tan(t - \pi/6)$, and inductively $x_n = \tan(t - (n-1)\pi/6), n \geq 1$. Since the tangent is periodic of period π, we obtain $x_n = x_{n+6}$, which shows that the sequence has period 6.

We conclude our discussion with a more difficult example.

Let $a, b, c, x, y, z \geq 0$. Prove that

$$(a^2 + x^2)(b^2 + y^2)(c^2 + z^2) \geq (ayz + bzx + cxy - xyz)^2.$$

Let $a = x \tan \alpha$, $b = y \tan \beta$, and $c = z \tan \gamma$ for some $\alpha, \beta, \gamma \in [0, \frac{\pi}{2})$. We have

$$1 \geq \cos(x + y + z)^2 = (\sin x \sin(y + z) - \cos x \cos(y + z))^2$$
$$= (\sin x \sin y \cos z + \sin x \cos y \sin z + \cos x \sin y \sin z - \cos x \cos y \cos z)^2.$$

Now, dividing through by $\cos^2 \alpha \cos^2 \beta \cos^2 \gamma$, we find that

$$\sec^2 \alpha \sec^2 \beta \sec^2 \gamma \geq (\tan \alpha + \tan \beta + \tan \gamma - 1)^2,$$

which implies that

$$(x^2 \tan^2 \alpha + x^2)(y^2 \tan^2 \beta + y^2)(z^2 \tan^2 \gamma + z^2)$$
$$\geq x^2 y^2 z^2 (\tan \alpha + \tan \beta + \tan \gamma - 1)^2.$$

Therefore

$$(a^2 + x^2)(b^2 + y^2)(c^2 + z^2) \geq (ayz + bzx + cxy - xyz)^2.$$

And now some problems.

1. For what values of the real parameter a does there exist a real number x satisfying

$$\sqrt{1-x^2} \geq a - x?$$

2. Given four distinct numbers in the interval $(0,1)$, show that there exist two of them, x and y, such that

$$0 < x\sqrt{1-y^2} - y\sqrt{1-x^2} < \frac{1}{2}.$$

3. Prove that among any four different real numbers, there are two a and b such that

$$\frac{1+ab}{\sqrt{1+a^2}\cdot\sqrt{1+b^2}} > \frac{1}{2}$$

4. Solve the equation

$$x^2 + (4x^3 - 3x)^2 = 1$$

in real numbers.

5. Compute the integral

$$I = \int \sqrt{2 + \sqrt{2 + \cdots + \sqrt{2 + x}}}\,dx,$$

where the expression contains $n \geq 1$ square roots.

6. The sequence $\{x_n\}_n$ satisfies $\sqrt{x_{n+2}+2} \leq x_n \leq 2$ for all $n \geq 1$. Find all possible values of x_{1986}.

7. Find all real solutions to the system of equations

$$2x + x^2 y = y,$$
$$2y + y^2 z = z,$$
$$2z + z^2 x = x.$$

8. Find all real solutions to the system of equations

$$x_1 - \frac{1}{x_1} = 2x_2,$$
$$x_2 - \frac{1}{x_2} = 2x_3,$$
$$x_3 - \frac{1}{x_3} = 2x_4,$$
$$x_4 - \frac{1}{x_4} = 2x_1.$$

9. Prove that

$$-\frac{1}{2} \le \frac{(x+y)(1-xy)}{(1+x^2)(1+y^2)} \le \frac{1}{2}$$

for all $x, y \in \mathbf{R}$.

10. For each real number x, define the sequence $\{x_n\}_n$ recursively by $x_1 = x$, and

$$x_{n+1} = \frac{1}{1-x_n} - \frac{1}{1+x_n}$$

for all n. If $x_n = \pm 1$, then the sequence terminates (for x_{n+1} would be undefined). How many such sequences terminate after the eighth term?

11. A sequence of real numbers $a_1, a_2, a_3, \ldots$ has the property that $a_{k+1} = (ka_k + 1)/(k - a_k)$ for every positive integer k. Prove that this sequence contains infinitely many positive terms and infinitely many negative terms.

12. Given $1 \le a_1 \le a_2 \le \cdots \le a_n \le 1$, prove that

$$\sum_{i=1}^{n-1} \sqrt{1 - a_i a_{i+1} - \sqrt{(1 - a_i^2)(1 - a_{i+1}^2)}} < \frac{\pi\sqrt{2}}{2}.$$

13. Let $x_0 = 0$ and $x_1, x_2, \ldots, x_n > 0$, with $\sum_{k=1}^{n} x_k = 1$. Prove that

$$\sum_{k=1}^{n} \frac{x_k}{\sqrt{1 + x_0 + \cdots + x_{k-1}} \sqrt{x_k + \cdots + x_n}} < \frac{\pi}{2}.$$

14. Find all triples of numbers $x, y, z \in (0, 1)$, satisfying

$$x^2 + y^2 + z^2 + 2xyz = 1.$$

15. Let a, b, and c be given positive numbers. Determine all positive real numbers x, y, and z such that

$$x + y + z = a + b + c,$$
$$4xyz - (a^2 x + b^2 y + c^2 z) = abc.$$

Chapter 2

Algebra and Analysis

T. Andreescu and R. Gelca, *Mathematical Olympiad Challenges*, DOI: 10.1007/978-0-8176-4611-0_2,   39
© Birkhäuser Boston, a part of Springer Science+Business Media, LLC 2009

2.1   No Square Is Negative

In this section, we consider some applications of the simplest inequality in algebra:

$$x^2 \geq 0, \text{ for all } x \in \mathbf{R},$$

where equality holds if and only if $x = 0$. We start with an easy example.

Let x be a real number. Prove that $4x - x^4 \leq 3$.

This problem was posed to young students, who did not even skim through a calculus book. Rewrite the inequality as $x^4 - 4x + 3 \geq 0$, then complete a square to obtain $x^4 - 2x^2 + 1 + 2x^2 - 4x + 2 \geq 0$, that is, $(x - 1)^2 + 2x^2 - 4x + 2 \geq 0$. This is the same as

$$(x^2 - 1)^2 + 2(x - 1)^2 \geq 0,$$

which is clearly true.

The second example appeared at the Romanian Mathematical Olympiad in 1981, proposed by T. Andreescu.

Determine whether there exists a one-to-one function $f : \mathbf{R} \to \mathbf{R}$ with the property that for all x,

$$f(x^2) - (f(x))^2 \geq \frac{1}{4}.$$

We will show that such functions do not exist. The idea is extremely simple: look at the two numbers that are equal to their squares, namely $x = 0$ and $x = 1$, for which $f(x)$ and $f(x^2)$ are equal. Plugging $x = 0$ into the relation, we obtain

$$f(0) - (f(0))^2 \geq \frac{1}{4}.$$

Moving everything to the right side yields

$$0 \geq \left(f(0) - \frac{1}{2}\right)^2.$$

This implies $f(0) = \frac{1}{2}$. Similarly, plugging $x = 1$ we obtain $f(1) = \frac{1}{2}$, which is the same as $f(0)$, and so f cannot be one-to-one.

We list below a number of problems that can be solved by applying similar ideas.

1. The sum of n real numbers is zero and the sum of their pairwise products is also zero. Prove that the sum of the cubes of the numbers is zero.

2. Let a, b, c, and d be real numbers. Prove that the numbers $a - b^2$, $b - c^2$, $c - d^2$, and $d - a^2$ cannot all be larger than $\frac{1}{4}$.

3. Let x, y, z be positive real numbers less than 4. Prove that among the numbers

$$\frac{1}{x} + \frac{1}{4-y}, \frac{1}{y} + \frac{1}{4-z}, \frac{1}{z} + \frac{1}{4-x}$$

there is at least one that is greater than or equal to 1.

4. Find all real solutions to the system of equations

$$x + y = \sqrt{4z - 1},$$
$$y + z = \sqrt{4x - 1},$$
$$z + x = \sqrt{4y - 1}.$$

5. Let x, y be numbers in the interval $(0, 1)$ with the property that there exists a positive number a different from 1 such that

$$\log_x a + \log_y a = 4\log_{xy} a.$$

Prove that $x = y$.

6. Find all real triples (x, y, z) that satisfy $x^4 + y^4 + z^4 - 4xyz = -1$.

7. Find all triples of real numbers x, y, z satisfying

$$2xy - z^2 \geq 1,$$
$$z - |x + y| \geq -1.$$

8. Show that if $x^4 + ax^3 + 2x^2 + bx + 1$ has a real solution, then $a^2 + b^2 \geq 8$.

9. Let a, b, and c be real numbers such that $a^2 + c^2 \leq 4b$. Prove that for all $x \in \mathbf{R}$, $x^4 + ax^3 + bx^2 + cx + 1 \geq 0$.

10. Prove that for all real numbers x, y, z, the following inequality holds:

$$x^2 + y^2 + z^2 - xy - yz - xz \geq \frac{3}{4}(x - y)^2.$$

11. Find the real numbers $x_1, x_2, \ldots, x_n$ satisfying

$$\sqrt{x_1 - 1^2} + 2\sqrt{x_2 - 2^2} + \cdots + n\sqrt{x_n - n^2} = \frac{1}{2}(x_1 + x_2 + \cdots + x_n).$$

12. (a) Let a, b, c be nonnegative real numbers. Prove that

$$ab + bc + ca \geq \sqrt{3abc(a + b + c)}.$$

(b) Let a, b, c be nonnegative real numbers such that $a + b + c = 1$. Prove that

$$a^2 + b^2 + c^2 + \sqrt{12abc} \leq 1.$$

13. Determine $f : \mathbf{N} \to \mathbf{R}$ such that for all k, m, n, one has

$$f(km) + f(kn) - f(k)f(mn) \geq 1.$$

14. Let a, b, c be the edges of a right parallelepiped and d its diagonal. Prove that

$$a^2 b^2 + b^2 c^2 + c^2 a^2 \geq abcd\sqrt{3}.$$

15. If $a_1, a_2, \ldots, a_n$ are real numbers, show that

$$\sum_{i=1}^{n} \sum_{j=1}^{n} ij \cos(a_i - a_j) \geq 0.$$

2.2   Look at the Endpoints

This section is about inequalities that are proved by using the fact that certain real functions reach their extrema at the endpoints of the interval of definition. Two kinds of functions are considered:

- linear functions, which have both extrema at the endpoints of their domain, and

- convex functions, whose maximum is attained on the boundary of the domain.

The main idea is to view an expression as a linear or convex function in each of the variables separately and use this to bound the expression from above or below.

It is important to remark that a linear function can have interior extrema, but only if the slope is zero, in which case the extrema are attained at the enpoints as well. We exemplify these ideas with a problem that appeared at a Romanian Team Selection Test for the International Mathematical Olympiad in 1980.

Given a positive number a, find the maximum of

$$\sum_{k=1}^{n} (a - a_1)(a - a_2) \cdots (a - a_{k-1}) a_k (a - a_{k+1}) \cdots (a - a_n),$$

where a_1, a_2, ..., a_n range independently over the interval $[0, a]$.

For an index k, fix $a_1, a_2, \ldots, a_{k-1}, a_{k+1}, \ldots, a_n$, and think of the given expression as a function in a_k. This function is linear, hence its maximum on the interval $[0, a]$ is attained at one of the endpoints. Repeating the argument for each variable, we conclude that the expression reaches its maximum for a certain choice of $a_k = 0$ or a, $k = 1, 2, \ldots, n$.

If all a_k's are equal to 0, or if two or more a_k's are equal to a, the sum is 0. If one a_k is a and the others are 0, the expression is equal to a^n; hence this is the desired maximum.

Here is an example that appeared at the W.L. Putnam Mathematical Competition, which we solve using convex functions.

Let n be a natural number and let $x_i \in [0,1]$, $i = 1,2,\ldots,n$. Find the maximum of the sum $\sum_{i<j} |x_i - x_j|$.

Note that for a fixed a, the function $f(x) = |x - a|$ is convex. Thus, if we keep $x_2, x_3, \ldots, x_n$ fixed, the expression is a convex function in x_1, being a sum of convex functions. In order to maximize it, one must choose x_1 to be one of the endpoints of the interval. The same argument applied to the other numbers leads to the conclusion that the maximum of the sum is obtained when all x_i's are either 0 or 1. Assume that p of them are 0, and $n - p$ are 1. The sum is then equal to $p(n - p)$. Looking at this value as a function of p, we deduce that when n is even, the maximum is attained for $p = n/2$ and is equal to $n^2/4$, and when n is odd, the maximum is attained for $p = (n \pm 1)/2$ and is equal to $(n^2 - 1)/4$.

Here are more problems of this kind.

1. Let $0 \leq a,b,c,d \leq 1$. Prove that
$$(1-a)(1-b)(1-c)(1-d) + a + b + c + d \geq 1.$$

2. The nonnegative numbers a,b,c,A,B,C, and k satisfy $a + A = b + B = c + C = k$. Prove that
$$aB + bC + cA \leq k^2.$$

3. Let $0 \leq x_k \leq 1$ for all $k = 1,2,\ldots,n$. Prove that
$$x_1 + x_2 + \cdots + x_n - x_1 x_2 \cdots x_n \leq n - 1.$$

4. Find the maximum value of the sum
$$S_n = a_1(1 - a_2) + a_2(1 - a_3) + \cdots + a_n(1 - a_1),$$
where $\frac{1}{2} \leq a_i \leq 1$ for every $i = 1,2,\ldots,n$.

5. Let $n \geq 2$ and $0 \leq x_i \leq 1$ for all $i = 1,2,\ldots,n$. Show that
$$(x_1 + x_2 + \cdots + x_n) - (x_1 x_2 + x_2 x_3 + \cdots + x_n x_1) \leq \left\lfloor \frac{n}{2} \right\rfloor$$
and determine when there is equality.

6. Let $a_1, a_2, \ldots, a_{19}$ be real numbers from the interval $[-98, 98]$. Determine the minimum value of $a_1 a_2 + a_2 a_3 + \cdots + a_{19} a_1$.

7. Prove that for numbers a,b,c in the interval $[0,1]$,
$$\frac{a}{b+c+1} + \frac{b}{c+a+1} + \frac{c}{a+b+1} + (1-a)(1-b)(1-c) \leq 1.$$

8. If $a,b,c,d,e \in [p,q]$ with $p > 0$, prove that
$$(a+b+c+d+e)\left(\frac{1}{a} + \frac{1}{b} + \frac{1}{c} + \frac{1}{d} + \frac{1}{e}\right) \leq 25 + 6\left(\sqrt{\frac{p}{q}} - \sqrt{\frac{q}{p}}\right)^2.$$

9. Prove that if $1 \leq x_k \leq 2$, $k = 1, 2, \ldots, n$, then

$$\left(\sum_{k=1}^{n} x_k \right) \left(\sum_{k=1}^{n} \frac{1}{x_k} \right)^2 \leq n^3.$$

10. Show that for all real numbers $x_1, x_2, \ldots, x_n$,

$$\sum_{i=1}^{n} \sum_{j=1}^{n} |x_i + x_j| \geq n \sum_{i=1}^{n} |x_i|.$$

11. Prove that $x^2 + y^2 + z^2 \leq xyz + 2$ where $x, y, z \in [0, 1]$.

12. Prove that the area of a triangle lying inside the unit square does not exceed $1/2$.

13. Let $P = A_1 A_2 \cdots A_n$ be a convex polygon. For each side $A_i A_{i+1}$, let T_i be the triangle of largest area having $A_i A_{i+1}$ as a side and another vertex of the polygon as its third vertex. Let S_i be the area of T_i, and S the area of the polygon. Prove that $\sum S_i \geq 2S$.

14. Show that if $x, y, z \in [0, 1]$, then $x^2 + y^2 + z^2 \leq x^2 y + y^2 z + z^2 x + 1$.

15. The numbers $x_1, x_2, \ldots, x_{1997}$ belong to the interval $\left[-\frac{1}{\sqrt{3}}, \sqrt{3} \right]$ and sum up to $-318\sqrt{3}$. Find the greatest possible value of $x_1^{12} + x_2^{12} + \cdots + x_{1997}^{12}$.

2.3   Telescopic Sums and Products in Algebra

We have already seen in the previous chapter that many sums can be easily computed by telescoping them, that is, by putting them in the form

$$\sum_{k=2}^{n} [F(k) - F(k-1)].$$

Indeed, in such a sum the $F(k)$'s, for k between 2 and $n-1$, cancel out, giving the answer $F(n) - F(1)$. The reader might notice the similarity between this method of summation and the fundamental theorem of calculus and conclude that what we do is find a discrete antiderivative for the terms of the sum.

A simple example employing the method of telescopic summation is the following.

Compute $\sum_{k=1}^{n} k! \cdot k$.

If we write $k! \cdot k = k! \cdot (k+1-1) = (k+1)! - k!$, then the sum becomes $\sum_{k=1}^{n} [(k+1)! - k!]$, which, after cancellations, is equal to $(n+1)! - 1$.

Evaluate the sum

$$\sum_{k=1}^{n} \frac{1}{(k+1)\sqrt{k} + k\sqrt{k+1}}.$$

The idea is to rationalize the denominator. We have

$$((k+1)\sqrt{k} - k\sqrt{k+1})((k+1)\sqrt{k} + k\sqrt{k+1})$$
$$= k(k+1)^2 - (k+1)k^2 = k(k+1).$$

The sum becomes

$$\sum_{k=1}^{n} \frac{(k+1)\sqrt{k} - k\sqrt{k+1}}{k(k+1)} = \sum_{k=1}^{n} \left(\frac{1}{\sqrt{k}} - \frac{1}{\sqrt{k+1}} \right) = 1 - \frac{1}{\sqrt{n+1}}.$$

And now a problem from the Leningrad Mathematical Olympiad.

Prove that for all positive integers n,

$$n - 1 < \frac{1}{\sqrt{1} + \sqrt{2}} + \frac{3}{\sqrt{2} + \sqrt{5}} + \cdots + \frac{2n-1}{\sqrt{(n-1)^2 + 1} + \sqrt{n^2 + 1}} < n.$$

In order to prove this double inequality, note that

$$\frac{2k-1}{\sqrt{(k-1)^2 + 1} + \sqrt{k^2 + 1}} = \frac{(2k-1)\left(\sqrt{k^2 + 1} - \sqrt{(k-1)^2 + 1}\right)}{k^2 + 1 - (k-1)^2 - 1}$$
$$= \sqrt{k^2 + 1} - \sqrt{(k-1)^2 + 1}.$$

Hence

$$\frac{1}{\sqrt{1} + \sqrt{2}} + \frac{3}{\sqrt{2} + \sqrt{5}} + \cdots + \frac{2n-1}{\sqrt{(n-1)^2 + 1} + \sqrt{n^2 + 1}}$$
$$= \sqrt{2} - 1 + \sqrt{5} - \sqrt{2} + \cdots + \sqrt{n^2 + 1} - \sqrt{(n-1)^2 + 1},$$

which telescopes to $\sqrt{n^2 + 1} - 1$. The double inequality

$$n - 1 < \sqrt{n^2 + 1} - 1 < n$$

is easy and is left to the reader.

The last example comes from the 2006 Bulgarian Mathematical Olympiad.

Find all pairs of polynomials P(x) and Q(x) such that for all x that are not roots of Q(x),

$$\frac{P(x)}{Q(x)} - \frac{P(x+1)}{Q(x+1)} = \frac{1}{x(x+2)}.$$

For the solution, write the equation from the statement as

$$\frac{P(x)}{Q(x)} - \frac{P(x+1)}{Q(x+1)} = \frac{2x+1}{2x(x+1)} - \frac{2(x+1)+1}{2(x+1)(x+2)}.$$

Choose a positive integer n and let x be large enough, then add the equalities

$$\frac{P(x+k)}{Q(x+k)} - \frac{P(x+k+1)}{Q(x+k+1)}$$
$$= \frac{2(x+k)+1}{2(x+k)(x+k+1)} - \frac{2(x+k+1)+1}{2(x+k+1)(x+k+2)}$$

for $k = 0, 1, \ldots, n-1$. The sums on both sides telescope to give

$$\frac{P(x)}{Q(x)} - \frac{P(x+n)}{Q(x+n)} = \frac{2x+1}{x(x+1)} - \frac{2(x+n)+1}{2(x+n)(x+n+1)}.$$

As the right-hand side converges when $n \to \infty$, so does $\frac{P(x+n)}{Q(x+n)}$. Thus by passing to the limit, we obtain

$$\frac{P(x)}{Q(x)} = \frac{2x+1}{2x(x+1)} + C,$$

where C is a constant. Of course, this holds for infinitely many x, hence for any x that is not a root of $Q(x)$. We obtain the simpler functional equation

$$2x(x+1)P(x) = (2x+1+2Cx(x+1))Q(x).$$

Note that $2x+1$ has no common factors with $x(x+1)$. Hence the general solution to this equation consists of polynomials of the form $P(x) = (2x+1+2Cx(x+1))R(x)$ and $Q(x) = 2x(x+1)R(x)$ where $R(x)$ is an arbitrary polynomial. Any such $P(x)$ and $Q(x)$ satisfy the original equation.

A similar method can be used for computing products. In this case, one writes the expression as a product of fractions whose numerators and denominators cancel out alternately to leave the numerator of the first fraction and the denominator of the last, or vice versa. Here is an example.

Prove that

$$\prod_{n=2}^{\infty} \left(1 - \frac{1}{n^2}\right) = \frac{1}{2}.$$

Truncating the product and writing each factor as a fraction, we get

$$\prod_{n=2}^{N}\left(1 - \frac{1}{n^2}\right) = \prod_{n=2}^{N}\left(1 - \frac{1}{n}\right)\left(1 + \frac{1}{n}\right) = \prod_{n=2}^{N}\frac{n-1}{n} \prod_{n=2}^{N}\frac{n+1}{n}$$
$$= \frac{1}{N} \cdot \frac{N+1}{2} = \frac{N+1}{2N}.$$

Letting N tend to infinity, we get that the product is equal to $\frac{1}{2}$.

We invite the reader to solve the following problems.

1. Compute $\sum_{k=1}^{n} k!(k^2 + k + 1)$.

2. Let $a_1, a_2, \ldots, a_n$ be an arithmetic progression with common difference d. Compute

$$\sum_{k=1}^{n} \frac{1}{a_k a_{k+1}}.$$

3. Evaluate the sum

$$\sum_{k=1}^{\infty} \frac{6^k}{(3^k - 2^k)(3^{k+1} - 2^{k+1})}.$$

4. The sequence $\{x_n\}_n$ is defined by $x_1 = \frac{1}{2}$, $x_{k+1} = x_k^2 + x_k$. Find the greatest integer less than

$$\frac{1}{x_1 + 1} + \frac{1}{x_2 + 1} + \cdots + \frac{1}{x_{100} + 1}$$

5. Let F_n be the Fibonacci sequence ($F_1 = 1, F_2 = 1, F_{n+1} = F_n + F_{n-1}$). Evaluate

(a) $\sum_{n=2}^{\infty} \frac{F_n}{F_{n-1}F_{n+1}}$;

(b) $\sum_{n=2}^{\infty} \frac{1}{F_{n-1}F_{n+1}}$.

6. Compute the sum

$$\sqrt{1 + \frac{1}{1^2} + \frac{1}{2^2}} + \sqrt{1 + \frac{1}{2^2} + \frac{1}{3^2}} + \cdots + \sqrt{1 + \frac{1}{1999^2} + \frac{1}{2000^2}}.$$

7. Prove the inequality

$$\frac{1}{\sqrt{1} + \sqrt{3}} + \frac{1}{\sqrt{5} + \sqrt{7}} + \cdots + \frac{1}{\sqrt{9997} + \sqrt{9999}} > 24.$$

8. Let $1 \leq m < n$ be two integers. Prove the double inequality

$$2(\sqrt{n+1} - \sqrt{m}) < \frac{1}{\sqrt{m}} + \frac{1}{\sqrt{m+1}} + \cdots + \frac{1}{\sqrt{n-1}} + \frac{1}{\sqrt{n}} < 2(\sqrt{n} - \sqrt{m-1}).$$

9. Let

$$a_k = \frac{k}{(k-1)^{4/3} + k^{4/3} + (k+1)^{4/3}}.$$

Prove that $a_1 + a_2 + \cdots + a_{999} < 50$.

10. Prove the inequality

$$\sum_{n=1}^{\infty} \frac{1}{(n+1)\sqrt{n}} < 2.$$

11. Evaluate the sum

$$\sum_{n=0}^{\infty} \frac{1}{F_{2^n}},$$

where F_m is the mth term of the Fibonacci sequence.

12. Prove that

$$\prod_{n=2}^{\infty} \frac{n^3 - 1}{n^3 + 1} = \frac{2}{3}.$$

13. Compute the product

$$\prod_{n=0}^{\infty} \left(1 + \frac{1}{2^{2^n}}\right).$$

14. Let $L_1 = 2$, $L_2 = 1$, and $L_{n+2} = L_{n+1} + L_n$, for $n \geq 1$, be the Lucas sequence. Prove that

$$\prod_{k=1}^{m} L_{2^k+1} = F_{2^{m+1}},$$

where $\{F_n\}_n$ is the Fibonacci sequence.

2.4   On an Algebraic Identity

The polynomial $X^2 + 1$ is irreducible over $\mathbf{R}$, but, of course, $X^4 + 1$ is not. To factor it, complete the square then view the result as a difference of two squares:

$$X^4 + 1 = (X^4 + 2X^2 + 1) - 2X^2 = (X^2 + 1)^2 - (\sqrt{2}X)^2$$
$$= (X^2 + \sqrt{2}X + 1)(X^2 - \sqrt{2}X + 1).$$

From this we can derive the Sophie Germain identity

$$X^4 + 4Y^4 = (X^2 + 2XY + 2Y^2)(X^2 - 2XY + 2Y^2),$$

with an alternative version

$$X^4 + \frac{1}{4}Y^4 = \left(X^2 + XY + \frac{1}{2}Y^2\right)\left(X^2 - XY + \frac{1}{2}Y^2\right).$$

Knowing this identity can be useful in some situations. Here are two examples.

Evaluate the sum

$$\sum_{k=1}^{n} \frac{4k}{4k^4 + 1}.$$

By using the above identity, we obtain

$$\sum_{k=1}^{n} \frac{4k}{4k^4 + 1} = \sum_{k=1}^{n} \frac{(2k^2 + 2k + 1) - (2k^2 - 2k + 1)}{(2k^2 + 2k + 1)(2k^2 - 2k + 1)}$$

$$= \sum_{k=1}^{n} \left(\frac{1}{2k^2 - 2k + 1} - \frac{1}{2(k+1)^2 - 2(k+1) + 1} \right)$$

$$= 1 - \frac{1}{2n^2 + 2n + 1},$$

and we are done.

Given an $n \times n$ matrix A with the property that $A^3 = 0$, prove that the matrix $M = \frac{1}{2}A^2 + A + I_n$ is invertible.

Indeed, the inverse of this matrix is $\frac{1}{2}A^2 - A + I_n$, since the product of the two gives $\frac{1}{4}A^4 + I_n$, which is equal, by hypothesis, to the identity matrix. Of course, the conclusion can also be derived from the fact that the eigenvalues of M are nonzero, which is a direct consequence of the spectral mapping theorem. Or, we can notice that $I_n + A + A^2/2 = e^A$, and its inverse is $e^{-A} = I - A + A^2/2$.

We give below more examples that make use of the Sophie Germain identity.

1. Prove that for every integer $n > 2$, the number $2^{2^n - 2} + 1$ is not a prime number.

2. Prove that any sequence satisfying the recurrence relation $x_{n+1} + x_{n-1} = \sqrt{2}x_n$ is periodic.

3. Compute the sum

$$\sum_{k=1}^{n} \frac{k^2 - \frac{1}{2}}{k^4 + \frac{1}{4}}.$$

4. Evaluate

$$\frac{(1^4 + \frac{1}{4})(3^4 + \frac{1}{4}) \cdots ((2n-1)^4 + \frac{1}{4})}{(2^4 + \frac{1}{4})(4^4 + \frac{1}{4}) \cdots ((2n)^4 + \frac{1}{4})}.$$

5. Show that there are infinitely many positive integers a such that for any n, the number $n^4 + a$ is not prime.

6. Show that $n^4 + 4^n$ is prime if and only if $n = 1$.

7. Consider the polynomial $P(X) = X^4 + 6X^2 - 4X + 1$. Prove that $P(X^4)$ can be written as the product of two polynomials with integer coefficients, each of degree greater than 1.

8. Prove that for any integer n greater than 1, $n^{12} + 64$ can be written as the product of four distinct positive integers greater than 1.

9. Let m and n be positive integers. Prove that if m is even, then

$$\sum_{k=0}^{m} (-4)^k n^{4(m-k)}$$

is not a prime number.

10. Find the least positive integer n for which the polynomial

$$P(x) = x^{n-4} + 4n$$

can be written as a product of four non-constant polynomials with integer coefficients.

2.5   Systems of Equations

For this section, we have selected non-standard systems of equations. The first example involves just algebraic manipulations.

Prove that the only positive solution of

$$x + y^2 + z^3 = 3,$$
$$y + z^2 + x^3 = 3,$$
$$z + x^2 + y^3 = 3$$

is $(x, y, z) = (1, 1, 1)$.

From the difference of the first two equations, we obtain that

$$x(1 - x^2) + y(y - 1) + z^2(z - 1) = 0.$$

From the difference of the last two equations, we obtain that

$$y(1 - y^2) + z(z - 1) + x^2(x - 1) = 0.$$

Multiplying this equation by z and subtracting it from the one above yields

$$x(x - 1)(1 + x + xz) = y(y - 1)(1 + z + yz).$$

Similarly

$$y(y - 1)(1 + y + yx) = z(z - 1)(1 + x + zx).$$

From the last two relations, it follows that if x, y, and z are positive, then x, y, and z are all equal to 1, all less than 1, or all greater than 1. The last two possibilities are excluded, for $x + y^2 + z^3 = 3$, and the result follows.

The second example is from the 1996 British Mathematical Olympiad.

Find all solutions in positive real numbers a, b, c, d to the following system:

$$a + b + c + d = 12,$$
$$abcd = 27 + ab + ac + ad + bc + bd + cd.$$

Using the Arithmetic Mean–Geometric Mean (AM–GM) inequality in the second equation, we obtain

$$abcd \geq 27 + 6\sqrt{abcd}.$$

Moving everything to the left and factoring the expression, viewed as a quadratic polynomial in $\sqrt{abcd}$, yields

$$(\sqrt{abcd} + 3)(\sqrt{abcd} - 9) \geq 0.$$

This implies $\sqrt{abcd} \geq 9$, which combined with the first equation of the system gives

$$\sqrt[4]{abcd} \geq \frac{a + b + c + d}{4}.$$

The AM–GM inequality implies that $a = b = c = d = 3$ is the only solution.

And now a problem with a surprising solution from the 1996 Vietnamese Mathematical Olympiad.

Find all positive real numbers x and y satisfying the system of equations

$$\sqrt{3x}\left(1 + \frac{1}{x + y}\right) = 2,$$
$$\sqrt{7y}\left(1 - \frac{1}{x + y}\right) = 4\sqrt{2}.$$

It is natural to make the substitution $\sqrt{x} = u$, $\sqrt{y} = v$. The system becomes

$$u\left(1 + \frac{1}{u^2 + v^2}\right) = \frac{2}{\sqrt{3}},$$
$$v\left(1 - \frac{1}{u^2 + v^2}\right) = \frac{4\sqrt{2}}{\sqrt{7}}.$$

But $u^2 + v^2$ is the square of the absolute value of the complex number $z = u + iv$. This suggests that we add the second equation multiplied by i to the first one. We obtain

$$u + iv + \frac{u - iv}{u^2 + v^2} = \left(\frac{2}{\sqrt{3}} + i\frac{4\sqrt{2}}{\sqrt{7}}\right).$$

The quotient $(u - iv)/(u^2 + v^2)$ is equal to $\bar{z}/|z|^2 = \bar{z}/(z\bar{z}) = 1/z$, so the above equation becomes

$$z + \frac{1}{z} = \left(\frac{2}{\sqrt{3}} + i\frac{4\sqrt{2}}{\sqrt{7}}\right).$$

Hence z satisfies the quadratic equation

$$z^2 - \left(\frac{2}{\sqrt{3}} + i\frac{4\sqrt{2}}{\sqrt{7}}\right)z + 1 = 0$$

with solutions

$$\left(\frac{1}{\sqrt{3}} \pm \frac{2}{\sqrt{21}}\right) + i\left(\frac{2\sqrt{2}}{\sqrt{7}} \pm \sqrt{2}\right),$$

where the signs $+$ and $-$ correspond.

This shows that the initial system has the solutions

$$x = \left(\frac{1}{\sqrt{3}} \pm \frac{2}{\sqrt{21}}\right)^2, \quad y = \left(\frac{2\sqrt{2}}{\sqrt{7}} \pm \sqrt{2}\right)^2,$$

where the signs $+$ and $-$ correspond.

The systems below are to be solved in real numbers, unless specified otherwise.

1. Solve the system of equations

$$x + \log(x + \sqrt{x^2 + 1}) = y,$$
$$y + \log(y + \sqrt{y^2 + 1}) = z,$$
$$z + \log(z + \sqrt{z^2 + 1}) = x.$$

2. Solve the system

$$\log(2xy) = \log x \log y,$$
$$\log(yz) = \log y \log z,$$
$$\log(2zx) = \log z \log x.$$

3. Solve the system of equations

$$xy + yz + zx = 12,$$
$$xyz = 2 + x + y + z$$

in positive real numbers x, y, z.

4. Find all real solutions to the system of equations

$$\frac{4x^2}{4x^2+1} = y,$$
$$\frac{4y^2}{4y^2+1} = z,$$
$$\frac{4z^2}{4z^2+1} = x.$$

5. Find $ax^5 + by^5$ if the real numbers a, b, x, and y satisfy the system of equations

$$ax + by = 3,$$
$$ax^2 + by^2 = 7,$$
$$ax^3 + by^3 = 16,$$
$$ax^4 + by^4 = 42.$$

6. Find all solutions to the system of equations

$$6(x - y^{-1}) = 3(y - z^{-1}) - 2(z - x^{-1}) = xyz - (zyz)^{-1}$$

in nonzero real numbers x, y, z.

7. Find all integer solutions to the system

$$3 = x + y + z = x^3 + y^3 + z^3.$$

8. Solve the system

$$x + \frac{2}{x} = 2y,$$
$$y + \frac{2}{y} = 2z,$$
$$z + \frac{2}{z} = 2x.$$

9. Solve the system of equations

$$(x + y)^3 = z,$$
$$(y + z)^3 = x,$$
$$(z + x)^3 = y.$$

10. Solve the system

$$x^2 - |x| = |yz|,$$
$$y^2 - |y| = |zx|,$$
$$z^2 - |z| = |xy|.$$

11. Find the solutions to the system of equations

$$x + \lfloor y \rfloor + \{z\} = 1.1,$$
$$\lfloor x \rfloor + \{y\} + z = 2.2,$$
$$\{x\} + y + \lfloor z \rfloor = 3.3,$$

where $\lfloor\ \rfloor$ and $\{\ \}$ denote respectively the greatest integer and fractional part functions.

12. For a given complex number a, find the complex solutions to the system

$$(x_1 + x_2 + x_3)x_4 = a,$$
$$(x_1 + x_2 + x_4)x_3 = a,$$
$$(x_1 + x_3 + x_4)x_2 = a,$$
$$(x_2 + x_3 + x_4)x_1 = a.$$

13. Find all real numbers a for which there exist nonnegative real numbers x_1, x_2, x_3, x_4, x_5 satisfying the system

$$\sum_{k=1}^{5} kx_k = a,$$

$$\sum_{k=1}^{5} k^3 x_k = a^2,$$

$$\sum_{k=1}^{5} k^5 x_k = a^3.$$

14. Solve the system of equations

$$x^3 - 9(y^2 - 3y + 3) = 0,$$
$$y^3 - 9(z^2 - 3z + 3) = 0,$$
$$z^3 - 9(x^2 - 3x + 3) = 0.$$

15. Solve the system

$$ax + by = (x - y)^2,$$
$$by + cz = (y - z)^2,$$
$$cz + ax = (z - x)^2,$$

where $a, b, c > 0$.

16. Let a, b, c be positive real numbers, not all equal. Find all solutions to the system of equations

$$x^2 - yz = a,$$
$$y^2 - zx = b,$$
$$z^2 - xy = c,$$

in real numbers x, y, z.

The second example is a problem from the 2007 USA Team Selection Test for the IMO, proposed by K. Kedlaya.

Let n be a positive integer and let $a_1 \leq a_2 \leq \cdots \leq a_n$ and $b_1 \leq b_2 \leq \cdots \leq b_n$ be two nondecreasing sequences of real numbers such that

$$a_1 + \cdots + a_i \leq b_1 + \cdots + b_i \quad \text{for every } i = 1, 2, \ldots, n-1$$

and

$$a_1 + \cdots + a_n = b_1 + \cdots + b_n.$$

Suppose that for every real number m, the number of pairs (i, j) with $a_i - a_j = m$ equals the number of pairs (k, l) with $b_k - b_l = m$. Prove that $a_i = b_i$ for all $i = 1, 2, \ldots, n$.

From the statement we can deduce immediately that

$$\sum_{1 \leq i < j \leq n} (a_i - a_j) = \sum_{1 \leq k < l \leq n} (b_k - b_l)$$

and

$$\sum_{1 \leq i < j \leq n} (a_i - a_j)^2 = \sum_{1 \leq k < l \leq n} (b_k - b_l)^2$$

since we are summing the same numbers. We will give two solutions, one based on the first equality, one based on the second.

First solution: Applying the Abel summation formula to the sequences $s_i = a_1 + a_2 + \cdots + a_{n-i}$ and $t_i = 1$, $i = 1, 2, \ldots, n-1$, we can write

$$2 \sum_{i=1}^{n-1} (a_1 + \cdots + a_i) = 2(n-1)a_1 + 2(n-2)a_2 + \cdots + 2a_{n-1}$$

$$= (n-1)a_1 + (n-3)a_2 + \cdots + (1-n)a_n + (n-1) \sum_{i=1}^{n} a_i$$

$$= (n-1) \sum_{i=1}^{n} a_i + \sum_{1 \leq i < j \leq n} (a_i - a_j).$$

By the same argument

$$2 \sum_{i=1}^{n-1} (b_1 + \cdots + b_i) = (n-1) \sum_{i=1}^{n} b_i + \sum_{1 \leq k < l \leq n-1} (b_1 + \cdots + b_i).$$

In both identities the right-hand sides are equal, hence

$$\sum_{i=1}^{n-1} (a_1 + \cdots + a_i) = \sum_{i=1}^{n-1} (b_1 + \cdots + b_i).$$

Consequently, each of the inequalities $a_1 + \cdots + a_i \leq b_1 + \cdots + b_i$ for $i = 1, \ldots, n-1$ must be an equality. Combining with $a_1 + \cdots + a_n = b_1 + \cdots + b_n$ we deduce that $a_i = b_i$ for all $i = 1, \ldots, n$, as desired.

Second solution: As announced, we will use the second equality. Expanding both sides, we obtain

$$(n-1)\sum_{i=1}^{n} a_i^2 + 2\sum_{1\le i<j\le n} a_i a_j = (n-1)\sum_{i=1}^{n} b_i^2 + 2\sum_{1\le k<l\le n} b_k b_l.$$

Also, squaring both sides of the equation $a_1 + \cdots + a_n = b_1 + \cdots + b_n$ yields

$$\sum_{i=1}^{n} a_i^2 + 2\sum_{1\le i<j\le n} a_i a_j = \sum_{i=1}^{n} b_i^2 + 2\sum_{1\le k<l\le n} b_k b_l.$$

Subtracting the second equality from the first and dividing by $n-2$ we deduce that

$$\sum_{i=1}^{n} a_i^2 = \sum_{i=1}^{n} b_i^2.$$

Of course, this does not work if $n = 2$, but in that case the problem is easy. So let us continue working under the assumption that $n > 2$. By the Cauchy–Schwarz inequality,

$$\left(\sum_{i=1}^{n} b_i^2\right)^2 = \left(\sum_{i=1}^{n} a_i^2\right)\left(\sum_{i=1}^{n} b_i^2\right) \ge \left(\sum_{i=1}^{n} a_i b_i\right)^2.$$

which yields

$$\sum_{i=1}^{n} b_i^2 \ge \left|\sum_{i=1}^{n} a_i b_i\right| \ge \sum_{i=1}^{n} a_i b_i.$$

We now set $s_i = a_1 + \cdots + a_i$ and $t_i = b_1 + \cdots + b_i$ for every $1 \le i \le n$. Using the Abel summation formula we can write

$$\sum_{i=1}^{n} a_i b_i = s_1 b_2 + (s_2 - s_1)b_2 + \cdots + (s_n - s_{n-1})b_n$$

$$= s_1(b_1 - b_2) + s_2(b_2 - b_3) + \cdots + s_{n-1}(b_{n-1} - b_n) + s_n b_n.$$

By the given conditions $s_i \le t_i$ and $b_i - b_{i+1} \le 0$ for every $1 \le i \le n-1$, and $s_n = t_n$. It follows that

$$\sum_{i=1}^{n} a_i b_i \ge t_1(b_1 - b_2) + t_2(b_2 - b_3) + \cdots + t_{n-1}(b_{n-1} - b_n) + t_n b_n$$

$$= t_1 b_1 + (t_2 - t_1)b_2 + \cdots + (t_n - t_{n-1})b_n = \sum_{i=1}^{n} b_i^2.$$

Hence we have equality in Cauchy–Schwarz, and also all of the above inequalities are equalities. In particular $s_i = t_i$ for $i = 1, \ldots, n$, and by subtraction we obtain $a_i = b_i$ for all $i = 1, \ldots, n$, completing the proof.

The following problems can also be solved using the above summation formula.

1. Using the Abel summation formula, compute the sums
   (a) $1 + 2q + 3q^2 + \cdots + nq^{n-1}$.
   (b) $1 + 4q + 9q^2 + \cdots + n^2 q^{n-1}$.

2. The numbers $a_1 \geq a_2 \geq \cdots \geq a_n > 0$ and $b_1 \geq b_2 \geq \cdots \geq b_n > 0$ satisfy $a_1 \geq b_1$, $a_1 + a_2 \geq b_1 + b_2, \ldots, a_1 + a_2 + \cdots + a_n \geq b_1 + b_2 + \cdots + b_n$. Prove that for every positive integer k,
   $$a_1^k + a_2^k + \cdots + a_n^k \geq b_1^k + b_2^k + \cdots + b_n^k.$$

3. Let a, b, c, and d be nonnegative numbers such that $a \leq 1$, $a + b \leq 5$, $a + b + c \leq 14$, $a + b + c + d \leq 30$. Prove that
   $$\sqrt{a} + \sqrt{b} + \sqrt{c} + \sqrt{d} \leq 10.$$

4. Let $a_1, a_2, \ldots, a_n$ be nonnegative numbers such that $a_1 a_2 \cdots a_k \geq \frac{1}{(2k)!}$ for all k. Prove that
   $$a_1 + a_2 + \cdots + a_n > \frac{1}{n+1} + \frac{1}{n+2} + \cdots + \frac{1}{2n}.$$

5. Let $x_1, x_2, \ldots, x_n$ and $y_1 \geq y_2 \geq \cdots \geq y_n$ be two sequences of positive numbers such that $x_1 \geq y_1$, $x_1 x_2 \geq y_1 y_2$, $\ldots$, $x_1 x_2 \cdots x_n \geq y_1 y_2 \cdots y_n$. Prove that $x_1 + x_2 + \cdots + x_n \geq y_1 + y_2 + \cdots + y_n$.

6. Let $\{a_n\}_n$ be a sequence of positive numbers such that for all n, $\sum_{k=1}^{n} a_k \geq \sqrt{n}$.
   Prove that for all n,
   $$\sum_{k=1}^{n} a_k^2 \geq \frac{1}{4}\left(1 + \frac{1}{2} + \cdots + \frac{1}{n}\right).$$

7. Let $\phi : \mathbf{N} \to \mathbf{N}$ be an injective function. Prove that for all $n \in \mathbf{N}$
   $$\sum_{k=1}^{n} \frac{\phi(k)}{k^2} \geq \sum_{k=1}^{n} \frac{1}{k}.$$

8. Let $a_1 + a_2 + a_3 + a_4 + \cdots$ be a convergent series. Prove that the series $a_1 + a_2/2 + a_3/3 + a_4/4 + \cdots$ is also convergent.

9. Let $x_1, x_2, \ldots, x_n, y_1, y_2, \ldots, y_n$ be positive real numbers such that
   (i) $x_1 y_1 < x_2 y_2 < \cdots < x_n y_n$,
   (ii) $x_1 + x_2 + \cdots + x_k \geq y_1 + y_2 + \cdots + y_k$, $1 \leq k \leq n$.

   (a) Prove that
   $$\frac{1}{x_1} + \frac{1}{x_2} + \cdots + \frac{1}{x_n} \leq \frac{1}{y_1} + \frac{1}{y_2} + \cdots + \frac{1}{y_n}.$$

(b) Let $A = \{a_1, a_2, \ldots, a_n\}$ be a set of positive integers such that for every distinct subsets B and C of A, $\sum_{x \in B} x \neq \sum_{x \in C} x$. Prove that

$$\frac{1}{a_1} + \frac{1}{a_2} + \cdots + \frac{1}{a_n} < 2.$$

10. The sequence $u_1, u_2, \ldots, u_n, \ldots$ is defined by

$$u_1 = 2 \text{ and } u_n = u_1 u_2 \cdots u_{n-1} + 1, \quad n = 2, 3, \ldots.$$

Prove that for all possible integers n, the closest underapproximation of 1 by n Egyptian fractions is

$$1 - \frac{1}{u_1 u_2 \cdots u_n}.$$

2.8 $x + 1/x$

The nth Chebyshev polynomial (of the first kind) is usually defined as the polynomial expressing $\cos nx$ in terms of $\cos x$. Closely related is the polynomial $P_n(X)$ that expresses $2\cos nx$ in terms of $2\cos x$. This polynomial can be obtained by writing $x^n + x^{-n}$ in terms of $x + x^{-1}$. Indeed, if $x = \cos t + i\sin t$, then $x + x^{-1} = 2\cos t$, while by the de Moivre formula $x^n + x^{-n} = 2\cos nt$. Note that the sum-to-product formula

$$\cos(n+1)x + \cos(n-1)x = 2\cos x \cos nx,$$

allows us to prove by induction that $P_n(X)$ has integer coefficients, and we can easily compute $P_0(X) = 2$, $P_1(X) = X$, $P_2(X) = X^2 - 2$, $P_3(X) = X^3 - 3X$.

The fact that $x^n + x^{-n}$ can be written as a polynomial with integer coefficients in $x + x^{-1}$ for all n can also be proved inductively using the identity

$$x^{n+1} + \frac{1}{x^{n+1}} = \left(x + \frac{1}{x}\right)\left(x^n + \frac{1}{x^n}\right) - \left(x^{n-1} + \frac{1}{x^{n-1}}\right).$$

Let us apply this fact to the following problem.

Prove that for all n, the number $\sqrt[n]{\sqrt{3} + \sqrt{2}} + \sqrt[n]{\sqrt{3} - \sqrt{2}}$ is irrational.

Because $x^n + x^{-n}$ can be written as a polynomial with integer coefficients in $x + x^{-1}$, if $x + x^{-1}$ were rational, then so would be $x^n + x^{-n}$. If $x = \sqrt[n]{\sqrt{3} + \sqrt{2}}$, then $x^{-1} = \sqrt[n]{\sqrt{3} - \sqrt{2}}$; hence $x^n + x^{-n} = \sqrt{3} + \sqrt{2} + \sqrt{3} - \sqrt{2} = 2\sqrt{3}$, which is irrational. It follows $x + x^{-1}$ must be irrational, too.

Here is another example.

Let r be a positive real number such that

$$\sqrt[4]{r} - \frac{1}{\sqrt[4]{r}} = 14.$$

Prove that

$$\sqrt[6]{r} + \frac{1}{\sqrt[6]{r}} = 6.$$

Squaring the relation from the statement, we obtain

$$\sqrt[3]{r} + \frac{1}{\sqrt[3]{r}} - 2 = 196,$$

hence

$$\sqrt[3]{r} + \frac{1}{\sqrt[3]{r}} = 198.$$

Let

$$\sqrt[6]{r} + \frac{1}{\sqrt[6]{r}} = x.$$

Then using the above recursive relation, we obtain

$$\sqrt{r} + \frac{1}{\sqrt{r}} = x^3 - 3x.$$

Hence $x^3 - 3x - 198 = 0$. Factoring the left-hand side as $(x - 6)(x^2 + 6x + 33)$, we see that the equation has the unique positive solution $x = 6$, as desired.

Note that the recursive relation for calculating $x^n + \frac{1}{x^n}$ from $x + \frac{1}{x}$ can be generalized to

$$u^{n+1} + v^{n+1} = (u + v)(u^n + v^n) - uv(u^{n-1} + v^{n-1})$$

for any two numbers u and v, and $n \in \mathbf{N}$. Another way to obtain this identity is to substitute x by u/v in the above and then multiply out by the common denominator.

And now the problems.

1. Prove that for all odd numbers n, $x^n - x^{-n}$ can be written as a polynomial in $x - x^{-1}$.

2. Prove that if $x, y > 0$ and for some $n \in \mathbf{N}$, $x^{n-1} + y^{n-1} = x^n + y^n = x^{n+1} + y^{n+1}$, then $x = y$.

3. Given that

$$x + x^{-1} = \frac{1 + \sqrt{5}}{2},$$

find $x^{2000} + x^{-2000}$.

4. If z is a complex number satisfying $|z^3 + z^{-3}| \leq 2$, show that $|z + z^{-1}| \leq 2$.

5. Prove that $\cos 1°$ is irrational.

6. Let a, b, c be real numbers such that $\max\{|a|, |b|, |c|\} > 2$ and $a^2 + b^2 + c^2 - abc = 4$. Prove that there exist real numbers u, v, w such that

$$a = u + \frac{1}{u}, b = v + \frac{1}{v}, c = w + \frac{1}{w},$$

where $uvw = 1$.

7. Show that for $x > 1$,

$$\frac{x - x^{-1}}{1} < \frac{x^2 - x^{-2}}{2} < \frac{x^3 - x^{-3}}{3} < \cdots < \frac{x^n - x^{-n}}{n} < \cdots.$$

8. Evaluate $\lim_{n \to \infty}\{(\sqrt{2} + 1)^{2n}\}$ where $\{a\}$ denotes the fractional part of a, i.e., $\{a\} = a - \lfloor a \rfloor$.

9. Prove that for all positive integers n, the number $\lfloor (1 + \sqrt{3})^{2n+1} \rfloor$ is divisible by 2^{n+1} but not by 2^{n+2}.

10. Show that if $\cos a + \sin a$ is rational for some a, then for any positive integer n, $\cos^n a + \sin^n a$ is rational.

11. Prove that for any $n \in \mathbf{N}$, there exists a rational number a_n such that the polynomial $X^2 + (1/2)X + 1$ divides $X^{2n} + a_n X^n + 1$.

12. Let a, b, c be real numbers, with ac and b rational, such that the equation $ax^2 + bx + c = 0$ has a rational solution r. Prove that for every positive integer n, there exists a rational number b_n for which r^n is a solution to the equation $a^n x^2 + b_n x + c^n = 0$.

13. If A is an $n \times n$ matrix with real entries and there exists $a \in [-2, 2]$ such that $A^2 - aA + I_n = 0_n$, prove that for every natural number m, there exists a unique $a_m \in [-2, 2]$ such that $A^{2m} - a_m A^m + I_n = 0_n$.

2.9   Matrices

The problems in this section can be solved using properties of matrices that do not involve the row–column structure. An example is the following:

Prove that if A and B are $n \times n$ matrices, then

$$\det(I_n - AB) = \det(I_n - BA).$$

For the solution, let us assume first that B is invertible. Then $I_n - AB = B^{-1}(I_n - BA)B$, and hence

$$\det(I_n - AB) = \det(B^{-1})\det(I_n - BA)\det B = \det(I_n - BA).$$

If B is not invertible, consider instead the matrix $B_x = xI_n + B$. Since $\det(xI_n + B)$ is a polynomial in x, the matrices B_x are invertible for all but finitely many values of x. Thus we can use the first part of the proof and conclude that $\det(I_n - AB_x) = \det(I_n - B_xA)$ for all except finitely many values of x. But these two determinants are polynomials in x, which are equal for infinitely many values of x, so they must always be equal. In particular, for $x = 0$, $\det(I_n - AB) = \det(I_n - BA)$.

As a consequence, we see that if $I_n - AB$ is invertible, then $I_n - BA$ is also invertible. Here is a direct proof of this implication. If V is the inverse of $I_n - AB$, then $V(I_n - AB) = I_n$, hence $VAB = V - I_n$. We have

$$
\begin{aligned}
(I_n + BVA)(I_n - BA) &= I_n - BA + BVA - BVABA \\
&= I_n - BA + BVA - B(V - I_n)A = I_n,
\end{aligned}
$$

hence $I_n + BVA$ is the inverse of $I_n - BA$.

1. Let A, B be two square matrices such that $A + B = AB$. Prove that A and B commute.

2. Prove that if A is a 5×4 matrix and B is a 4×5 matrix, then

$$
\det(AB - I_5) + \det(BA - I_4) = 0.
$$

3. Let X, Y, Z be $n \times n$ matrices such that

$$
X + Y + Z = XY + YZ + ZX.
$$

Prove that the equalities

$$
\begin{aligned}
XYZ &= XZ - ZX, \\
YZX &= YX - XY, \\
ZXY &= ZY - YZ,
\end{aligned}
$$

are equivalent.

4. Show that for any two $n \times n$ matrices A and B, the following identity holds:

$$
\det(I_n - AB) = \det \begin{bmatrix} I_n & A \\ B & I_n \end{bmatrix}.
$$

5. Let A be an $n \times n$ matrix such that $A^n = \alpha A$ where α is a real number different from 1 and -1. Prove that the matrix $A + I_n$ is invertible.

6. If A and B are different matrices satisfying $A^3 = B^3$ and $A^2B = B^2A$, find $\det(A^2 + B^2)$.

7. Prove that if A is an $n \times n$ matrix with real entries, then

$$
\det(A^2 + I_n) \geq 0.
$$

8. Show that if A and B are $n \times n$ matrices with real entries and $AB = 0_n$, then $\det(I_n + A^{2p} + B^{2q}) \geq 0$ for any positive integers p and q.

9. Let A, B, C be $n \times n$ real matrices that are pairwise commutative and $ABC = 0_n$. Prove that

$$\det(A^3 + B^3 + C^3)\det(A + B + C) \geq 0.$$

10. Let p and q be real numbers such that $x^2 + px + q \neq 0$ for every real number x. Prove that if n is an odd positive integer, then

$$X^2 + pX + qI_n \neq 0_n$$

for all real matrices X of order $n \times n$.

11. Let A and B be two $n \times n$ matrices and let $C = AB - BA$. Show that if C commutes with both A and B, then there exists an integer m such that $C^m = 0_n$.

2.10   The Mean Value Theorem

Rolle's theorem states that if a function $f : [a,b] \to \mathbf{R}$ is continuous on $[a,b]$, differentiable on (a,b), and such that $f(a) = f(b)$, then there exists a point $c \in (a,b)$ with $f'(c) = 0$. This is a consequence of the fact that f has an extremum in the interior of the interval, and that the derivative vanishes at extremal points.

An important corollary of Rolle's theorem is the mean value theorem, due to Lagrange:

Let $f : [a,b] \to \mathbf{R}$ be a function that is continuous on $[a,b]$ and differentiable on (a,b). Then there exists $c \in (a,b)$ such that

$$\frac{f(b) - f(a)}{b - a} = f'(c).$$

A more general form of this theorem is due to Cauchy:

Let $f, g : [a,b] \to \mathbf{R}$ be two functions, continuous on $[a,b]$ and differentiable on (a,b). Then there exists $c \in (a,b)$ such that

$$(f(b) - f(a))g'(c) = (g(b) - g(a))f'(c).$$

This follows by applying Rolle's theorem to the function

$$H(x) = (f(b) - f(a))(g(x) - g(a)) - (g(b) - g(a))(f(x) - f(a)).$$

The mean value theorem is the special case where $g(x) = x$.

Let us apply the mean value theorem to solve a problem from the 1978 Romanian Mathematical Olympiad, proposed by S. Rădulescu.

It can be easily seen that $2^0 + 5^0 = 3^0 + 4^0$ and $2^1 + 5^1 = 3^1 + 4^1$. Are there other real numbers x such that

$$2^x + 5^x = 3^x + 4^x?$$

Rewrite the equation as

$$5^x - 4^x = 3^x - 2^x.$$

Now view x as a constant, and the numbers $2, 3, 4, 5$ as points where we evaluate the function $f(t) = t^x$. On the intervals $[2,3]$ and $[4,5]$, this function satisfies the hypothesis of the mean value theorem; hence there exist $t_1 \in (2,3)$ and $t_2 \in (4,5)$ with $xt_1^{x-1} = 5^x - 4^x$ and $xt_2^{x-1} = 3^x - 2^x$ (note that the derivative of f is $f'(t) = xt^{x-1}$).

It follows that $t_1^{x-1} = t_2^{x-1}$, and hence $(t_1/t_2)^{x-1} = 1$. The numbers t_1 and t_2 are distinct, since they lie in disjoint intervals, so this equality cannot hold. Thus there are no other real numbers x besides 0 and 1 that satisfy this equation.

1. Prove that if the real numbers $a_0, a_1, \ldots, a_n$ satisfy

$$\sum_{i=0}^{n} \frac{a_i}{i+1} = 0,$$

then the equation $a_n x^n + a_{n-1} x^{n-1} + \cdots + a_1 x + a_0 = 0$ has at least one real root.

2. One can easily check that $2^2 + 9^2 = 6^2 + 7^2$ and $1^3 + 12^3 = 9^3 + 10^3$. But do there exist distinct pairs (x,y) and (u,v) of positive integers such that the equalities

$$x^2 + y^2 = u^2 + v^2$$
$$x^3 + y^3 = u^3 + v^3$$

hold simultaneously?

3. Let p be a prime number, and let a, b, c, d be distinct positive integers such that $a^p + b^p = c^p + d^p$. Prove that

$$|a - c| + |b - d| \geq p.$$

4. Let a, b be two positive numbers, and let $f : [a,b] \to \mathbf{R}$ be a continuous function, differentiable on (a,b). Prove that there exists $c \in (a,b)$ such that

$$\frac{1}{a-b}(af(b) - bf(a)) = f(c) - cf'(c).$$

5. Let $f : [a,b] \to \mathbf{R}$ a continuous positive function, differentiable on (a,b). Prove that there exists $c \in (a,b)$ such that

$$\frac{f(b)}{f(a)} = e^{(b-a)\frac{f'(c)}{f(c)}}.$$

6. Let $f, g : [a,b] \to \mathbf{R}$ be two continuous functions, differentiable on (a,b). Assume in addition that g and g' are nowhere zero on (a,b) and that $f(a)/g(a) = f(b)/g(b)$. Prove that there exists $c \in (a,b)$ such that

$$\frac{f(c)}{g(c)} = \frac{f'(c)}{g'(c)}.$$

7. Let $f : [a,b] \to \mathbf{R}$ be a continuous function, differentiable on (a,b) and nowhere zero on (a,b). Prove that there exists $\theta \in (a,b)$ such that

$$\frac{f'(\theta)}{f(\theta)} = \frac{1}{a-\theta} + \frac{1}{b-\theta}.$$

8. Let $f : \mathbf{R} \to \mathbf{R}$ be a differentiable function. Assume that $\lim_{x \to \infty} f(x) = a$ for some real number a and that $\lim_{x \to \infty} x f'(x)$ exists. Evaluate this limit.

9. Compute the limit

$$\lim_{n \to \infty} \sqrt{n} \left[\left(1 + \frac{1}{n+1} \right)^{n+1} - \left(1 + \frac{1}{n} \right)^{n} \right].$$

10. Let $f : [0,1] \to \mathbf{R}$ be a continuous function, differentiable on $(0,1)$, with the property that there exists $a \in (0,1]$ such that $\int_0^a f(x)dx = 0$. Prove that

$$\left| \int_0^1 f(x)dx \right| \le \frac{1-a}{2} \sup_{x \in (0,1)} |f'(x)|.$$

Can equality hold?

11. Suppose that $f : [0,1] \to \mathbf{R}$ has a continuous derivative and that $\int_0^1 f(x)\,dx = 0$. Prove that for every $\alpha \in (0,1)$,

$$\left| \int_0^\alpha f(x)dx \right| \le \frac{1}{8} \max_{0 \le x \le 1} |f'(x)|.$$

12. The function $f : \mathbf{R} \to \mathbf{R}$ is differentiable and

$$f(x) = f\left(\frac{x}{2} \right) + \frac{x}{2} f'(x)$$

for every real number x. Prove that f is a linear function, that is, $f(x) = ax + b$ for some $a, b \in \mathbf{R}$.

Chapter 3

Number Theory and Combinatorics

T. Andreescu and R. Gelca, *Mathematical Olympiad Challenges*, DOI: 10.1007/978-0-8176-4611-0_3,   69
© Birkhäuser Boston, a part of Springer Science+Business Media, LLC 2009

3.1   Arrange in Order

This section is about a problem-solving technique that although simple, can be very powerful. As the title says, the idea is to arrange some objects in increasing or decreasing order. Here is an example.

Given 7 distinct positive integers that add up to 100, *prove that some three of them add up to at least* 50.

For the proof let $a < b < c < d < e < f < g$ be these numbers. We will show that $e + f + g \geq 50$. If $e > 15$, this is straightforward, since $e + f + g \geq 16 + 17 + 18 = 51$. If $e \leq 15$, then $a + b + c + d \leq 14 + 13 + 12 + 11 = 50$; hence $e + f + g = 100 - a - b - c - d \geq 50$.

The second problem comes from combinatorial geometry.

Given $2n + 2$ *points in the plane, no three collinear, prove that two of them determine a line that separates n of the points from the other n.*

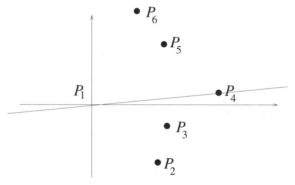

Figure 3.1.1

Imagine the points lying on a map, and choose the westmost point, say P_1, as one of the two that will determine the line (there are at most two westmost points, choose any of them). Place a Cartesian system of coordinates with the origin at P_1, the x-axis in the direction west–east, and the y-axis in the direction south–north. Order the rest of the points in an increasing sequence $P_2, P_3, \ldots, P_{2n+2}$ with respect to the oriented angles that $P_1 P_i$ form with the x-axis (see Figure 3.1.1). This is possible because no three points are collinear and the angles are between $-90°$ and $90°$. If we choose $P_1 P_{n+2}$ to be the line, then $P_2, P_3, \ldots, P_{n+1}$ lie inside the angle formed by $P_1 P_{n+2}$ and the negative half of the y-axis, and $P_{n+3}, P_{n+4}, \ldots, P_{2n+2}$ lie inside the angle formed by $P_1 P_{n+2}$ and the positive half of the y-axis, so the two sets of points are separated by the line $P_1 P_{n+2}$, which shows that P_1 and P_{n+2} have the desired property.

The following problems are left to the reader.

1. Prove that the digits of any six-digit number can be permuted in such a way that the sum of the first three digits of the new number differs by at most 9 from the sum of the remaining digits.

2. The unit cube

$$C = \{(x,y,z) \mid 0 \le x,y,z \le 1\}$$

is cut along the planes $x = y$, $y = z$, and $z = x$. How many pieces are there?

3. Show that if $2n + 1$ real numbers have the property that the sum of any n is less than the sum of the remaining $n + 1$, then all these numbers are positive.

4. Consider seven distinct positive integers not exceeding 1706. Prove that there are three of them, say a, b, c, such that $a < b + c < 4a$.

5. Let a be the least and A the largest of n distinct positive integers. Prove that the least common multiple of these numbers is greater than or equal to na and that the greatest common divisor is less than or equal to A/n.

6. Consider $2n$ distinct positive integers a_1, a_2, ..., a_{2n} not exceeding n^2 ($n > 2$). Prove that some three of the differences $a_i - a_j$ are equal.

7. Given $2n + 3$ points in the plane, no three collinear and no four on a circle, prove that there exists a circle containing three of the points such that exactly n of the remaining points are in its interior.

8. Given $4n$ points in the plane, no three collinear, show that one can form n non-intersecting (not necessarily convex) quadrilateral surfaces with vertices at these points.

9. Given 69 distinct positive integers not exceeding 100, prove that one can choose four of them a, b, c, d such that $a < b < c$ and $a + b + c = d$. Is this statement true for 68 numbers?

10. Prove that from any 25 distinct positive numbers, one can choose two whose sum and difference do not coincide with any of the remaining 23.

11. In a 10×10 table are written the integers from 1 to 100. From each row we select the third largest number. Show that the sum of these numbers is not less than the sum of the numbers in some row.

12. Given n positive integers, consider all possible sums formed by one or more of them. Prove that these sums can be divided into n groups such that in each group the ratio of the largest to the smallest does not exceed 2.

3.2 Squares and Cubes

In this section, we have selected problems about perfect squares and cubes. They are to be solved mainly through algebraic manipulations. The first problem was given at the 29th International Mathematical Olympiad in 1988, proposed by Germany.

Show that if a,b are integers such that $\frac{a^2+b^2}{1+ab}$ is also an integer, then $\frac{a^2+b^2}{1+ab}$ is a perfect square.

Set $q = \frac{a^2+b^2}{1+ab}$. We want to show that among all pairs of nonnegative integers (α, β) with the property that $\frac{\alpha^2+\beta^2}{1+\alpha\beta} = q$ and $\alpha \geq \beta$, the one with $\alpha + \beta$ minimal has $\beta = 0$. Then we would have $q = \alpha^2$, a perfect square.

Thus let (α, β), $\alpha \geq \beta \geq 0$, be a pair that minimizes $\alpha + \beta$, and suppose that $\beta > 0$. The relation $q = \frac{x^2+\beta^2}{1+x\beta}$ is equivalent to the quadratic equation in x, $x^2 - \beta q x + \beta^2 - q = 0$. This has one root equal to α, and since the roots add up to βq, the other root is equal to $\beta q - \alpha$. Let us show that $0 \leq \beta q - \alpha < \alpha$, which will contradict the minimality of (α, β).

We have

$$\beta q - \alpha + 1 = \frac{\beta^3 - \alpha + \alpha\beta + 1}{1 + \alpha\beta} > 0,$$

and hence $\beta q - \alpha \geq 0$. Also, from

$$q = \frac{\alpha^2 + \beta^2}{1 + \alpha\beta} < \frac{2\alpha^2}{\alpha\beta} = \frac{2\alpha}{\beta}$$

we deduce $\beta q - \alpha < \alpha$. Thus the minimal pair is of the form $(\alpha, 0)$, therefore $q = \alpha^2$.

Here is an example involving squares and cubes, published in *Revista Matematică din Timişoara (Timişoara's Mathematics Gazette)*.

Prove that for every positive integer m, there is a positive integer n such that $m + n + 1$ is a perfect square and $mn + 1$ is a perfect cube.

The solution is straightforward. If we choose $n = m^2 + 3m + 3$, then $m + n + 1 = (m+2)^2$ and $mn + 1 = (m+1)^3$.

1. Find all triples of positive integers (x, y, z) for which

$$x^2 + y^2 + z^2 + 2xy + 2x(z-1) + 2y(z+1)$$

    is a perfect square.

2. Show that each of the numbers

$$\frac{107811}{3}, \quad \frac{110778111}{3}, \quad \frac{111077781111}{3}, \dots$$

    is the cube of a positive integer.

3. Prove that the product of four consecutive positive integers cannot be a perfect square.

4. Which are there more of among the natural numbers from 1 to 1,000,000, inclusive: numbers that can be represented as the sum of a perfect square and a (positive) perfect cube, or numbers that cannot be?

5. Prove that the equation $x^2 + y^2 + 1 = z^2$ has infinitely many integer solutions.

6. Let a and b be integers such that there exist consecutive integers c and d for which $a - b = a^2c - b^2d$. Prove that $|a - b|$ is a perfect square.

7. Let $k_1 < k_2 < k_3 < \cdots$ be positive integers, no two consecutive, and let $s_m = k_1 + k_2 + \cdots + k_m$ for $m = 1, 2, 3, \ldots$. Prove that for each positive integer n, the interval $[s_n, s_{n+1})$ contains at least one perfect square.

8. Let $\{a_n\}_{n \geq 0}$ be a sequence such that $a_0 = a_1 = 5$ and

$$a_n = \frac{a_{n-1} + a_{n+1}}{98}$$

for all positive integers n. Prove that

$$\frac{a_n + 1}{6}$$

is a perfect square for all nonnegative integers n.

9. Prove that for any integers a, b, c, there exists a positive integer n such that $\sqrt{n^3 + an^2 + bn + c}$ is not an integer.

10. Prove that there is a perfect cube between n and $3n$ for any integer $n \geq 10$.

11. Consider the operation of taking a positive integer and writing it to its left. For example, starting with $n = 137$, we get 137137. Does there exist a positive integer for which this operation gives a perfect square?

12. Determine all pairs of positive integers (a, b) such that

$$\frac{a^2}{2ab^2 - b^3 + 1}$$

is a positive integer.

13. Find the maximal value of $m^2 + n^2$ if m and n are integers between 1 and 1981 satisfying $(n^2 - mn - m^2)^2 = 1$.

14. Prove that if n is a positive integer such that the equation

$$x^3 - 3xy^2 + y^3 = n$$

has a solution in integers (x, y), then it has at least three such solutions. Show that the equation has no integer solutions when $n = 2891$.

15. (a) Prove that if there exists a triple of positive integers (x, y, z) such that

$$x^2 + y^2 + 1 = xyz,$$

then $z = 3$.
(b) Find all such triples.

3.3   Repunits

We call repunits the numbers that contain only the digit 1 in their writing, namely numbers of the form $11\ldots1$.

Here is a modified version of a problem from the 2005 Bulgarian Mathematical Olympiad.

Prove that the equation

$$x^2 + 2y^2 + 98z^2 = \underbrace{111\cdots1}_{666\ times}$$

does not have integer solutions.

First solution: Assume to the contrary that an integer solution (x,y,z) does exist. The repunit on the right can be factored as

$$111111 \cdot (1 + 10^6 + \cdots + 10^{6\cdot110}) = \frac{10^6 - 1}{9} \cdot (1 + 10^6 + \cdots + 10^{6\cdot110}).$$

By Fermat's little theorem $10^6 - 1$ is divisible by 7, so it is natural to take residues modulo 7. As the residues of a square modulo are $0, 1, 2$, and 4, and as 98 is divisible by 7, we see that the only possibility is that both x and y are divisible by 7. In this case, the left-hand side is divisible by 7^2. But the repunit on the right equals $7 \cdot 15873 \cdot (1 + 10^6 + \cdots + 10^{6\cdot110})$, whose second factor is 4 modulo 7, while its third factor is, by the same Fermat's little theorem, a sum of 111 ones, which is not divisible by 7. Hence the right-hand side is not divisible by 7^2, a contradiction. Therefore the solution does not exist.

Second solution: Because 1000 is divisible by 8, the size of the number on the right is greatly reduced when working modulo 8. The equation becomes

$$x^2 + 2y^2 + 2z^2 \equiv 111 (\mathrm{mod}\ 8),$$

or

$$x^2 + 2y^2 + 2z^2 \equiv 7 (\mathrm{mod}\ 8).$$

The residue of a square modulo 8 can be $0, 1$, or 4, and the residue of the double of a square can be 0 or 2. A case check shows that the residue class 7 cannot be obtained by adding a $0, 1$, or 4 to a 0 or 2 and another 0 or 2. The conclusion follows.

The problem below appeared in the Russian journal *Quantum*.

Find all quadratic polynomials with integer coefficients that transform repunits into repunits.

Suppose $f(x) = ax^2 + bx + c$ is a quadratic polynomial. The condition desired is that for every m, there is some n with $f((10^m - 1)/9) = (10^n - 1)/9$. Let

$$g(x) = 9f\left(\frac{x-1}{9}\right) + 1 = \frac{a}{9}x^2 + \left(b - \frac{2a}{9}\right)x + \left(9c + 1 - b + \frac{a}{9}\right).$$

Then the equivalent condition for g is that for every m, there is an n with $g(10^m) = 10^n$, i.e., g transforms powers of 10 to powers of 10. This condition translates into

$$10^{-2m}g(10^m) = 10^{n-2m} = \frac{a}{9} + \left(b - \frac{2a}{9}\right)10^{-m} + \left(9c + 1 - b + \frac{a}{9}\right)10^{-2m}.$$

The first equality shows that $10^{-2m}g(10^m)$ is a power of 10. Letting m tend to infinity in the second equality, we see that $10^{-2m}g(10^m)$ converges to $a/9$, hence $a/9 = 10^k$ for some k and $10^{-2m}g(10^m) = 10^k$ for all sufficiently large m. Hence

$$\left(b - \frac{2a}{9}\right)10^{-m} + \left(9c + 1 - b + \frac{a}{9}\right)10^{-2m} = 0$$

for all sufficiently large m, and we see $b - \frac{2a}{9} = 9c + 1 - b + \frac{a}{9} = 0$. It follows that $b = 2 \cdot 10^k$ and $c = (10^k - 1)/9$.

Conversely, for these polynomials we have $g(x) = 10^k x^2$, which clearly transforms powers of 10 into powers of 10.

Here are more problems.

1. (a) Prove that any repunit in base 5 with an even number of digits is the product of two consecutive positive integers.

   (b) Prove that any repunit in base 9 is a triangular number.

2. Prove that

$$\underbrace{11\ldots1}_{2n \text{ times}} = \underbrace{22\ldots2}_{n \text{ times}} + (\underbrace{333\ldots3}_{n \text{ times}})^2.$$

3. Find the least repunit divisible by 19.

4. Show that the number of digits of a prime repunit is prime. Is the converse true?

5. Is the number

$$\underbrace{111\ldots1}_{81 \text{ times}}$$

   divisible by 81?

6. (a) Prove that $\underbrace{11\ldots1}_{n \text{ times}}$ is divisible by 41 if and only if n is divisible by 5.

   (b) Prove that $\underbrace{11\ldots1}_{n \text{ times}}$ is divisible by 91 if and only if n is divisible by 6.

7. Prove that a repunit greater than 1 cannot be the square of any integer.

8. Show that for any number n that ends in 1, 3, 7, or 9, there exists a repunit that is divisible by n.

9. Prove that there exists an infinite sequence of repunits with any two terms coprime.

10. Prove that for infinitely many n, there exists an n-digit number, not containing 0, divisible by the sum of its digits.

11. Find the n-digit approximation of the number $\sqrt{11\ldots1}$, provided that the repunit has $2n$ digits.

12. Find all polynomials with real coefficients that map repunits to repunits.

3.4   Digits of Numbers

Encoding data in the digits of numbers is the idea that governs the world of computers. There are some mathematical problems that can be solved by extracting information out of the digits of the expansion of numbers in a certain base. The first example is solved by looking at the binary expansion of numbers.

Let $f : \mathbf{N} \to \mathbf{N}$ be a function satisfying $f(1) = 1$, $f(2n) = f(n)$, and $f(2n+1) = f(2n) + 1$, for all positive integers n. Find the maximum of $f(n)$ when $1 \leq n \leq 1994$.

Let us find an explicit formula for f. Since the recurrence relation involves multiplications by 2 and additions of 1, a good idea is to write numbers in base 2. Then multiplication by $2 = 10_2$ adds a zero at the end of the number. We compute $f(10_2) = 1$, $f(11_2) = 2$, $f(110_2) = 2$. Let us show by induction that $f(n)$ is equal to the number of 1's in the binary expansion of n. If n is even, that is, $n = 10_2 \cdot m$, then $f(m) = f(10_2 \cdot m)$ by definition, and since m and $10_2 \cdot m$ have the same number of 1's in their binary expansions, the conclusion follows from the induction hypothesis. If n is odd, $n = 10_2 \cdot m + 1$, then $f(n) = f(m) + 1$, and since n has one more 1 in its binary expansion than m, the property follows again from the induction hypothesis.

The problem reduces to finding the largest number of 1's that can appear in the binary expansions of numbers less than 1994. Since $1994 < 2^{11} - 1$, $f(n)$ can be at most 10. We have $f(1023) = f(1111111111_2) = 10$; hence the required maximum is 10.

The second problem appeared at the 1995 Chinese Mathematical Olympiad. Its solution naturally leads to the consideration of both binary and ternary expansions of numbers.

Suppose that $f : \mathbf{N} \to \mathbf{N}$ satisfies $f(1) = 1$ and for all n,
(a) $3f(n)f(2n+1) = f(2n)(1 + 3f(n))$,
(b) $f(2n) < 6f(n)$.
Find all solutions to the equation $f(k) + f(m) = 293$.

As before, we want to find an explicit form for f. Since $3f(n)$ and $3f(n) + 1$ are relatively prime, condition (a) implies that $f(2n)$ is divisible by $3f(n)$. This combined with (b) gives that $f(2n) = 3f(n)$. Using (a) again, we have $f(2n+1) = 3f(n) + 1$.

The two relations we obtained suggest that f acts in the following way. For a number n written in base 2, $f(n)$ has the same digits but is read in base 3. For example, $f(7) = f(111_2) = 111_3 = 13$. That this is so can be proved as before by induction.

Since $293 = 101212_3$, the number of solutions to the equation

$$f(k) + f(m) = 293$$

is the same as the number of ways of writing 101212_3 as a sum of two numbers whose ternary expansions contain only 0's and 1's. Note that when two such numbers are added there is no digit transfer, so the numbers must both contain a 0 in the second position and a 1 in the fourth and sixth. In the first, third, and fifth positions, one of the numbers must have a 0 and the other a 1. Altogether there are eight solutions.

We conclude the introduction with a theorem of W. Sierpiński from real analysis. We recall that a function $f : \mathbf{R} \to \mathbf{R}$ is said to have the intermediate value property if for any $a, b \in \mathbf{R}$ and any λ between $f(a)$ and $f(b)$, there exists c between a and b with $f(c) = \lambda$. Sierpiński's result states the following.

Every function $f : \mathbf{R} \to \mathbf{R}$ can be written as the sum of two functions satisfying the intermediate value property.

The proof is constructive. The idea is to read from the decimals of numbers information used in the construction. For simplicity, we will do the proof for functions defined on the open unit interval. The case of the function defined on the whole real line follows by composing with a continuous bijection from the interval to the line.

Let $x \in (0, 1)$. We want to define the values of the functions $g(x)$ and $h(x)$ that add up to $f(x)$, based on the decimal expansion $x = 0.x_1 x_2 x_3 \ldots$. To this end, we define the sequences $a_n^x = x_{2n-1}$, $n \geq 1$, and $b_n^x = x_{2n}$, $n \geq 1$. Let

$$A = \{x, \text{ there exists } n_x \text{ such that } a_n^x = 0 \text{ for } n \geq n_x\};$$
$$B = \{x, \text{ there exists } n_x \text{ such that } a_n^x = 2 \text{ for } n \geq n_x\}.$$

Note that A and B do not intersect. Now we use the sequence $\{a_n^x\}_n$ to determine the sign of $g(x)$ and $h(x)$, the number of digits of their integer parts, and the truncation of the sequence $\{b_n^x\}_n$, from which we read all the digits of $g(x)$ and $h(x)$. Define $g : A \to \mathbf{R}$ by $g(x) = (-1)^{a_{n-1}^x} b_n^x b_{n+1}^x \ldots b_m^x . b_{m+1}^x b_{m+2}^x \ldots$ if $a_{n-1}^x \neq 1$, $a_n^x = \cdots = a_m^x = 1$, and $a_k^x = 0$ for $k > m$. Also define $h : B \to \mathbf{R}$, by $h(x) = (-1)^{a_{n-1}^x} b_n^x b_{n+1}^x \ldots b_m^x . b_{m+1}^x b_{m+2}^x \cdots$ if $a_{n-1}^x \neq 1$, $a_n^x = \cdots = a_m^x = 1$, and $a_k^x = 2$ for $k > m$. On B let $g(x) = f(x) - h(x)$ and on $\mathbf{R} - (A \cup B)$ let $g(x) = 0$. Finally, on $\mathbf{R} - B$ let $h(x) = f(x) - g(x)$. It follows that $f(x) = g(x) + h(x)$, for all x.

By examining the definitions of g and h, we see that what really matters in computing their values is how the decimal expansion of x looks at its far end. To be more explicit, for any two numbers $y, z \in (0, 1)$, one can find x between y and z such that for some integers $n \leq m$, $a_{n-1}^x \neq 1$, $a_n^x = \cdots = a_m^x = 1$, and a_k^x equal to 0 (or 2) for $k > m$, and such that b_k^x for $k \geq n$ are any digits we desire. It follows that the restrictions of g and h to any interval are surjective. Therefore, g and h satisfy trivially the intermediate value property.

The problems below can be solved by looking at the expansion in an appropriately chosen base.

1. Determine $f : \mathbf{N} \to \mathbf{R}$ such that $f(1) = 1$ and

$$f(n) = \begin{cases} 1 + f\left(\frac{n-1}{2}\right), & n \text{ odd}, \\ 1 + f\left(\frac{n}{2}\right), & n \text{ even}. \end{cases}$$

2. Prove that for any natural number n,

$$\left\lfloor \frac{n+2^0}{2^1} \right\rfloor + \left\lfloor \frac{n+2^1}{2^2} \right\rfloor + \left\lfloor \frac{n+2^2}{2^3} \right\rfloor + \cdots = n.$$

3. Let $\{x_n\}_n$ be a sequence defined recursively by $x_0 = 0$, $x_1 = 1$, and $x_{n+1} = (x_n + x_{n-1})/2$. Prove that the sequence is convergent and find its limit.

4. The sequence $\{a_n\}_n$ is defined by $a_0 = 0$ and

$$a_n = a_{\lfloor \frac{n}{2} \rfloor} + \left\lfloor \frac{n}{2} \right\rfloor, \quad \text{for } n \geq 1.$$

Prove that the sequence $\left\{ \frac{a_n}{n} \right\}_n$ converges and find its limit.

5. A biologist watches a chameleon. The chameleon catches flies and rests after each catch. The biologist notices that:
   (i) The first fly is caught after a resting period of 1 minute.
   (ii) The resting period before the $2m$th fly is caught is the same as the resting period before catching the mth fly and 1 minute shorter than the resting period before the $(2m+1)$st fly.
   (iii) When the chameleon stops resting, he catches a fly instantly.
   (a) How many flies were caught by the chameleon before his first resting period of 9 consecutive minutes?
   (b) After how many minutes will the chameleon catch his 98th fly?
   (c) How many flies have been caught by the chameleon after 1999 minutes have passed?

6. For a natural number k, let $p(k)$ denote the least prime number that does not divide k. If $p(k) > 2$, define $q(k)$ to be the product of all primes less than $p(k)$; otherwise let $q(k) = 1$. Consider the sequence

$$x_0 = 1, \quad x_{n+1} = \frac{x_n p(x_n)}{q(x_n)} \quad n = 0, 1, 2, \ldots.$$

Determine all natural numbers n such that $x_n = 111111$.

7. Let $f : \mathbf{N} \to \mathbf{N}$ be defined by $f(1) = 1$, $f(3) = 3$, and, for all $n \in \mathbf{N}$,
   (a) $f(2n) = f(n)$,
   (b) $f(4n+1) = 2f(2n+1) - f(n)$,
   (c) $f(4n+3) = 3f(2n+1) - 2f(n)$.
   Find the number of $n \leq 1988$ for which $f(n) = n$.

8. Let the sequence a_n, $n = 1, 2, 3, \ldots$, be generated as follows: $a_1 = 0$, and for $n \geq 1$,

$$a_n = a_{\lfloor n/2 \rfloor} + (-1)^{n(n+1)/2}.$$

(a) Determine the maximum and the minimum value of a_n over $n \leq 1996$, and find all $n \leq 1996$ for which these extreme values are attained.

(b) How many terms a_n, $n \leq 1996$, are equal to 0?

9. Prove that there exists a continuous function from the interval $[0, 1]$ onto the square $[0, 1] \times [0, 1]$.

10. Find all increasing functions $f : \mathbf{N} \to \mathbf{N}$ with the property that $f(f(n)) = 3n$ for all n.

11. Prove that there exist functions $f : [0, 1] \to [0, 1]$ satisfying the intermediate value property such that the equation $f(x) = x$ has no solution.

12. Determine whether there exists a set X of integers with the property that for any integer n, there is exactly one solution of $a + 2b = n$ with $a, b \in X$.

13. Let $p > 2$ be a prime. The sequence $\{a_n\}_n$ is defined by $a_0 = 0$, $a_1 = 1$, $a_2 = 2$, $\ldots$, $a_{p-2} = p - 2$, and for all $n \geq p - 1$, a_n is the least integer greater than a_{n-1} that does not form an arithmetic sequence of length p with any of the preceding terms. Prove that for all n, a_n is the number obtained by writing n in base $p - 1$ and reading the result in base p.

3.5   Residues

Below we have listed problems that can be solved by looking at residues. We have avoided the usual reductions modulo 2, 3, or 4 and consider more subtle situations. Before we proceed with the examples, let us mention two results that will be needed in some of the solutions.

Fermat's little theorem. Let p be a prime number and n a positive integer. Then

$$n^p - n \equiv 0 (\mathrm{mod}\ p).$$

Euler's theorem. Let $n > 1$ be an integer and a an integer coprime with n. Then

$$a^{\phi(n)} \equiv 1 (\mathrm{mod}\ n),$$

where $\phi(n)$ is the number of positive integers smaller than n and coprime to n.

We begin with a problem that appeared in *Kvant (Quantum)*, proposed by S. Konyagin.

Denote by $S(m)$ the sum of the digits of the positive integer m. Prove that there does not exist a number N such that $S(2^n) < S(2^{n+1})$ for all $n \geq N$.

The idea is to consider residues modulo 9. This is motivated by the fact that a number is congruent modulo 9 to the sum of its digits.

Thus let us assume that there exists a number N such that $S(2^n)$ is increasing for $n \geq N$. The residues of $S(2^n)$ modulo 9 repeat periodically and are $1,2,4,8,7,5$. If $S(2^n)$ is increasing, then in an interval of length 6 it increases at least by $(2-1) + (4-2) + (8-4) + (7-8+9) + (5-7+9) + (1-5+9) = 27$. Here we have used the fact that if a residue is less than the previous one, we must add a multiple of 9, since the difference between any two consecutive sums is assumed to be positive. Therefore, $S(2^{n+6}) \geq S(2^n) + 27$ for $n \geq N$.

Also, 2^n has $\lfloor n \log 2 \rfloor + 1$ digits, and hence $S(2^n) \leq 9(n \log 2 + 1) \leq 3n + 9$. It follows that for $k \geq 0$,

$$S(2^N) + 27k \leq S(2^{N+6k}) \leq 3(N + 6k) + 9.$$

This implies that

$$9k \leq 3N - S(2^N) + 9$$

for all $k \in \mathbf{N}$, which is a contradiction, since the right side is bounded. Thus $S(2^n)$ does not eventually become increasing.

To illustrate the use of the criterion for divisibility by 11, we use a short listed problem from the 44th International Mathematical Olympiad, 2003.

Let $N(k)$ be the number of integers n, $0 \leq n \leq 10^k$ whose digits can be permuted in such a way that they yield an integer divisible by 11. Prove that $N(2m) = 10N(2m-1)$ for every positive integer m.

As mentioned, the solution uses the fact that a number whose base 10 representation is $n = a_{k-1}a_{k-2} \ldots a_0$ is divisible by 11 if and only if the alternating sum of its digits $a_0 - a_1 + a_2 - \cdots \pm a_{k-1}$ is divisible by 11. Thus the problem is about k-tuples $(a_0, a_1, \cdots, a_{k-1})$, with $a_i \in \{0, 1, \ldots, 9\}$ that can be permuted such that the alternated sum becomes divisible by 11. Let $A(k)$ be the set of such k-tuples. We are to show that $A(2m)$ has 10 times more elements than does $A(2m-1)$.

Note that if $(a_0, a_1, \ldots, a_{2m-1})$ is a $2m$-tuple having a permutation with the alternated sum divisible by 11, then for any positive integer s, both $(a_0 + s, a_1 + s, \ldots, a_{2m-1} + s)$ and $(sa_0, sa_1, \ldots, sa_{2m-1})$ have the same property.

Thus let us consider $(j, a_1, a_2, \ldots, a_{2m-1}) \in A(2m)$. Let also l be the unique integer in $\{1, 2, \ldots, 10\}$ such that $(j+1)l \equiv 1 \pmod{11}$. The $2m-1$-tuple $((j+1)\,l-1, (a_1+1)l-1, (a_2+1)l-1, \ldots (a_{2m-1}+1)l-1)$ has a permutation whose alternated sum is divisible by 11, therefore $(0, (a_1+1)l-1, (a_2+1)l-1, \ldots (a_{2m-1}+1)l-1)$ has the same property. Thus if we let b_i be the residue of $(a_i+1)l-1$ modulo 11, then $(b_1, b_2, \ldots, b_{2m-1})$ is in $A(2m-1)$. Note that $(a_i+1)l-1$ is not congruent to 10 modulo 11, therefore the b_i's are indeed digits.

Conversely, for each $j \in \{0, 1, \ldots, 9\}$ and $(b_1, b_2, \ldots, b_{2m-1})$ in $A(2m-1)$, one can reconstruct $(j, a_1, a_2, \ldots, a_{2m-1})$ by simply letting a_i be the remainder of $(b_i + 1)(j+1) - 1$ modulo 11. Thus each element in $A(2m-1)$ is associated to a 10-element subset of $A(2m)$. It follows that $A(2m)$ has 10 times more elements than does $A(2m-1)$, as desired.

We continue with an example that is solved working modulo 31.

The numbers 31, 331, 3331 *are prime, but there exist numbers of this form that are composite. Prove that indeed, there exist infinitely many numbers of this form that are composite.*

Denote by a_n the number $33\ldots 31$ with n occurrences of the digit 3. Then $3a_n + 7 = 10^{n+1}$. From Fermat's little theorem, $10^{30} \equiv 1 \pmod{31}$, so $10^{30k+2} \equiv 10^2 \pmod{31}$, for all $k \geq 0$. Hence

$$a_{30k+1} = \frac{10^{30k+2} - 7}{3} \equiv \frac{10^2 - 7}{3} \equiv 0 \pmod{31}.$$

We conclude that all numbers of the form a_{30k+1} are divisible by 31, and being greater than 31, they are composite.

In the next problem we denote by $\{x\}$ the fractional part of x, namely $\{x\} = x - \lfloor x \rfloor$.

Let p and q be relatively prime positive integers. Evaluate the sum

$$\left\{ \frac{p}{q} \right\} + \left\{ \frac{2p}{q} \right\} + \cdots + \left\{ \frac{(q-1)p}{q} \right\}.$$

The key idea is to look at the residues of the numerators modulo q. This is so, since if a and b are positive integers and r is the residue of a modulo b, then $\{a/b\} = r/b$. Indeed, if $a = bc + r$, then $a/b = c + r/b$, and since $r/b \in [0,1)$, it must be equal to the fractional part of a/b.

Let $r_1, r_2, \ldots, r_{q-1}$ be the residues of the numerators modulo q. Since p and q are relatively prime, these residues represent a permutation of the numbers $1, 2, \ldots, q-1$. Hence

$$\frac{r_1}{q} + \frac{r_2}{q} + \cdots + \frac{r_{q-1}}{q} = \frac{1}{q} + \frac{2}{q} + \cdots + \frac{q-1}{q} = \frac{q-1}{2}.$$

As a corollary, we obtain the the well-known reciprocity law: if p and q are relatively prime positive integers, then

$$\left\lfloor \frac{p}{q} \right\rfloor + \left\lfloor \frac{2p}{q} \right\rfloor + \cdots + \left\lfloor \frac{(q-1)p}{q} \right\rfloor = \frac{(p-1)(q-1)}{2}.$$

It is called a reciprocity law because the value of the expression on the left does not change when we switch p and q.

A list of interesting problems follows.

1. Let p be a prime number and w, n integers such that $2^p + 3^p = w^n$. Prove that $n = 1$.

2. Show that if for the positive integers m, n one has $\sqrt{7} - m/n > 0$, then

$$\sqrt{7} - \frac{m}{n} > \frac{1}{mn}.$$

3. Let n be an integer. Prove that if $2\sqrt{28n^2+1}+2$ is an integer, then it is a perfect square.

4. Prove that the system of equations

$$x^2 + 6y^2 = z^2,$$
$$6x^2 + y^2 = t^2,$$

has no nontrivial integer solutions.

5. Determine all possible values for the sum of the digits of a perfect square.

6. Find all pairs of positive integers (x, y) that satisfy

$$x^2 - y! = 2001.$$

7. Does there exist a power of 2 such that after permuting its digits, we obtain another power of 2? (When permuting the digits, you are not allowed to bring zeros to the front of the number.)

8. Let A be the sum of the digits of the number 4444^{4444} and B the sum of the digits of A. Compute the sum of the digits of B.

9. Prove that the equation $y^2 = x^5 - 4$ has no integer solutions.

10. Prove that 19^{19} cannot be written as the sum of a perfect cube and a perfect fourth power.

11. Prove that there are no integers x and y for which

$$x^2 + 3xy - 2y^2 = 122.$$

12. Let n be an integer greater than 1.
    (a) Prove that $1! + 2! + 3! + \cdots + n!$ is a perfect power if and only if $n = 3$.
    (b) Prove that $(1!)^3 + (2!)^3 + \cdots + (n!)^3$ is a perfect power if and only if $n = 3$.

13. Find, with proof, the least positive integer n for which the sum of the digits of $29n$ is as small as possible.

14. Find the fifth digit from the end of the number

$$5^{5^{5^{5^{5}}}}.$$

15. The sequence $\{a_n\}_{n \geq 0}$ is defined as follows: a_0 is a positive rational number smaller than $\sqrt{1998}$, and if $a_n = p_n/q_n$ for some relatively prime integers p_n and q_n, then

$$a_{n+1} = \frac{p_n^2 + 5}{p_n q_n}.$$

Prove that $a_n < \sqrt{1998}$ for all n.

16. Let k and n be two coprime natural numbers such that $1 \leq k \leq n-1$, and let $M - \{1, 2, \ldots, n-1\}$. Every element of M is colored with one of two colors such that
    (a) i and $n-i$ have the same color, for all $i \in M$,
    (b) for $i \in M$ and $i \neq k$, i and $|k-i|$ have the same color.
    Prove that all elements of M are colored with the same color.

3.6 Diophantine Equations with the Unknowns as Exponents

The equations included in this section have in common the fact that the unknowns are positive integers that appear as exponents. In some of the more standard problems, one can consider residue classes modulo the base of one of the exponentials to derive properties of the exponents, such as their parity. These can then be used to factor expressions and use some algebraic trick.

Let us consider the following example.

Find the positive integer solutions to the equation

$$3^x + 4^y = 5^z.$$

First solution: Looking at the residues mod 4, we find that $3^x \equiv 1 \pmod 4$. This implies that $x = 2x_1$ for some integer x_1. Also, $5^z \equiv 1 \pmod 3$, and hence z is even, say $z = 2z_1$. We obtain

$$4^y = (5^{z_1} + 3^{x_1})(5^{z_1} - 3^{x_1}).$$

Hence $5^{z_1} + 3^{x_1} = 2^s$ and $5^{z_1} - 3^{x_1} = 2^t$, with $s > t$ and $s + t = 2y$. Solving for 5^{z_1} and 3^{x_1}, we get

$$5^{z_1} = 2^{t-1}(2^{s-t} + 1) \text{ and } 3^{x_1} = 2^{t-1}(2^{s-t} - 1).$$

Since the left side of both equalities is odd, t must be equal to 1. Denoting $s - t$ by u, we obtain the equation $3^{x_1} = 2^u - 1$. Let us solve it.

Taking everything mod 3, we see that u must be even, say $u = 2u_1$. Factoring the right side and repeating the argument yields

$$2^{u_1} + 1 = 3^\alpha \text{ and } 2^{u_1} - 1 = 3^\beta$$

for some α and β. But this gives $3^\alpha - 3^\beta = 2$, and hence $\alpha = 1$ and $\beta = 0$. Consequently, $u_1 = 1$, $u = 2$, and the unique solution is $x = y = z = 2$.

Second solution: There is a completely different approach to this problem. We start with the same observation that $x = 2x_1$ and $z = 2z_1$, with x_1 and z_1 integers, and then write the equation as

$$(3^{x_1})^2 + (2^y)^2 = (5^{z_1})^2.$$

It follows that the numbers $X = 3^{x_1}$, $Y = 2^y$, $Z = 5^{z_1}$ satisfy the Pythagorean equation

$$X^2 + Y^2 = Z^2.$$

Note that X, Y, Z are pairwise coprime and that Y is even. From the general theory of the Pythagorean equation, we know that there exist coprime positive integers u and v such that $X = u^2 - v^2$, $Y = 2uv$, $Z = u^2 + v^2$, that is,

$$3^{x_1} = u^2 - v^2, \quad 2^y = 2uv, \quad 5^{z_1} = u^2 + v^2.$$

From the second equality, we see that u and v are both powers of 2, and from the first equality, we find that $v = 1$ (because a power of 3 cannot be divisible by a power of 2). Therefore $u = 2^{y-1}$, and we obtain the system

$$3^{x_1} = 2^y - 1$$
$$5^{z_1} = 2^y + 1.$$

Taking the second equation modulo 4, we see that y is even, say $y = 2y_1$. But then

$$3^{x_1} = (2^{y_1} - 1)(2^{y_1} + 1),$$

and therefore $2^{y_1} - 1$ and $2^{y_1} + 1$ are both powers of 3. The only powers of 3 that lie this close to each other are 1 and 3. Thus $y_1 = 1$, and then $x_1 = z_1 = 1$, and finally $x = y = z = 2$, as desired.

And now a problem from a 2005 Romanian Team Selection Test for the International Mathematical Olympiad that has more to it than just the simple application of residues.

Find the positive integer solutions to the equation

$$3^x = 2^x y + 1.$$

Rewrite the equation as

$$3^x - 1 = 2^x y.$$

We can now infer that x cannot exceed the exponent of 2 in the prime number factorization of $3^x - 1$. Let us estimate how large this exponent is. Note that modulo 4 the number $3^n - 1$ is congruent to 2 if n is odd and to 0 if n is even. This observation shows that it is worth factoring x as $2^m(2n+1)$, with m and n nonnegative integers. We can then write

$$3^x - 1 = 3^{2^m(2n+1)} - 1 = (3^{2n+1})^{2^m} - 1$$
$$= (3^{2n+1} - 1)(3^{2n+1} + 1) \prod_{k=1}^{m-1} \left((3^{2n+1})^{2^k} + 1 \right).$$

Taken modulo 8, the first two factors are 2, respectively 4. Hence they contribute a factor of 2^3 to the product $3^x - 1$. Each of the other factors contributes a factor of 2.

Consequently, the exponent of 2 in $3^x - 1$ is $m+2$. We obtain the inequality $x \leq m+2$, which translates to

$$2^m(2n+1) \leq (m+2).$$

In particular, $2^m \leq m+2$, which restricts us to $m = 0, 1, 2$. An easy case check yields the solutions to the original equation $(x,y) = (1,1)$, $(x,y) = (2,2)$, $(x,y) = (4,5)$.

Similar arguments can be used to solve the following problems.

1. Find all positive integers x and y that satisfy the equation

$$|3^x - 2^y| = 1.$$

2. Find all positive integers x and y satisfying the equation

$$3^x - 2^y = 7.$$

3. Find all positive integers x, y, z, t, and n satisfying

$$n^x + n^y + n^z = n^t.$$

4. (a) Find all nonnegative integer solutions to the equation

$$3^x - y^3 = 1.$$

(b) Find all pairs of nonnegative integers x and y that solve the equation

$$p^x - y^p = 1,$$

where p is a given odd prime.

5. Find all positive integers n for which

$$1^n + 9^n + 10^n = 5^n + 6^n + 11^n.$$

6. Show that the equation

$$2^x - 1 = z^m$$

has no integer solutions if $m > 1$.

7. Find the positive integer solutions to the equation

$$5^x + 12^y = z^2.$$

8. Find the positive integers x, y, z satisfying

$$1 + 2^x 3^y = z^2.$$

9. Find all nonnegative integer solutions to the equation

$$5^x 7^y + 4 = 3^z.$$

10. Find all nonnegative integer solutions to the equation

$$3^x - 2^y = 19^z.$$

11. Find all nonnegative integer solutions of the equation

$$2^x 3^y - 5^z 7^w = 1.$$

12. Show that the equation

$$x^x + y^y = z^z$$

does not admit positive integer solutions.

13. Solve in positive integers

$$x^{x^{x^x}} = (19 - y^x) y^{x^y} - 74.$$

14. Find all positive integer solutions to the equation

$$x^y - y^x = 1.$$

3.7 Numerical Functions

This section contains functional equations for functions having as domain and range the set $\mathbf{N} = \{1, 2, \ldots, n\}$ of positive integers or the set $\mathbf{N}_0 = \{0, 1, 2, \ldots, n\}$ of nonnegative integers. In each solution we use clever manipulations of the given equation, combined with properties of the set of positive integers such as the fact that each of its subsets is bounded from below and has a least element and that every positive integer has a unique decomposition into a product of primes.

We start with an example from B.J. Venkatachala's book *Functional Equations, A Problem Solving Approach.*

Find all functions $f : \mathbf{N} \to \mathbf{N}$ such that
(a) $f(n)$ is a square for each $n \in \mathbf{N}$;
(b) $f(m+n) = f(m) + f(n) + 2mn$, for all $m, n \in \mathbf{N}$.

The functional equation (b) shows that for any $m, n \in \mathbf{N}$, $f(m+n) > f(n)$, so f is strictly increasing. Using the condition (a), we deduce that $f(n) \geq n^2$ for all $n \in \mathbf{N}$. Considering the function $h : \mathbf{N} \to \mathbf{N}_0$, $h(n) = f(n) - n^2$, we obtain from (b) that h satisfies the functional equation $h(m+n) = h(m) + h(n)$, so $h(n) = nh(1)$ for all $n \in \mathbf{N}$. Consequently, $f(n) - n^2 = n(f(1) - 1)$, and so $f(n) = n^2 + nf(1) - n$ for all $n \in \mathbf{N}$. For each prime number p, $f(p) = p^2 + pf(1) - p$ is on the one hand a perfect square and on the other hand divisible by p. This number must therefore be divisible by p^2, and so for each prime number p, $f(1) - 1$ is divisible by p. This can only happen if $f(1) = 1$, and so the only function that satisfies the conditions from the statement is $f : \mathbf{N} \to \mathbf{N}$, $f(n) = n^2$.

The second example is a problem submitted by the United States for the International Mathematical Olympiad in 1997, proposed by the authors of this book.

Let $\mathbf{N}$ be the set of positive integers. Find all functions $f : \mathbf{N} \to \mathbf{N}$ such that

$$f^{(19)}(n) + 97f(n) = 98n + 232,$$

where $f^{(k)}(n) = f(f(\ldots(f(n))))$, k times.

Note that $232 = 2(19 + 97)$, so the function $f(n) = n + 2$ satisfies the condition of the problem. We will prove that this is the only solution. For the proof, we first determine the value of $f(n)$ for small n, then proceed by induction. Set $f(n) = n + 2 + a_n$. We want to show that $a_n = 0$ for $n \le 36 = 2 \cdot (19 - 1)$, and then use an inductive argument with step 36.

The condition from the statement implies $97f(n) < 98n + 232$; hence $f(n) \le n + 3$ for $n < 156$. By applying this inequality several times, we get $f^{(19)}(n) \le f^{(18)}(n + 3) \le \cdots \le n + 57$ for $n \le 102$.

In particular, for $n \le 36$ we have

$$97f(n) < f^{(19)}(n) + 97f(n) \le n + 57 + 97f(n).$$

Here we used the fact that $f^{(19)}(n)$ is positive. The above relation implies

$$97(n + 2 + a_n) < 98n + 232 \le n + 57 + 97(n + 2 + a_n).$$

The first inequality gives $97a_n - 38 < n$, and since $n \le 36$, we must have $a_n \le 0$. The second inequality implies $97a_n + 251 \ge 232$; hence $a_n \ge 0$. Therefore, for $n \le 36$, we have $f(n) = n + 2$.

Let us prove by induction that $f(n) = n + 2$. We have already proved the property for numbers less than or equal to 36. Let $k > 36$ and assume that the equality is true for all $m < k$. From the induction hypothesis, $f^{(18)}(k - 36) = f^{(17)}(k - 34) = \cdots = f(k - 2) = k$; hence

$$
\begin{aligned}
f(k) &= f(f^{(18)}(k - 36)) = f^{(19)}(k - 36) \\
&= 98(k - 36) + 232 - 97f(k - 36).
\end{aligned}
$$

Again, by the induction hypothesis, $f(k - 36) = k - 34$, which gives $f(k) = 98k - 98 \cdot 36 + 2 \cdot 97 + 2 \cdot 19 - 97k + 97 \cdot 34 = k + 2$, and we are done.

The third example was communicated to us by I. Boreico.

Find all increasing functions $f : \mathbf{N} \to \mathbf{N}$ such that the only natural numbers that are not in the image of f are those of form $f(n) + f(n + 1)$, $n \in \mathbf{N}$.

Let us assume first that f grows as a linear function, say $f(x) \approx cx$. In this case, let us compute the actual value of c. If $f(n) = m$, then there are exactly $m - n$ positive integers up to m that are not values of f. Therefore we conclude that they are exactly $f(1) + f(2), \ldots, f(m-n) + f(m-n+1)$. Hence $f(m-n) + f(m-n+1) < m < f(m-n+1) + f(m-n+2)$. Now as $f(x) \approx cx$, we conclude $m \approx cn$. Hence $2c(m-n) \approx m$ or $2c(c-1)n \approx cn$, which means $2c - 2 = 1$, so $c = \frac{3}{2}$.

Next, let us make the assumption that $f(x) = \lfloor \frac{3}{2}x + a \rfloor$ for some a. We want to determine a. Clearly, $f(1) = 1, f(2) = 2$ as $1, 2$ must necessarily belong to the range of f. Then 3 does not belong to the range of f, hence $f(3) \geq 4$. Thus $f(2) + f(3) \geq 6$, hence 4 belongs to the range of f and $f(3) = 4$. Similarly $f(4) = 5, f(5) = 7$, and so on. Hence $\lfloor \frac{3}{2} + a \rfloor = 1, \lfloor 3 + a \rfloor = 2$, which implies that $a \in [-\frac{1}{2}, 0)$. And we see that for any a, b in this interval, $\lfloor \frac{3}{2}x + a \rfloor = \lfloor \frac{3}{2}x + b \rfloor$. Thus we can assume $a = -\frac{1}{2}$ and infer that $f(n) = \lfloor \frac{3n-1}{2} \rfloor$. Let us prove that this is so. First, we wish to show that $\lfloor \frac{3n-1}{2} \rfloor$ satisfies the condition from the statement. Indeed, $\lfloor \frac{3n-1}{2} \rfloor + \lfloor \frac{3(n+1)-1}{2} \rfloor = \lfloor \frac{3n-1}{2} \rfloor + 1 + \lfloor \frac{3n}{2} \rfloor = 3n$ by Hermite's identity. We need to show that number of the form $3k$ are not values of $f(n) = \lfloor \frac{3n-1}{2} \rfloor$. Indeed, if $n = 2k$, then $\lceil \frac{3n-1}{2} \rceil = 3k - 1$, and if $n = 2k + 1$, then $\lceil \frac{3n-1}{2} \rceil = 3k + 1$, and the conclusion follows.

The fact that $f(n) = \lfloor \frac{3n-1}{2} \rfloor$ stems now from the inductive assertion that f is unique. Indeed, if we have determined $f(1), f(2), \ldots, f(n-1)$, then we have determined all of $f(1) + f(2), f(2) + f(3), \ldots, f(n-2) + f(n-1)$. Then $f(n)$ must be the least number that is greater than $f(n-1)$ and not among $f(1) + f(2), f(2) + f(3), \ldots, f(n-2) + f(n-1)$. This is because if m is this number and $f(n) \neq m$, then $f(n) > m$, and then m does not belong either to the range of f or to the set $\{f(n) + f(n+1) | n \in \mathbf{N}\}$, a contradiction. Hence $f(n)$ is computed uniquely from the previous values of f and thus f is unique.

Below are listed more numerical functional equations.

1. Find all functions $f : \mathbf{N} \to \mathbf{N}$ satisfying $f(1) = 1$ and $f(m+n) = f(m) + f(n) + mn$ for all $m, n \in \mathbf{N}$.

2. Find all surjective functions $f : \mathbf{N_0} \to \mathbf{N_0}$ with the property that for all $n \geq 0$,

$$f(n) \geq n + (-1)^n.$$

3. Find all functions $f : \mathbf{N} \to \mathbf{N}$ with the property that $f(f(m) + f(n)) = m + n$, for all m and n.

4. Prove that there are no functions $f : \mathbf{N} \to \mathbf{N}$ such that the function $g : \mathbf{N} \to \mathbf{Z}$, $g(n) = f(3n+1) - n$ is increasing and the function $h : \mathbf{N} \to \mathbf{Z}$, $h(n) = f(5n+2) - n$ is decreasing.

5. Find all pairs of functions $f, g : \mathbf{N_0} \to \mathbf{N_0}$ satisfying

$$f(n) + f(n + g(n)) = f(n+1).$$

6. Find all functions $f : \mathbf{N_0} \to \mathbf{N_0}$ such that

$$f(m + f(n)) = f(f(m)) + f(n) \quad \text{for all} \quad m, n \geq 0.$$

7. Prove that there is no function $f : \mathbf{N} \to \mathbf{N}$ with the property that

$$6f(f(n)) = 5f(n) - n \quad \text{for all } n \in \mathbf{N}.$$

8. Let $f : \mathbf{N} \to \mathbf{N}$ be such that

$$f(n+1) > f(f(n)) \quad \text{for all } n \in \mathbf{N}.$$

Prove that $f(n) = n$ for all $n \in \mathbf{N}$.

9. Find all functions $f : \mathbf{N} \to \mathbf{N}$ with the property that for all $n \in \mathbf{N}$,

$$\frac{1}{f(1)f(2)} + \frac{1}{f(2)f(3)} + \cdots + \frac{1}{f(n)f(n+1)} = \frac{f(f(n))}{f(n+1)}.$$

10. Prove that there exists no function $f : \mathbf{N} \to \mathbf{N}$ such that

$$f(f(n)) = n + 1987 \quad \text{for all } n \in \mathbf{N}.$$

11. Find all functions $f : \mathbf{N_0} \to \mathbf{N_0}$ satisfying the following two conditions:
    (a) For any $m, n \in \mathbf{N_0}$,

    $$2f(m^2 + n^2) = (f(m))^2 + (f(n))^2.$$

    (b) For any $m, n \in \mathbf{N_0}$ with $m \geq n$,

    $$f(m^2) \geq f(n^2).$$

12. Let $f : \mathbf{N} \to \mathbf{N}$ be a strictly increasing function such that $f(2) = 2$ and $f(mn) = f(m)f(n)$ for every relatively prime pair of natural numbers m and n. Prove that $f(n) = n$ for every positive integer n.

13. Find a bijective function $f : \mathbf{N_0} \to \mathbf{N_0}$ such that for all m, n,

$$f(3mn + m + n) = 4f(m)f(n) + f(m) + f(n).$$

14. Determine whether there exists a function $f : \mathbf{N} \to \mathbf{N}$ such that

$$f(f(n)) = n^2 - 19n + 99,$$

for all positive integers n.

15. A function $f : \mathbf{N} \to \mathbf{N}$ is such that for all $m, n \in \mathbf{N}$, the number $(m^2 + n)^2$ is divisible by $(f(m))^2 + f(n)$. Prove that $f(n) = n$ for each $n \in \mathbf{N}$.

16. Find all increasing functions $f : \mathbf{N_0} \to \mathbf{N_0}$ satisfying $f(2) = 7$ and $f(mn) = f(m) + f(n) + f(m)f(n)$, for all $m, n \in \mathbf{N_0}$.

3.8   Invariants

A central concept in mathematics is that of an invariant. Invariants are quantities that do not change under specific transformations and thus give obstructions to transforming one object into another. A good illustration of how invariants can be used is the following problem.

Is it possible to start with a knight at some corner of a chessboard and reach the opposite corner passing once through all squares?

As a matter of fact, it is possible to perform all these moves if you drop the restriction on where to end. For the solution, define for each finite path of moves the invariant equal to the color of the knight's last square. Since at each move the color of the square on which the knight rests changes, the invariant depends on the parity of the number of moves. It is equal to the color of the initial square for paths of even length and is equal to the opposite color for paths of odd length.

If a path with the desired property existed, then its length would be 63, which is odd. Thus the initial and the final square would have opposite colors and so could not be opposite corners.

Figure 3.8.1

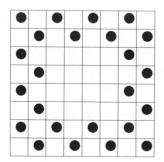

Figure 3.8.2

We now present a classic example from group theory. Recall that checkers can be moved only diagonally, and they jump over a piece into the square immediately after, which must be empty. A piece that we jump over is removed (see Figure 3.8.1).

Given the configuration from Figure 3.8.2, show that there is no sequence of moves that leaves only one piece on the board.

The proof uses the Klein four group. It is the group of the symmetries of a rectangle and consists of four elements $a, b, c,$ and e with e the identity, subject to the relations $ab = c$, $bc = a$, $ca = b$, $a^2 = b^2 = c^2 = e$. If we color the chessboard as in Figure 3.8.3, then the product of the colors of squares that contain checkers is invariant under moves. The initial configuration has product equal to e; hence the final configuration must also have product equal to e, so it contains at least 2 checkers.

b	a		c		b
	c	b		a		c
b	a		c		b
	c	b		a		c
b	a		c		b
	c	b		a		c
b	a		c		h
	c		b	a		c

Figure 3.8.3

And here is a third example, of a different algebraic flavor, which was given at a training test during the Mathematical Olympiad Summer Program in 2006, proposed by R. Gelca.

To a polynomial $P(x) = ax^3 + bx^2 + cx + d$, of degree at most 3, one can apply two operations: (a) switch simultaneously a and d, respectively b and c, (b) translate the variable x to $x + t$, where t is a real number. Can one transform, by a successive application of these operations, the polynomial $P_1(x) = x^3 + x^2 - 2x$ into $P_2(x) = x^3 - 3x - 2$?

The answer is no. Associate to each polynomial P the set of its roots in the complex plane, together with the point at infinity in case the degree of the polynomial is strictly less than 3, and denote by $I(P)$ the number of elements in this set. Here multiplicities should be ignored. It is easy to see that $I(P)$ is invariant under these operations. As $P_1(x) = x(x - 1)(x + 2)$, and $P_2(x) = (x + 1)^2(x - 2)$, we find that $I(P_1) = 3 \neq 2 = I(P_2)$, and so the two polynomials cannot be transformed one into the other.

And now a problem from a training test from the Mathematical Olympiad Summer Program in 2007, proposed by R. Gelca and Z. Feng.

Consider the 7×7 array $a_{ij} = (i^2 + j)(i + j^2)$, $1 \leq i, j \leq 7$. We are allowed to perform the following operation: choose an arbitrary integer (positive or negative) and seven entries of the array, one from each column, then add the chosen integer to each chosen entry in the array. Determine if it is possible that after applying this operation finitely many times, we can transform the given array into an array all of whose rows are arithmetic progressions?

The answer is no. Here is a simplified version of the original argument due to Z. Sunic. We concentrate on the 3×7 subarray a_{ij}, $1 \leq i \leq 3$, $1 \leq j \leq 6$. The invariant that provides the obstruction is the residue modulo 3 of the sum of the entries of this subarray. The given operation does not change this residue. Originally, it is

$$\sum_{i=1}^{3}\sum_{j=1}^{6}(i^2+j)(i+j^2) = \sum_{i=1}^{3}\sum_{j=1}^{6}i^3 + \sum_{i=1}^{3}\sum_{j=1}^{3}j^3 + \sum_{i=1}^{3}i\sum_{j=1}^{6}j + \sum_{i=1}^{6}i^2\sum_{j=1}^{6}j^2$$

$$\equiv 0 + \left(\frac{3\cdot 4\cdot 7}{6}\right)\cdot\left(\frac{6\cdot 7\cdot 13}{6}\right) \equiv 2(\text{mod }3).$$

However, if all rows are arithmetic progressions, then the sum is congruent to 0 modulo 3. We therefore conclude that the original configuration cannot be transformed into one in which all rows are arithmetic progressions. Note that the argument fails if we work with the entire 7×7 array!

We let the reader find the appropriate invariants for the problems below.

1. Is it possible to move a knight on a 5×5 chessboard so that it returns to its original position after having visited each square of the board exactly once?

2. Prove that if we remove two opposite unit square corners from the usual 8×8 chessboard, the remaining part cannot be covered with 2×1 dominoes.

3. Can one cover an 10×10 chessboard with 25 pieces like the one described in Figure 3.8.4?

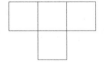

Figure 3.8.4

4. In the squares of a 3×3 chessboard are written the signs $+$ and $-$ as shown in Figure 3.8.5(a). Consider the operations in which one is allowed to simultaneously change all signs on some row or column. Can one change the given configuration to the one in Figure 3.8.5(b) by applying such operations finitely many times?

Figure 3.8.5

5. On every square of a 1997×1997 board is written either $+1$ or -1. For every row we compute the product R_i of all numbers written in that row, and for every column we compute the product C_i of all numbers written in that column. Prove that $\sum_{i=1}^{1997}(R_i + C_i)$ is never equal to zero.

6. There is one stone at each vertex of a square. We are allowed to change the number of stones according to the following rule: We may take away any number of stones from one vertex and add twice as many stones to the pile at one of the adjacent vertices. Is it possible to get 1989, 1988, 1990, and 1989 stones at consecutive vertices after a finite number of moves?

7. Given a circle of n lights, exactly one of which is initially on, it is permitted to change the state of a bulb, provided that one also changes the state of every dth bulb after it (where d is a divisor of n strictly less than n), provided that before the move, all these n/d bulbs were in the same state as another. For what values of n is it possible to turn all the bulbs on by making a sequence of moves of this kind?

8. Given a stack of $2n + 1$ cards, we can perform the following two operations:
(a) Put the first k at the end, for any k.
(b) Put the first n in order in the spaces between the other $n + 1$.
Prove that we have exactly $2n(2n + 1)$ distinct configurations.

9. Three piles of stones are given. Sisyphus carries the stones one by one from one pile to another. For each transfer of a stone, he receives from Zeus a number of coins equal to the number of stones from the pile from which the stone is drawn minus the number of stones in the recipient pile (with the stone Sisyphus just carried not counted). If this number is negative, Sisyphus pays back the corresponding amount (the generous Zeus allows him to pay later if he is broke). At some point, all stones have been returned to their piles. What is the maximum possible income for Sisyphus at this moment?

10. Starting at $(1, 1)$, a stone is moved in the coordinate plane according to the following rules:
(a) From any point (a, b), the stone can move to $(2a, b)$ or $(a, 2b)$.
(b) From any point (a, b), the stone can move to $(a - b, b)$ if $a > b$ or to $(a, b - a)$ if $a < b$.
For which positive integers x, y can the stone be moved to (x, y)?

11. In the sequence $1, 0, 1, 0, 1, 0, 3, 5, 0, \ldots$, each term starting with the seventh is equal to the last digit of the sum of the preceding six terms. Prove that this sequence does not contain six consecutive terms equal to $0, 1, 0, 1, 0, 1$, respectively.

12. Prove that a 4×11 rectangle cannot be covered with L-shaped 3×2 pieces (Figure 3.8.6).

Figure 3.8.6

13. At a round table are 1994 girls playing a game with a deck of n cards. Initially, one girl holds all the cards. At each turn, if at least one girl holds at least two cards, one of these girls must pass a card to each of her two neighbors. The game ends when and only when each girl is holding at most one card.
    (a) Prove that if $n \geq 1994$, then the game cannot end.
    (b) Prove that if $n < 1994$, then the game must end.

14. A solitaire game is played on an $m \times n$ regular board, using mn markers that are white on one side and black on the other. Initially, each square of the board contains a marker with its white side up, except for one corner square, which contains a marker with its black side up. In each move, one may take away one marker with its black side up but must then turn over all markers that are in squares having an edge in common with the square of the removed marker. Determine all pairs (m,n) of positive integers such that all markers can be removed from the board.

15. Three numbers are written on a blackboard. We can choose one of them, say c, and replace it by $2a + 2b - c$, where a and b are the other two numbers. Can we reach the triple $11, 12, 13$ from the triple $20, 21, 24$?

16. The sides of a polygon are colored in three colors: red, yellow, and blue. Initially, their colors read in the clockwise direction: red, blue, red, blue, ..., red, blue, yellow. We are allowed to change the color of a side, but we have to ensure that no two adjacent sides are colored by the same color. Is it possible that after some operations, the colors of the sides are in the clockwise order: red, blue, red, blue, ..., red, yellow, blue?

3.9 Pell Equations

The equation $u^2 - Dv^2 = 1$, with D a positive integer that is not a perfect square, is called a Pell equation. Already known from ancient times in connection with the famous "cattle problem" of Archimedes, it was brought to the attention of the mathematicians of modern times by Fermat in 1597. Euler, who also studied it, attributed the equation to Pell, whence the name. The first complete solution was given by Lagrange in 1766.

To find all positive integer solutions to this equation, one first determines a minimal solution (i.e., the solution (u,v) for which $u + v\sqrt{D}$ is minimal and $(u,v) \neq (1,0)$) and then computes all other solutions recursively. There is a general method of determining the minimal solution, which we will only sketch, since in all problems below the minimal solution is not difficult to guess. The method works as follows. If we consider the continued fraction expansion

$$\sqrt{D} = a_1 + \cfrac{1}{a_2 + \cfrac{1}{a_3 + \cdots}},$$

then the sequence $a_1, a_2, a_3 \ldots$ is periodic starting with the second term. Let $a_2, a_3, \ldots, a_m$ be a cycle of minimal period. Define the positive integers p_m and q_m to be

13. Prove that there exist infinitely many positive integers n such that $n^2 + 1$ divides $n!$.

14. Prove that the equation

$$x^2 + y^2 + z^2 + 2xyz = 1$$

has infinitely many integer solutions.

15. Does there exist a row in Pascal's triangle containing four distinct elements a, b, c, d such that $b = 2a$ and $d = 2c$?

16. Show that for infinitely many positive integers n, we can find a triangle with integer sides whose semiperimeter divided by its inradius is n.

3.10 Prime Numbers and Binomial Coefficients

Let a and b be nonnegative integers. Here and henceforth, the binomial coefficient $\binom{a}{b}$ will denote the number of b element subsets of an a-element set. In other words, it is equal to $(a(a-1)\cdots(a-b+1))/(b!)$ if $0 < b \le a$, to 0 if $0 \le a < b$, and to 1 if $b = 0$, $a \ge 0$.

The problems in this section are based on the following observation. If p is a prime number and $0 < k < p$, then p divides $\binom{p}{k}$. Indeed, $\binom{p}{k} = p!/(k!(p-k)!)$, and we see that p appears in the numerator but not in the denominator. The p in the numerator does not cancel out, being a prime. This fact can be generalized to the following theorem of Lucas.

Let p be a prime, and let n, k be nonnegative integers with base p representations

$$n = n_0 + n_1 p + \cdots + n_t p^t, \quad k = k_0 + k_1 p + \cdots + k_t p^t.$$

Then

$$\binom{n}{k} \equiv \binom{n_0}{k_0}\binom{n_1}{k_1}\cdots\binom{n_t}{k_t} \pmod{p}.$$

The exponent t in both representations may be considered the same because one of them can always be amended by leading zeros. For the proof, note that $(a+b)^{p^r} \equiv a^{p^r} + b^{p^r} \pmod{p}$, for any $r \ge 0$. We have

$$(1+X)^n \equiv (1+X)^{n_0 + n_1 p + \cdots + n_t p^t} \equiv (1+X)^{n_0}(1+X)^{pn_1}\cdots(1+X)^{p^t n_t}$$

$$\equiv (1+X)^{n_0}(1+X^p)^{n_1}\cdots(1+X^{p^t})^{n_t}$$

$$\equiv \left(\sum_{i=0}^{n_0}\binom{n_0}{i_0}X^{i_0}\right)\left(\sum_{i_1=0}^{n_1}\binom{n_1}{i_1}X^{pi_1}\right)\cdots\left(\sum_{i_t=0}^{n_t}\binom{n_t}{i_t}X^{p^t i_t}\right) \pmod{p}.$$

Since k has a unique expansion in base p, $k = k_0 + k_1 p + \cdots + k_t p^t$, the coefficient of X^k in the last product is equal to $\binom{n_0}{k_0}\binom{n_1}{k_1}\cdots\binom{n_t}{k_t}$, and the congruence is proved.

Here is an easy application.

Let p be a prime number, m a positive integer, and $k > 1$ an integer that is not divisible by p. Prove that $\binom{kp^m}{p^m}$ is not divisible by p.

Let $k = k_0 + k_1 p + \cdots + k_t p^t$ be the base p representation of k. Then $kp^m = k_0 p^m + k_1 p^{m+1} + \cdots + k_t p^{m+t}$, so by Lucas's theorem

$$\binom{kp^m}{p^m} \equiv \binom{k_0}{1}\binom{k_1}{0} \cdots \binom{k_t}{0} \equiv \binom{k_0}{1} \pmod{p}.$$

Since k is not divisible by p, $\binom{k_0}{1} = k_0$ is not divisible by p, and the conclusion follows.

The following problems are either direct applications of Lucas's theorem or can be solved using similar ideas.

1. For a given n, find the number of odd elements of the form $\binom{n}{k}$, $k = 0, 1, \ldots, n$.

2. Let p be a prime number. Prove that the number of binomial coefficients $\binom{n}{0}$, $\binom{n}{1}$, $\ldots$, $\binom{n}{n}$ that are multiples of p is equal to

$$(n+1) - (n_1 + 1)(n_2 + 1) \cdots (n_m + 1),$$

   where $n_1, n_2, \ldots, n_m$ are the digits of the base p expansion of n.

3. Prove that the numbers $\binom{2^n}{k}$, $k = 1, 2, \ldots, 2^n - 1$, are all even and that exactly one of them is not divisible by 4.

4. Let p be a prime number. Prove that

$$\binom{p^n}{p} \equiv p^{n-1} \pmod{p^n}$$

   for all positive integers n.

5. Prove that the number of binomial coefficients that give the residue 1 when divided by 3 is greater than the number of binomial coefficients that give the residue 2 when divided by 3.

6. Show that if p is prime and $0 \le m < n < p$, then

$$\binom{np+m}{mp+n} \equiv (-1)^{m+n+1} p \pmod{p^2}.$$

7. Suppose p is an odd prime. Prove that

$$\sum_{j=0}^{p} \binom{p}{j}\binom{p+j}{j} \equiv 2^p + 1 \pmod{p^2}.$$

8. Prove that for any prime p, the number $\binom{2p}{p} - 2$ is divisible by p^2.

9. Prove that for any prime $p > 3$, the number $\binom{2p-1}{p-1} - 1$ is divisible by p^3.

10. For each polynomial $P(X)$ with integer coefficients, let us denote the number of odd coefficients by $w(P)$. For $i = 0, 1, \ldots$, let $Q_i(X) = (1+X)^i$. Prove that if $0 \le i_1 < i_2 < \cdots < i_n$ are integers, then $w(Q_{i_1} + Q_{i_2} + \cdots + Q_{i_n}) \ge w(Q_{i_1})$.

11. Let p be an odd prime and n a positive integer. Prove that $p+1$ divides n if and only if

$$\sum_{k \equiv j (\mathrm{mod}\ p-1)} \binom{n}{k} (-1)^{(k-j)/p-1} \equiv 0 \pmod{p}$$

for every $j \in \{1, 3, 5, \ldots, p-2\}$.

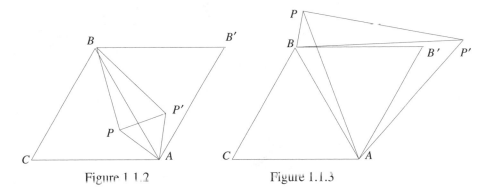

Figure 1.1.2   Figure 1.1.3

6. Let us assume first that PC is the hypotenuse. If P is inside the triangle, then as in the solution to the previous problem we conclude that $PP'B$ is a right triangle if and only if $\angle APB = 150°$. If P is outside the triangle, as shown in Figure 1.1.3, then $\angle BPP' = 90°$ if and only if $\angle BPA = 30°$, because $\angle BPA = \angle BPP' - \angle APP' = \angle BPP' - 60°$. Thus the locus is a circle with center outside the triangle ABC in which AB determines an arc of $60°$. The cases where PA, respectively PB, are hypotenuses give two more circles congruent to this one, corresponding to the sides BC and CA.

7. This is the inverse construction to the one described in the first part of the book, and we expect again some $60°$ rotations to be helpful. If we let $A = X$, $P = Y$, and $P' = Z$, then C can be obtained by rotating P' around P clockwise through $60°$ (see Figure 1.1.1 from the introductory part of Section 1.1 of this chapter). Point B is then obtained by rotating C around A clockwise through $60°$.

8. Let O be the center of the three circles and ABC the desired equilateral triangle. By Pompeiu's theorem, the radii OA, OB, OC must satisfy the inequalities $PA \le PB + PC$, $PB \le PA + PC$, $PC \le PA + PB$. Hence these are necessary conditions for the construction to be possible. We will show how to construct the equilateral triangle if these three conditions are simultaneously satisfied.

Let us examine Figure 1.1.4, in which we assume that the triangle ABC has already been constructed. Let A' and O' be the images through the $60°$ rotation of A respectively O around C. As seen before, the triangle OAO' is the Pompeiu triangle of the point O with respect to the equilateral triangle ABC. The Pompeiu triangle might be degenerate.

Here is the construction. Choose A on one of the circles, then construct the Pompeiu triangle OAO'. It is constructible with a straightedge and a compass because its sides are the radii of the three circles. Note that even if in the begining the common center of the circles is not specified, there is a standard construction that produces it (see for example the last problem in Section 1.6). The point C is obtained by intersecting a second of the three concentric circles with the circle of the same radius centered at O'. Finally, B is obtained by intersecting the third circle with the circle of center A and radius AC.

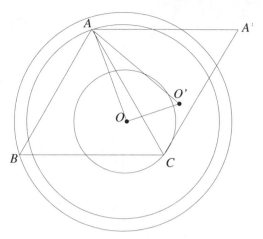

Figure 1.1.4

9. Rotate triangle ZMY through $60°$ counterclockwise about Z to ZNW (see Figure 1.1.5). First note that triangles ZMN and ZYW are equilateral; hence $MN = ZM$ and $YW = YZ$. Now $\angle XMN$ and $\angle MNW$ are straight angles, both being $120° + 60°$, so $XW = XM + YM + ZM$. On the other hand, as in the solution to Problem 7, when constructing backwards the triangle ABC from triangle XYZ, we can choose $A = W$ and $C = X$. Then the side length of the equilateral triangle is XW, which is equal to $XM + YM + ZM$.

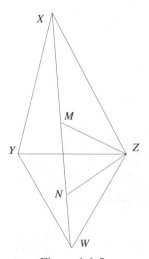

Figure 1.1.5

(A version of this problem appeared at the USAMO in 1974)

10. *First solution:* We prove that the locus is the empty set, one circle, or two circles centered at the centroid O of the equilateral triangle ABC. For simplicity, assume that the area of the equilateral triangle is 1, so its side length is $2/\sqrt[4]{3}$.

For a triangle XYZ, denote by $S[XYZ]$ the signed area of this triangle (by definition, $S[XYZ]$ is the area of triangle XYZ if the triangle is positively oriented, namely if when going from X to Y, Y to Z, and Z to X we turned counterclockwise, and the negative of the area otherwise). Recall Figure 1.1.1 from the introduction, with A' and P' the images of A and P through the $60°$ clockwise rotation around C. Then APP' is a triangle with side lengths PA, PB, and PC.

We set $S[PBC] = k_1$, $S[PCA] = k_2$, and $S[PAB] = k_3$ and do all computations in terms of these three numbers (i.e., we use barycentric coordinates). Triangles APB and $A'P'A$ are congruent and so are triangles $A'P'C$ and APC, hence

$$S[APP'] = 2S[ABC] = 2k_3 - k_1 - k_2 - S[P'PC] = 1 - k_3 - S[P'PC].$$

Let us compute $S[P'PC]$ in terms of k_1, k_2, and k_3. Note that since we rotate clockwise, the triangle $P'PC$ is positively oriented. The triangle being equilateral, it suffices to compute the length of one of its sides, say PC. Let $\angle PCB = \alpha$ and use the area of $S[PBC]$ and $S[APC]$ to get $\sqrt[4]{3}k_1 = PC\sin\alpha$ and $\sqrt[4]{3}k_2 = PC \cdot \sin(60° - \alpha)$.

The subtraction formula for sine combined with the first equality transforms the second into

$$\sqrt[4]{3}k_2 = PC\sin 60° \cos\alpha - PC\cos 60° \sin\alpha = PC\frac{\sqrt{3}}{2}\cos\alpha - \frac{\sqrt[4]{3}}{2}k_1.$$

Hence $PC\cos\alpha = (2k_2 + k_1)/\sqrt[4]{3}$. We obtain

$$PC^2 = PC^2\sin\alpha^2 + PC^2\cos^2\alpha = k_1^2\sqrt{3} + (2k_2 + k_1)^2/\sqrt{3}$$
$$= \frac{4\sqrt{3}}{3}(k_1^2 + k_1k_2 + k_2^2).$$

It follows that the area of $P'PC$ is $k_1^2 + k_1k_2 + k_2^2$. Hence

$$S[APP'] = 1 - k_3 - (k_1^2 + k_1k_2 + k_2^2) = (k_1 + k_2 + k_3)^2 - k_3(k_1 + k_2 + k_3)$$
$$- (k_1^2 + k_1k_2 + k_2^2) = k_1k_2 + k_1k_3 + k_2k_3.$$

Next, we express the distance from P to the center O of the triangle in terms of k_1, k_2, and k_3. For this we use a vectorial approach. First note that

$$\overrightarrow{OP} = k_1\overrightarrow{OA} + k_2\overrightarrow{OB} + k_3\overrightarrow{OC}.$$

Indeed, the equality holds for P at one of the vertices or at the center of the triangle, and since both sides are linear in P, the equality holds everywhere.

By squaring, the relation becomes

$$OP^2 = \overrightarrow{OP} \cdot \overrightarrow{OP} = (k_1^2 OA^2 + k_2^2 OB^2 + k_3^2 OC^2 + 2k_1k_2\overrightarrow{OA} \cdot \overrightarrow{OB}$$
$$+ 2k_1k_3\overrightarrow{OA} \cdot \overrightarrow{OC} + 2k_2k_3\overrightarrow{OB} \cdot \overrightarrow{OC}).$$

Since the angles $\angle AOB$, $\angle AOC$, and $\angle BOC$ are all equal to $120°$, and $OA = OB = OC$, this is further equal to

$$(k_1^2 + k_2^2 + k_3^2 - k_1 k_2 - k_1 k_3 - k_2 k_3)OA^2$$
$$= ((k_1 + k_2 + k_3)^2 - 3(k_1 k_2 + k_1 k_3 + k_2 k_3))OA^2$$
$$= (1 - 3(k_1 k_2 + k_1 k_3 + k_2 k_3))OA^2.$$

Hence the condition for the area of triangle APP' to be constant is the same as for the distance OP to be constant. If the area is some $k < \frac{1}{3}$, the locus consists of two circles, one inside and one outside the circumcircle. If the area is $\frac{1}{3}$, the locus consists of a circle and O. Finally, if the constant area is greater than $\frac{1}{3}$, then the locus is a circle.

Second solution: We work in complex coordinates and use the fact that if a triangle in the plane has vertices at 0, $s_1 + it_1$, $s_2 + it_2$, then its area is $|s_1 t_2 - s_2 t_1|/2$.

Place the unit circle on the complex plane so that A, B, C correspond to the complex numbers $1, \omega, \omega^2$, where $\omega = e^{2\pi i/3}$, and let P correspond to the complex number x. The distances PA, PB, and PC are then $|x - 1|, |x - \omega|, |x - \omega^2|$. The identity

$$(x - 1) + \omega(x - \omega) + \omega^2(x - \omega^2) = 0$$

implies that the sides of the Pompeiu triangle, as vectors, correspond to the complex numbers $x - 1, \omega(x - \omega), \omega^2(x - \omega^2)$. Translating this triangle such that one vertex is at the origin, we obtain a triangle with vertices 0, $x - 1$, and $\omega(x - \omega)$. By the above-mentioned formula, the area of this triangle is

$$(\omega^2 - \omega)(x\bar{x} - 1) = i\sqrt{3}(|x|^2 - 1),$$

and this only depends on $|x|$, the distance from P to the origin. The conclusion follows.
(Z. Skopetz)

1.2   Cyclic Quadrilaterals

1. We argue on Figure 1.2.1. Denote by C the center of the square. Since $\angle NOM + \angle NCM = 90° + 90° = 180°$, it follows that the quadrilateral $MONC$ is cyclic. Thus $\angle MOC = \angle MNC = 45°$, which shows that C lies on the bisector of $\angle AOB$.

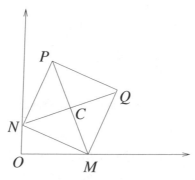

Figure 1.2.1

Conversely, for any C on the bisector, construct the squares $MONC$ and $MNPQ$. Since CMN is an isosceles right triangle, C is the center of $MNPQ$. Thus the locus is the bisector of $\angle AOB$.

(Romanian high school textbook)

2. Translate triangle DCP to triangle ABP' (see Figure 1.2.2). This way we obtain the quadrilateral $APBP'$, which is cyclic, since $\angle APB + \angle AP'B = 360° - \angle APD - \angle BPC = 180°$. Let Q be the intersection of AB and PP'. Then since the lines AD, PP', and BC are parallel, we have $\angle DAP + \angle BCP = \angle APQ + \angle QP'B$. The latter two angles have measures equal to half of the arcs $\overset{\frown}{AP'}$ and $\overset{\frown}{BP}$ of the circle circumscribed to the quadrilateral $APBP'$. On the other hand, the angle $\angle BQP$, which is right, is measured by half the sum of these two arcs. Hence $\angle DAP + \angle BCP = \angle BQP = 90°$.

(Mathematical Olympiad Summer Program, 1995)

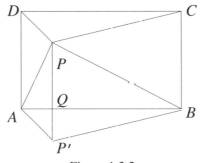

Figure 1.2.2

3. Without loss of generality, we may assume that P lies on the arc $\overset{\frown}{AB}$, not containing the points C and D (Figure 1.2.3). We have to prove that XY is perpendicular to ZW. This reduces to proving that the angles $\angle XYP$ and $\angle ZWP$ add up to $90°$. But $\angle XYP = \angle XBP$ and $\angle ZWP = \angle ZDP$ from the rectangles $XBYP$ and $ZDWP$ (which are cyclic quadrilaterals). It follows that

$$\angle XYP + \angle ZWP = \angle XBP + \angle ZDP = \frac{\overset{\frown}{AP}}{2} + \frac{\overset{\frown}{PC}}{2} = \frac{\overset{\frown}{AC}}{2} = 90°,$$

and we are done.

4. Let M and N be the projections of the vertex A onto the interior angle bisectors of $\angle B$ and $\angle C$, respectively, and let P and Q be the projections of A onto the exterior bisectors of the same angles (Figure 1.2.4). Let us prove that P lies on MN. Since the interior and the exterior bisectors of an angle of the triangle are orthogonal, the quadrilateral $APBM$ is a rectangle. Hence $\angle AMP = \angle ABP$. We have $\angle ABP = (180° - \angle B)/2 = \angle A/2 + \angle C/2$.

Denote by I the incenter. The quadrilateral $ANIM$ is cyclic, since it has two opposite right angles. Hence $\angle AMN = \angle AIN$. The angle $\angle AIN$ is exterior to triangle AIC; hence $\angle AIN = \angle A/2 + \angle C/2$. This shows that $\angle AMP = \angle A/2 + \angle C/2 = \angle AMN$, and thus M, N, P are collinear. The same argument shows that Q lies on MN, which solves the problem.

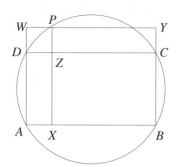

Figure 1.2.3

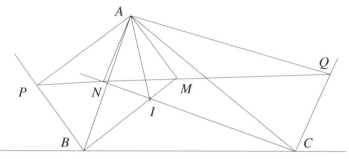

Figure 1.2.4

5. The quadrilateral that we want to prove is cyclic is a ratio-2 dilation of the one determined by the projections of P onto the sides, and thus it suffices to prove that the latter is cyclic. Let X, Y, Z, W be the projections of P onto the sides AB, BC, CD, and AD, respectively (Figure 1.2.5). The quadrilaterals $AXPW$, $BYPX$, $CZPY$, and $DWPZ$ are cyclic, since they all have a pair of opposite right angles.

Considering angles formed by a side and a diagonal, we get $\angle WAP = \angle WXP$, $\angle PXY = \angle PBY$, $\angle YZP = \angle YCP$, and $\angle PZW = \angle PDW$. In the triangles APD and BPC we have $\angle PAD + \angle PDA = 90°$ and $\angle PBC + \angle PCB = 90°$. Hence

$$\angle WXY + \angle WZY = \angle WXP + \angle PXY + \angle YZP + \angle PZW$$
$$= \angle WAP + \angle PDW + \angle PBY + \angle YCP$$
$$= 90° + 90° = 180°$$

which shows that the quadrilateral $XYZW$ is cyclic, and the problem is solved.
    (USAMO, 1993)

6. *First solution.* Let $R \in BQ$ be such that Q is between B and R and $QR = QC$ (see Figure 1.2.6). Since $\angle BQC$ is right, Q lies on the semicircle. Hence the quadrilateral $BAQC$ is cyclic, so $\angle AQC = 180° - \angle ABC = 135°$. It follows that $\angle AQR = 360° - 135° - 90° = 135°$. This implies that the triangles AQC and AQR are congruent, from

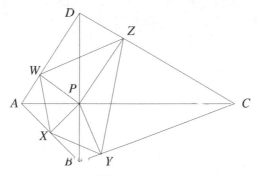

Figure 1.2.5

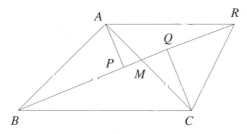

Figure 1.2.6

which $AR = AC = AB$. In the isosceles triangle ABR, AP is an altitude, so $BP = PR$, and since $PR = PQ + QC$, the conclusion follows.

Second solution. We use the construction from Figure 1.2.7, where S has been chosen on the line AP such that $AQCS$ is an isosceles trapezoid. Since $\angle BQC = 90°$, Q is on the semicircle, so the quadrilateral $ABCQ$ is cyclic. It follows that $\angle AQB = \angle ACB = 45°$. In the right triangle PAQ, $\angle PAQ = 45°$, which implies $\angle ASC = 45°$. Since $\angle ASC = \angle ABC$, the quadrilateral $ABSC$ is cyclic, so $\angle ASB = \angle ACB = 45°$. The triangle BPS is then isosceles; hence $BP = PS$. Finally, in the isosceles trapezoid $AQCS$, $AS = 2AP + QC$; hence $BP = PS = AP + QC = PQ + QC$.

(A modified version of a theorem due to Archimedes)

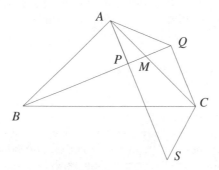

Figure 1.2.7

7. By the equality of $\angle EAF$ and $\angle FDE$, the quadrilateral $AEFD$ is cyclic (see Figure 1.2.8). Therefore, $\angle AEF + \angle FDA = 180°$. By the equality of $\angle BAE$ and $\angle CDF$, we have

$$\angle ADC + \angle ABC = \angle FDA + \angle CDF + \angle AEF - \angle BAE = 180°.$$

Hence the quadrilateral $ABCD$ is cyclic, so $\angle BAC = \angle BDC$. It follows that $\angle FAC = \angle BAC - \angle BAF = \angle BDC - \angle EDC = \angle EDB$.

(Russian Mathematical Olympiad, 1996)

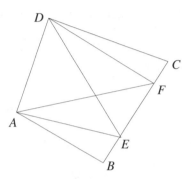

Figure 1.2.8

8. Since the measure of $\angle A$ is $60°$, it follows that the other two angles add up to $120°$. Hence $\angle IBC + \angle ICB = 60°$. This implies that $\angle B'IC' = \angle BIC = 120°$, and consequently, the quadrilateral $AB'IC'$ is cyclic, since two opposite angles add up to $180°$. It follows that $\angle IB'C' = \angle IAC' = 30°$ and $\angle IC'B' = \angle IAB' = 30°$, since AI is a bisector. Hence the triangle $IB'C'$ is isosceles; therefore, $IB' = IC'$.

9. The interior bisectors of $\angle A$ and $\angle B$ meet at I, while the exterior bisectors at the same angles meet at I_c, the excenter of the triangle opposite C. The interior and the exterior bisector of an angle are orthogonal, so the quadrilateral $AIBI_c$ has two opposite right angles. Hence A, B, I, and I_c lie on the circle with diameter II_c, which is then the circumcircle of the triangle ABI. Its center is the midpoint of II_c, and since C, I, and I_c are collinear, the conclusion follows.

10. Let P be such that $ADMP$ is a rectangle. On line AP, choose points Q and R such that $QBDA$ and $ADCR$ are rectangles (Figure 1.2.9). The points Q, B, and D lie on the circle of diameter AB; hence $ADEQ$ is a cyclic quadrilateral. Similarly, R, C, and D lie on the circle of diameter AC; hence $ADFR$ is a cyclic quadrilateral. The quadrilaterals $ADEQ$ and $ADFR$ share a side and have the same supporting lines for the other two sides. Since they are cyclic, the remaining two sides EQ and RF must be parallel. Thus E, Q, R, and F are the vertices of a trapezoid.

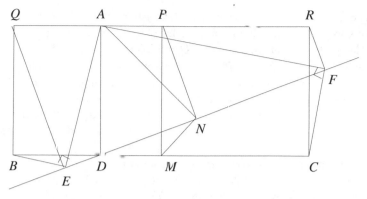

Figure 1.2.9

On the other hand, in the rectangle *QBCR*, *M* is the midpoint of *BC*, and *MP* is parallel to *QB*, so *P* is the midpoint of *QR*. Since *N* is the midpoint of *EF*, it follows that in the trapezoid *QEFR*, *NP* is parallel to *QE*. This implies that quadrilateral *ADNP* is cyclic, having sides parallel to the sides of *ADFR*. Moreover, *M* lies on the circumcircle of this quadrilateral, because the other three vertices of the rectangle *ADMP* lie on it. Hence *A*, *D*, *M*, *N* are on the circle with diameter *AM*, and consequently, $\angle ANM = \angle ADM = 90°$.

(Asian–Pacific Mathematical Olympiad, 1998; proposed by R. Gelca)

11. All reasoning is done on Figure 1.2.10. We begin with an observation that should illuminate us on how to proceed. The triangles *PDE* and *CFG* have parallel

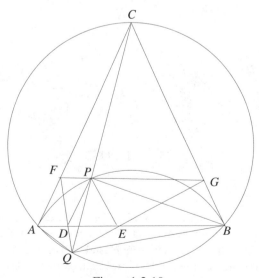

Figure 1.2.10

sides, thus either *DF*, *EG*, and *CP* are parallel, or they meet at a point, the common center of homothety. With this in mind we let Q be the intersection of *CP* with the circumcircle of *ABC* and prove that both *FD* and *EG* pass through Q.

For reasons of symmetry, it suffices to show that *FD* passes through Q. From the hypothesis and the fact that *AQBC* is cyclic, we obtain $\angle AQP = \angle ABC = \angle BAC = \angle PFC$. It follows that the quadrilateral *AQPF* is cyclic, and so $\angle FQP = \angle PAF$. As I is the incenter of *ABC*, $\angle IBA = \angle CBA/2 = \angle CAB/2 = \angle IAC$, so the circumcircle of *AIB* is tangent to *CA* at A. This implies that $\angle PAF = \angle DBP$ (they both subtend the same arc).

Using again the fact that *AQBC* is cyclic, we obtain $\angle QBD = \angle QCA = \angle QPD$, and so the quadrilateral *DQBP* is also cyclic. Hence $\angle DBP = \angle DQP$. Combining everything we obtained so far, we find that $\angle FQP = \angle PAF = \angle DBP = \angle DQP$. This implies that the lines *FQ* and *DQ* coincide, meaning that *FD* passes through Q. The problem is solved.

(Short list, 44th IMO, 2003)

12. The quadrilaterals *PN'NP'* and *APTN* are cyclic (see Figure 1.2.11), so $\angle AN'P' = \angle APN = \angle ATN$. Similarly $\angle CN'M' = \angle CMN = \angle CTN$. It follows that $\angle AN'P' + \angle CN'M' = \angle ATN + \angle CTN = \angle ATC = 120°$. Therefore, $\angle P'N'M' = 180° - 120° = 60°$. Similarly, $\angle N'M'P' = \angle N'P'M' = 60°$; thus the triangle *M'N'P'* is equilateral.

(Romanian Mathematical Olympiad, 1992; proposed by C. Cocea)

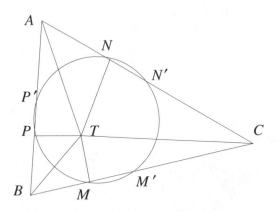

Figure 1.2.11

13. Let G be the point on *Oy* such that $OG = OA$, let F be the point of tangency of $\mathscr{C}$ to *Ox*, and let H be the center of $\mathscr{C}$. When F approaches A, *DE* approaches *GA*, so the fixed point of *DE* should lie on *AG*. Let P be the intersection of *DE* and *AG*.

Let us assume first that A is between F and O (Figure 1.2.12). Computing angles in terms of the arcs of the circle $\mathscr{C}$, we deduce that $\angle DEF = 90° + \frac{1}{2}\angle DOF$. Also, $\angle GAF$ is exterior to the isosceles triangle *OAG*; thus $\angle GAF = 90° + \frac{1}{2}\angle DOF$. Hence $\angle DEF = \angle GAF$, which implies that the quadrilateral *PAFE* is cyclic.

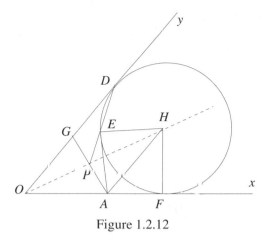

Figure 1.2.12

Using the orthogonality of the radius and the tangent at the point of tangency, we deduce that the quadrilateral *HFAF* has two opposite right angles, hence is cyclic. Therefore, *PAFH* is cyclic, too (since P and H lie both on the circumcircle of triangle *AEF*). It follows that $\angle HPA = \angle HFA = 90°$. But *HO* is orthogonal to *AG*, being a bisector in the isosceles triangle *OAG*. This implies that P is the midpoint of *AG*, so it does not depend on D.

If F is between A and O, then a similar computation shows that $\angle DEC = 90° - \frac{1}{2}\angle AOG = \angle GAF$. Thus the quadrilateral *PFAE* is cyclic. From here the proof proceeds as before.

(Gh. Ţiţeica, *Probleme de geometrie (Problems in geometry)*, Ed. Tehnică, Bucharest, 1929)

14. Let B_0 and B_1 be the intersections of the circumcircle of triangle $A_1A_4P_1$ with the lines A_0A_1 and A_3A_4 (see Figure 1.2.13). The quadrilateral $P_1B_1A_1A_4$ is cyclic; hence $\angle B_1P_1A_1 = \angle A_1A_4B_1$. Also, the quadrilateral $A_1A_2A_3A_4$ is cyclic; hence

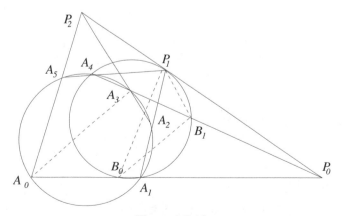

Figure 1.2.13

$\angle A_1A_4A_3 = \angle A_3A_2P_1$. Since $\angle A_1A_4A_3$ and $\angle A_1A_4B_1$ are in fact the same angle, it follows that $\angle B_1P_1A_1 = \angle A_3A_2P_1$; hence P_1B_1 and A_2P_2 are parallel. In a similar manner, one shows that B_0P_1 and A_0P_2, respectively B_1B_0 and A_0A_3, are parallel. Since the triangles $B_1B_0P_1$ and $A_3A_0P_2$ have parallel sides, they are perspective.

This means that A_0A_1, A_3A_4, and P_1P_2 intersect. It follows that P_0, P_1, and P_2 are collinear.

(Pascal's theorem, solution published by Jan van Yzeren in the American Mathematical Monthly, 1993)

1.3   Power of a Point

1. The three lines are the radical axes of the pairs of circles, and they intersect at the radical center.

2. Let AB, CD, and EF be the three chords passing through the point P. If we set $AP = a$, $BP = b$, $CP = c$, $DP = d$, $EP = e$, and $FP = f$, then writing the power of the point with respect to the circle, we obtain $ab = cd = ef$. Since the chords have equal lengths, we also have that $a + b = c + d = e + f$, and hence

$$\{a,b\} = \{c,d\} = \{e,f\}.$$

Without loss of generality, we may assume that $a = c = e$. The initial circle and the circle with center P and radius a have the common points A, C, and E; hence they must coincide. Thus P is the center of the circle.

3. We claim that if A and B minimize the product, then OAB is an isosceles triangle. Indeed, if A_1B_1 is another segment passing through P and with endpoints on the two rays, then the circle tangent to Ox and Oy at A respectively B cuts these segment at two interior points, which shows that $PA_1 \cdot PB_1$ is greater than the power of P with respect to the circle, and the latter is equal to $PA \cdot PB$.

(M. Pimsner and S. Popa, *Probleme de geometrie elementară (Problems in elementary geometry)*, Ed. Didactică şi Pedagogică, Bucharest, 1979)

4. Let MN be the diameter of the circle of intersection of the sphere with the plane, through the point P that lies at the intersection of AB with the plane. Set $AP = a$, $BP = b$, $MP = x$, and $NP = y$. Writing the power of the point P with respect to the great circle determined by AB and MN, we obtain $xy = ab$. We need to minimize the sum $x + y$ when the product xy is given. Using the AM–GM inequality, we conclude that the minimum of $x + y$ is $2\sqrt{ab}$ and is attained for $x = y = \sqrt{ab}$.

Now consider the points M and N in the plane through A and B that is perpendicular to $\mathscr{P}$, such that $M, N \in \mathscr{P}$ and $MP = NP = \sqrt{ab}$. Because $AP \cdot BP = MP \cdot NP$, it follows that A, B, M, N lie on a circle. Rotate this circle around a diameter perpendicular to $\mathscr{P}$ to obtain a sphere. The radius of this sphere is $\sqrt{ab}$, which shows that the minimal radius can be attained.

(Communicated by S. Savchev)

5. Let A'' be the second intersection of the line AH with the circumcircle. Writing the power of H with respect to the circumcircle we get $HA \cdot HA'' = R^2 - OH^2$.

Note that since $\angle BHC = 180° - \angle BAC$, A'' is symmetric to H with respect to the side BC. Thus $HA'' = 2HD$, where D is the intersection of AH with BC. Let us compute the lengths of AH and HD. If we denote the intersection of BH and AC by E, then in triangle ABE, $AE = AB\cos A = 2R\sin C\cos A$. In triangle AHE, $AH = AE/\sin C = 2R\cos A$ and $HD = AD - AH = 2R\sin B\sin C - 2R\cos A = 2R\sin B\sin C + 2R\cos(B+C) = 2R\cos B\cos C$. Hence $OH^2 = R^2 - HA \cdot HA'' = R^2 - 2R\cos A \cdot 4R\cos B\cos C = R^2 - 8R^2\cos A\cos B\cos C$.

If the triangle has one obtuse angle, then one has to take the right-hand side with opposite sign. This is because in this case, H lies outside the circumcircle, so its power with respect to the circumcircle is $OH^2 - R^2$.

Remark. Using the identity

$$\cos^2 A + \cos^2 B + \cos^2 C = 1 - 2\cos A\cos B\cos C,$$

which holds for the angles of a triangle, we can further transform the above formula into

$$OH^2 = R^2 - 4R^2(1 - \cos^2 A - \cos^2 B - \cos^2 C)$$

or

$$OH^2 = 9R^2 - (2R\sin A)^2 - (2R\sin B)^2 - (2R\sin C)^2.$$

By applying the law of cosines, we transform this into

$$OH^2 = 9R^2 - (a^2 + b^2 + c^2).$$

6. The lines AB, $A'M$, and $B'N$ are the radical axes of the three pairs of circles determined by ABA', ABB', and $A'B'C'$. Either they are parallel or they intersect at the radical center of the three circles. Similarly, BC, $B'P$, and $C'Q$ are the three radical axes of the circles BCB', BCC', and $A'B'C'$, and CA, $C'R$, and $A'S$ are the three radical axes of the circles ACC', ACA', and $A'B'C'$. Thus either case (ii) occurs, or these lines intersect respectively in three points A'', B'', and C''. This proves the first part.

The point A'' has the same power with respect to the circles ABB', ABA', and $A'B'C'$, that is, $A''A \cdot A''B = A''M \cdot A''A' = A''N \cdot A''B'$. It follows that A'' has the same power with respect to the circles ABC and $A'B'C'$, hence is on the radical axis of the two. Similarly, B'' and C'' are on the radical axis of the circle ABC and $A'B'C'$, thus the three points are collinear.

(Romanian IMO Team Selection Test, 1985)

7. In fact, we will prove that the circles are coaxial (have a common radical axis), which then proves the claim: If two of the circles meet, then the radical axis passes through the intersection; hence the third circle passes through the intersection as well.

Let H be the orthocenter of triangle ADE. As seen in the solution to Problem 5, the reflections of H across each side lie on the circumcircle. If A', D', and E' are the feet of the altitudes from A, D, and E, respectively, this means that $AH \cdot A'H$ equals half of the power of H with respect to the circumcircle, and similarly for the other altitudes. In particular,

$$AH \cdot A'H = DH \cdot D'H = EH \cdot E'H.$$

What about the power of H with respect to the circle with diameter AC? Since A' lies on this circle, the power is again $AH \cdot A'H$. By a similar argument and the above equality, H has the same power with respect to the circle with diameters AC, BD, and EF.

On the other hand, the same must be true for the orthocenters of the other three triangles formed by any three of the four lines AB, BC, CD, DA. Since these do not all coincide, the three circles must have a common radical axis.

(Hungarian Mathematical Olympiad, 1995)

8. Writing the power of A with respect to $\mathscr{C}_2$, we get $AE \cdot AF = AB^2$ (see Figure 1.3.1). On the other hand, $AD \cdot AC = (AB/2) \cdot 2AB = AB^2$. Hence $AE \cdot AF = AD \cdot AC$. This shows that triangles ADE and AFC (with the shared angle at A) are similar; thus $\angle AED = \angle ACF$, so $DEFC$ is cyclic.

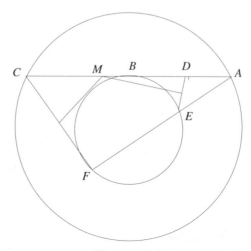

Figure 1.3.1

Since M is the intersection of the perpendicular bisectors of DE and CF, it must be the circumcenter of $DEFC$. Consequently, M also lies on the perpendicular bisector of CD. Since M is on AC, it must be the midpoint of CD. Hence $AM/MC = \frac{5}{3}$.

(USAMO, 1998; proposed by R. Gelca)

9. Let F and G be the feet of the altitudes from B and C, respectively (see Figure 1.3.2). The points F and G lie on the two circles, since the angles $\angle BFN$ and $\angle MGC$ are right. Let X and Y be the intersection of PH with these circles. The problem requires us to prove that $X = Y$.

Writing the power of H with respect to the circles of diameters BN, CM, and BC, we get $PH \cdot HX = CH \cdot HG = BH \cdot HF = PH \cdot HY$, where the middle

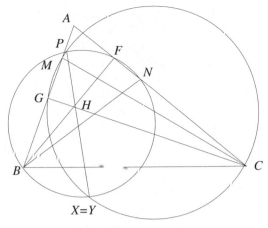

Figure 1.3.2

equality is proved as in Problem 7. This implies $HX = HY$; hence $X = Y$, and we are done.

(Leningrad Mathematical Olympiad, 1988)

10. Since AC and BD are altitudes in the triangle ABF (see Figure 1.3.3), E is the orthocenter of this triangle, so FE is perpendicular to AB. The triangles HEB and HAF are similar, so $HE/HA = HB/HF$. Hence $HE \cdot HF = HA \cdot HB$, which equals HG^2 (the power of H with respect to the circle), and the equivalence is now clear.

(Submitted by USA for the IMO, 1997; proposed by T. Andreescu)

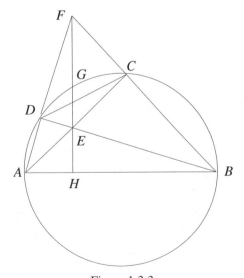

Figure 1.3.3

11. It suffices to show that A is on the radical axis of the circles circumscribed to triangles DPG and FPE, i.e., that A has the same power with respect to these two circles. Thus we try to compute the two powers. Denote by M the other point of intersection of AB with the circumcircle of the triangle DPG and by N the other point of intersection of AC with the circumcircle of the triangle FPE. The two powers are now $AD \cdot AM$ and $AE \cdot AN$. To prove that they are equal, it is enough to show that the points M, N, D, E lie on the same circle. If D is between A and M, then since $MDPG$ is cyclic, $\angle DMP = \angle DGP$. Also, DG and BC are parallel, so the latter angle is equal to $\angle BCP$; hence M, B, C, P are concyclic. If M is between A and D, then arguing similarly, $\angle DMP$ and $\angle BCP$ are supplementary, and this again implies M, B, C, P concyclic. Similarly, N, B, C, P are concyclic. Thus M, N, B, C, are concyclic, and since DE is parallel to BC, M, N, D, E are concyclic, and the proof is finished.

(Indian IMO Team Selection Test, 1995)

12. We argue on Figure 1.3.4. Because the lines DF and AC are parallel, it follows that $\angle DFA = \angle CAF$. On the other hand, $\angle DFA = \angle DEG$, since both angles subtend the arc $\overset{\frown}{DG}$. Thus $\angle CAF = \angle DEG$, and so the triangles AMG and EMA are similar. This implies that $AM/MG = EM/AM$, hence $AM^2 = MG \cdot ME$.

Writing the power of M with respect to the circle, we obtain $MG \cdot ME = MB \cdot MC$, hence

$$AM^2 = MB \cdot MC = (AB - AM)(AC - AM)$$
$$= AB \cdot AC - AM(AB + AC) + AM^2.$$

This yields $AM(AB + AC) = AB \cdot AC$, and the conclusion follows.

(Romanian Team Selection Test for the IMO, 2006)

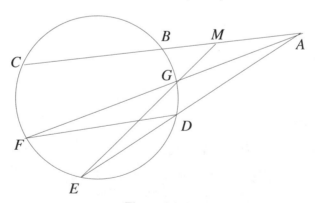

Figure 1.3.4

13. Writing the power of P with respect to the two circles, we get $BP \cdot PN = XP \cdot PY = CP \cdot PM$ (Figure 1.3.5). This implies that the quadrilateral $MBCN$ is cyclic, so $\angle MNB = \angle MCB$. Both triangles MAC and BND are right, since AC and BD are diameters; hence $\angle A = 90° - \angle MCA = 180° - \angle MND$. It follows that the quadrilateral $AMND$ is cyclic. The lines XY, AM, and DN are the radical axes of the circles of

diameter *AC*, *BD* and of the circumcircle of *AMND*; hence they are concurrent at the radical center of the circles.

(36th IMO, 1995; proposed by Bulgaria)

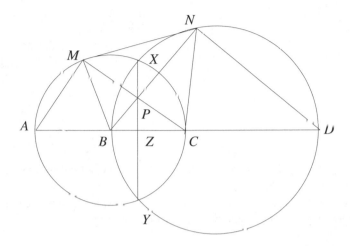

Figure 1.3.5

14. Let *L* be the intersection of *AD* and *CB*. We will proceed by proving that *K*, *L*, and *M* are collinear (Figure 1.3.6). In the cyclic quadrilateral *AOKC*, we have $\angle AKO = \angle ACO$. Similarly, in the cyclic quadrilateral *BOKD*, $\angle BKO = \angle BDO$. In the circle of diameter *AB*, $\angle ACO = 90° - \angle COA/2$, and $\angle BDO = 90° - \angle BOD/2$. By relating angles to arcs of the semicircle, we get

$$\angle AKB = \angle AKO + \angle OKB = 180° - \left(\frac{\angle COA}{2} + \frac{\angle BOD}{2} \right)$$

$$= 180° - \left(\frac{\overset{\frown}{CA}}{2} + \frac{\overset{\frown}{BD}}{2} \right) = 180° - \angle CLA = \angle ALB.$$

Therefore, $\angle AKB = \angle ALB$, so the quadrilateral *AKLB* is cyclic.

As a consequence, we get

$$\angle CKL = 360° - \angle AKL - \angle CKA$$
$$= (180° - \angle AKL) + (180° - \angle CKA)$$
$$= \angle LBA + (180° - \angle COA).$$

Referring to the arcs on the given semicircle, we obtain $180° - \angle COA = 180° - \overset{\frown}{AC}$ and $\angle LBA = \overset{\frown}{AC}/2$. We obtain $\angle CKL + \angle CDL = 180°$; hence the quadrilateral *CKLD* is cyclic.

The three radical axes of the circumcircles of *AKLB*, *CKLD*, and *ABCD* intersect at the radical center. These radical axes are *CD*, *AB*, and *KL*. Hence *M*, the intersection

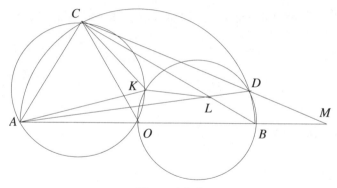

Figure 1.3.6

of the first two, lies also on the third. Thus we have reduced the problem to proving that LK is orthogonal to KO.

Since the angles $\angle AKL$ and $\angle ABL$ add up to $180°$, it suffices to prove that $\angle AKO + \angle ABL = 90°$. Using the cyclic quadrilateral $ACKO$, we write

$$\angle AKO + \angle ABL = \angle ACO + \angle CBA = \frac{180° - \angle COA}{2} + \frac{\overset{\frown}{AC}}{2}$$

$$= 90° - \frac{\overset{\frown}{AC}}{2} + \frac{\overset{\frown}{AC}}{2} = 90°,$$

and the problem is solved.

(Balkan Mathematical Olympiad, 1996)

15. We use directed angles modulo $180°$ and argue on Figure 1.3.7. From the cyclic quadrilaterals $AHFD$ and $BHFC$, we deduce that $\angle AHF = \angle ADF (\text{mod} 180°)$ and $\angle ADF = \angle FCB (\text{mod} 180°)$, and since $\angle ADF = \angle ADB = \angle ACB = \angle FHB$, it follows that

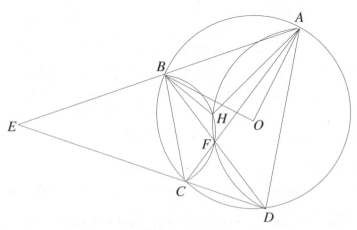

Figure 1.3.7

$\angle AHB = 2\angle ADB$. Let O be the circumcenter of $ABCD$. Then $\angle AOB = 2\angle ADB$, hence $\angle AOB = \angle AHB$. Thus O lies on the circumcircle of the triangle AHB and for similar reasons on the circumcircle of the triangle CHD. The radical axes of the circumcircles of AHB, CHD, and $ABCD$ intersect at one point; these lines are AB, CD, and HO. It follows that E, H, and O are collinear.

On the other hand, in the cyclic quadrilaterals $HOCD$ and $BHFC$, $\angle OHC = \angle ODC$, and $\angle CHF = \angle CBF$. Hence $\angle OHF = \angle OHC + \angle CHF = \angle ODC + \angle CBF = 90° - \angle CAD + \angle CBD$, so $\angle EHF = \angle OHF = 90°$, as desired.
(Bulgarian Mathematical Olympiad, 1996)

16. One can boost the intuition by considering the third circle, which we call ω_3, to be interior tangent to the given two circles ω_1 and ω_2, with the tangency points A respectively B (see Figure 1.3.8). Let P and Q be the intersections of AD and BD with ω_2 respectively ω_1. Writing the fact that D is on the radical axis of ω_1 and ω_2, we obtain $DA \cdot DP = DB \cdot DQ$, hence the triangles DAB and DQP are similar. It follows that $\angle DAB = \angle DQP$.

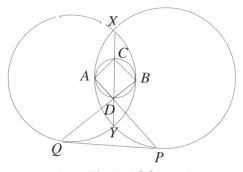

Figure 1.3.8

But note that $\angle DAB$ is equal to the angle made with DA by the tangent at D to ω_3. And this is mapped to the tangent to ω_2 at P by the homothety of center A that transforms ω_3 into ω_1. We conclude that line PQ is tangent to ω_2, and by a similar argument it is also tangent to ω_1. And so the four points required by the statement are the four tangency points of the two common tangents. Of course, one has to consider the other possible situations of the circle ω_3, but the same argument applies *mutatis mutandis*.
(Short list, 41st IMO, 2000)

1.4 Dissections of Polygonal Surfaces

1. Draw the line passing through the centers of the rectangles. As it cuts each rectangle in half, it divides the shaded region into parts of equal area.

Remark. This problem was given at a job interview for computer programmers.

2. Figure 1.4.1 describes three possible dissections.
(Russian Mathematical Olympiad, 1987–1988)

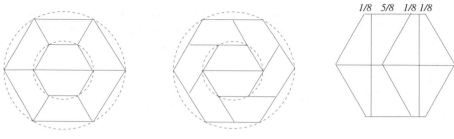

Figure 1.4.1

3. One of the many possible cuts and reassembling is shown in Figure 1.4.2.

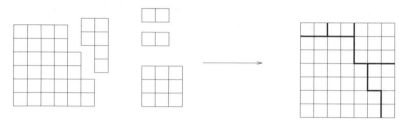

Figure 1.4.2

4. Figure 1.4.3 shows how to dissect a square into isosceles trapezoids and equilateral triangles and further how to dissect an equilateral triangle into three isosceles trapezoids.

(*Kvant (Quantum)*, proposed by V. Lev and A. Sivatski)

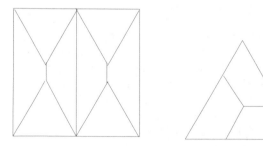

Figure 1.4.3

5. The right triangle whose legs are in the ratio 1:2 can be dissected into 5 equal right triangles as shown in Figure 1.4.4(a). The equilateral triangle can be dissected into 12 equal isosceles triangles as shown in Figure 1.4.4(b).

(Russian Mathematical Olympiad, 1988–1989)

6. The answer is affirmative! A dissection is provided in Figure 1.4.5.

   (P. Boychev)

a)

b)

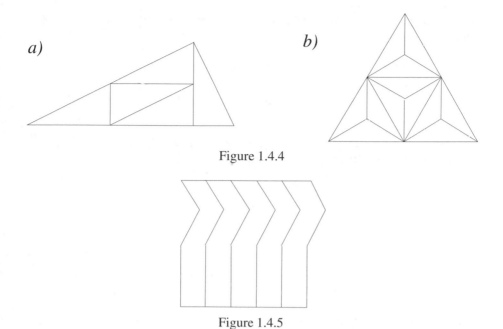

Figure 1.4.4

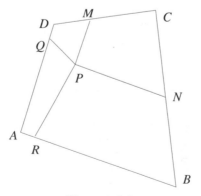

Figure 1.4.5

7. We will prove that there exists a dissection into 4 cyclic quadrilaterals, with one of the quadrilaterals an isosceles trapezoid. Since an isosceles trapezoid can be cut into an arbitrary number of isosceles trapezoids, this will prove the statement for all $n \geq 4$.

Let $ABCD$ be the quadrilateral (Figure 1.4.6). If it is a rectangle, the dissection is straightforward, if not, without loss of generality we may assume that $\angle D$ is obtuse. Choose P in the interior, M on CD, and N on BC such that $PNCM$ has sides parallel to the sides of $ABCD$. If P is chosen close enough to the side AD, then there exists a point Q on AD such that $PMDQ$ is an isosceles trapezoid. Note that $\angle PQA$ is acute.

Figure 1.4.6

If P is chosen close enough to the side AB, then one can find a point R_1 close to A such that $\angle PR_1A + \angle PQA > 180°$. Also, under the same condition, one can find a point R_2 close to B with $\angle PR_2A < 90°$; hence with $\angle PQA + \angle PR_2A < 180°$. A continuity argument yields $R \in AB$ with $\angle PQA + \angle PRA = 180°$; hence with $ARPQ$ cyclic. Finally, $\angle RPN = 360° - \angle RPQ - \angle QPM - \angle MPN = 360° - (180° - \angle A) - (180° - \angle D) - (180° - \angle C) = 180° - \angle B$, which shows that $RBNP$ is also cyclic. This gives the desired decomposition for $n = 4$. If $n > 4$ then dissect the isosceles trapezoid into isosceles trapezoids by $n - 4$ lines parallel to the bases.

(14th IMO, 1972)

8. We will prove inductively that a square can be divided into n squares for $n \geq 6$. The cases $n = 6, 7$, and 8 are described in Figure 1.4.7.

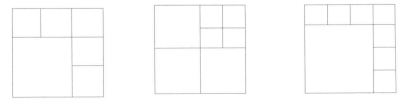

Figure 1.4.7

The conclusion now follows from an inductive argument, since if the property is true for a certain k, then by dividing one of the squares of the decomposition in four we add three more squares, so the property is true for $k + 3$ as well.

Let us show that one cannot divide a square into 5 squares. Suppose that such a dissection exists. Then each of the sides of the initial square is touched by at least two squares from the decomposition, and the squares that touch one side do not touch the opposite side. Hence there is one side of the square that is touched by exactly two squares of the decomposition. If the two squares are equal, then on top of them we have the other three squares, which are either equal, and consequently are of smaller size, as in Figure 1.4.8(a), or lie as in Figure 1.4.8(b). In the first case, the side of the initial square is, on the one hand, equal to twice the length of the side of one of the big squares, and on the other hand equal to the sum of the lengths of the sides of a small square and a big square, which is impossible.

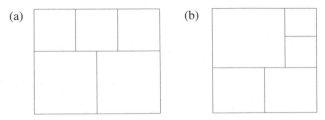

Figure 1.4.8

The second case reduces to the situation where two squares of different sizes touch a side. These two squares leave an L-shaped piece, which can be filled only by two squares equal to the smaller one and one square equal to the bigger one, and we end up again with the impossible configuration from Figure 1.4.8(a).

9. This problem is similar to the previous. Consider a cube C and let $P(n)$ be the proposition that C can be partitioned into n cubes.

If $P(k)$ is true for some k, then, by dividing one of the cubes of the partition into 8 cubes with planes that are parallel to the faces and that run through its center, it follows that $P(k+7)$ is also true. The problem reduces to examining $P(55)$, $P(56)$, $P(57)$, $P(58)$, $P(59)$, $P(60)$, and $P(61)$.

$P(55)$: Divide C into 27 cubes and four of these into 8 cubes each. This gives a division of C into $27 + 4 \cdot 7 = 55$ cubes.

$P(56)$. Divide C into 8 cubes, and four of these, determining a square on a face F of C, into 27 cubes each. Then consider 8 (of the 9) cubes having one ninth of F as base, formed by joining 8 little cubes into a single one. We have divided C into $8 + 4 \cdot 26 - 8 \cdot 7 = 56$ cubes.

$P(57)$: Divide C into 64 cubes; then join 8 of them to form a new one. This gives $64 - 7 = 57$.

$P(58)$: Divide C into 27 cubes and join 8 of them into a new one. Then do the same thing with two cubes of the partition. This produces $27 - 7 + 2 \cdot (26 - 7) = 58$ cubes.

$P(59)$: Divide C into 64 cubes and join 27 of them into a new one. Then divide 3 of the remaining cubes into 8 each. We have divided C into $64 - 26 + 3 \cdot 7 = 59$ cubes.

$P(60)$: Divide C into 8 cubes, then 2 of these into 27 each to obtain $8 + 2 \cdot 26 = 60$ cubes.

$P(61)$: Divide C into 27 cubes and join 8 of them into a new one. Then consider 4 of the remaining cubes that share a part of a face of C; let us call this part P, and divide them into 27 cubes each. Consider the 9 cubes having as one face the ninth part of P, obtained by joining 8 little cubes into a single one. We have divided C into $27 - 7 + 4 \cdot 26 - 9 \cdot 7 = 61$ cubes.

Since $P(55)$, $P(56)$, ..., $P(61)$ are true, by using an inductive argument with step 7 we get that $P(k)$ is true for all $k \geq 55$.

(Romanian IMO Team Selection Test, 1978)

10. We prove that only polygons having a center of symmetry satisfy this condition. To this end, let P be a polygon that can be dissected into parallelograms. Subdivide to assume that any two neighboring parallelograms share a full side. Start with one side. Then for any piece of that side, there is a well-defined path of parallelograms that ends when we get to another parallel section of the boundary of the same length. Since a convex polygon cannot have three parallel sides, this means we have another side parallel to the original of the same length.

This argument shows that the sides of P are parallel in pairs and congruent. This immediately implies that P has an even number of sides, and by convexity, the opposite sides are the ones that are parallel and congruent. Hence P has a center of symmetry (the midpoint of the great diagonals).

Conversely, let us prove by induction that any polygon with a center of symmetry can be dissected into parallelograms. For a quadrilateral (i.e., a parallelogram), the statement is obvious, and the induction step is described in Figure 1.4.9.

(*Kvant (Quantum)*)

11. We generalize the statement slightly by requiring that there are $2n$ or less points inside the polygon. Now we proceed by induction on n. For $n = 1$, choose the

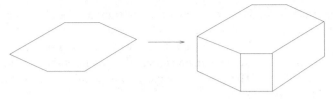

Figure 1.4.9

dissection determined by the line passing through both points (or through the point if there is only one).

Now assume that the property is true for all polygons with $2(n-1)$ interior points. Let us prove it for a polygon with $2n$ points. Fix a line d that does not intersect the polygon. Let A be the point that is closest to d and choose B among the remaining points such that the angle formed by the lines AB and d is minimal (it can eventually be zero). The line AB divides the polygon into two convex polygons P_1 and P_2 such that P_1 does not contain in its interior any of the $2n$ points, and P_2 contains in its interior at most $2n-2$ of the points. Applying the induction hypothesis, we deduce that P_2 can be divided into n polygons containing the remaining $2n-2$ points on their boundaries, and these polygons, together with P_1, provide the required dissection.

12. Like in the case of the previous problem, it is easier to prove the statement with a slightly weaker hypothesis, namely that only $\angle A$ is less than $90°$. We will show by induction on n that the triangle ABC can be dissected into n isosceles triangles whose equal sides are all equal to AC.

For $n=1$, the conclusion is obvious. Assume that it holds for $n-1$ and let us prove it for n. Choose $M \in AB$ such that $\angle CMA = \angle A$ (Figure 1.4.10). Then in triangle CMB,

$$\angle MCB = 180° - \angle CBM - \angle CMB = 180° - \angle A - (180° - n\angle A) = (n-1)\angle A;$$

hence by the induction hypothesis, this triangle can be decomposed into $n-1$ isosceles triangles with the equal sides equal to $MC = AC$. Adding triangle CAM to this dissection gives a dissection of triangle ABC.

(M. Pimsner and S. Popa, *Probleme de geometrie elementară (Problems in elementary geometry)*, Ed. Didactică şi Pedagogică, Bucharest, 1979)

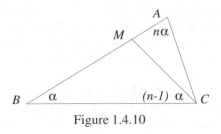

Figure 1.4.10

13. Color the squares of the board with pairs of letters as shown in Figure 1.4.11. On the board there are 15 squares colored by ac, 15 squares colored by ad, 15 squares

colored by bc, and 15 squares colored by bd. Call the colors ac and bd, respectively ad and bc complementary. In each of the tiles there are exactly two squares of the same color, and if a color shows up twice in a tile, its complementary color is absent.

ac	ad	ac	ad	ac	ad	ac	ad	ac	ad
bc	bd	bc	bd	bc	bd	bc	bd	bc	bd
ac	ad	ac	ad	ac	ad	ac	ad	ac	ad
bc	bd	bc	bd	bc	bd	bc	bd	bc	bd
ac	ad	ac	ad	ac	ad	ac	ad	ac	ad
bc	bd	bc	bd	bc	bd	bc	bd	bc	bd

Figure 1.4.11

Arguing by contradiction, let us assume that a dissection exists. Denote by k, l, m, n the number of tiles in which ac, ad, bd, respectively bc shows up twice. There are exactly $15 - 2k$ tiles in which ac shows up only once, and in none of these does bd (the complement of ac) show up twice. Hence in each of these $15 - 2k$ tiles there are either two squares colored by ad or two squares colored by bc. Moreover, any tile that contains two bc's or two ad's also contains one ac. Hence we can write $15 - 2k = m + l$. Similarly, we deduce $15 - 2l = n + k$, $15 - 2m = n + k$, $15 - 2n = m + l$. It follows that $n = k$ and $m = l$, and hence $15 = 2k + m + l = 2(k + m)$, which is impossible. We conclude that such a dissection does not exist.

(Romanian regional contest "Gh. Țiţeica," 1983)

14. Let n be the number of rectangles in the dissection. If n is a perfect square, say $n = k^2$, then we can rearrange the rectangles of the dissection in the usual $k \times k$ pattern.

If n is not a square, the ratio of similarity between the rectangles is $\sqrt{n}$, since the area of the initial rectangle is n times bigger than the area of a rectangle from the dissection. Denote the lengths of the sides of the big rectangle by A and B and the lengths of sides of the small rectangles by a and b, such that $A/a = B/b = \sqrt{n}$. If we look at the small rectangles that touch one of the sides of length A, we note that some share with this side their side of length a, and some their side of length b; thus there exist positive integers n_1 and n_2 such that $A = n_1 a + n_2 b$. Similarly, there exist positive integers n_3 and n_4 such that $B = n_3 a + n_4 b$. Hence a and b satisfy the system of equations

$$(n_1 - \sqrt{n})a + n_2 b = 0,$$
$$n_3 a + (n_4 - \sqrt{n})b = 0.$$

For this system to admit nontrivial solutions, its determinant must be 0. The determinant is equal to $-(n_1 + n_4)\sqrt{n} + n_1 n_4 + n - n_2 n_3$. Since n_1, n_2, n_3, and n_4 are nonnegative integers and $\sqrt{n}$ is irrational, the determinant can be zero only if $n_1 = n_4 = 0$. Hence the only possible dissection is when all sides a of small rectangles are parallel to the side B of the big rectangle.

(Gh. Eckstein; a version of this problem appeared also in *Kvant (Quantum)*, proposed by P. Pankov)

15. We will prove that the sum of the areas of the rectangles from the dissection is *n*. Consider an arbitrary dissection into finitely many parallelograms. Since we are interested only in a sum of areas, a finer dissection will have the same sum of areas of rectangles. Thus consider a dissection in which any two neighboring rectangles share a full side.

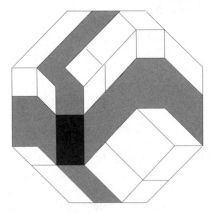

Figure 1.4.12

In such a dissection, each rectangle is the crossroad of two paths, each of which connects opposite sides, the sides connected by the first path being perpendicular to the sides connected by the second (Figure 1.4.12). Let L_1 and L_2 be two segments that lie on orthogonal sides. The intersection of the path starting at L_1 intersects the path starting at L_2 in exactly one rectangle, and the area of this rectangle is $L_1 L_2$. Thus if we add the areas of all rectangles that are crossroads of paths linking orthogonal pairs of sides, we get 1. On the other hand, there are $2n$ pairs of opposite sides, which paired with their orthogonals give *n* families of rectangles. The area of each family is 1; hence the total area of rectangles is *n*.

   (*Kvant (Quantum)*)

16. Figure 1.4.13 shows that the largest angle of the triangle can be 90° and 120°. Let us show that these are the only possible solutions.

Figure 1.4.13

Let $\alpha \leq \beta \leq \gamma$ be the angles of the triangle, and assume $\gamma \neq 90°$ and $\gamma \neq 120°$. There is only one triangle of the dissection containing the vertex at α.

Note that two triangles from the dissection cannot share a side and have two other sides on the same supporting line, for this would imply that the initial triangle has two supplementary angles.

Case 1. The triangle containing the vertex at α is as in Figure 1.4.14.

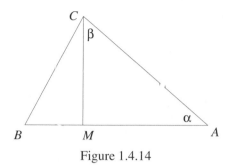

Figure 1.4.14

The $\angle MCA$ can only be equal to β, so $\angle BCA \geq \beta + \alpha$. Therefore $\angle BCA$ is obtuse and is equal to γ. The triangle BMC is decomposed into four triangles. At an interior vertex, all four triangles or only three of them meet. The first situation is clearly impossible, since this would imply that there are two triangles that share a side and have a vertex on one of the sides of triangle BMC.

On the sides of the triangle BMC can be at most one additional vertex. Indeed, since $\angle BMC = 180° - \gamma = \alpha + \beta$, there exists another edge originating at M. If there were two or more vertices on the sides of BMC, then, counting with repetitions, we would have at least five edges on the sides of BMC, plus 10 more (the one originating in M, plus two at each point on the sides of BMC, each counted twice, for they have a triangle on each side). From these, at most two could have been counted twice. This gives at least 13 sides of triangles. But there are only four triangles. This proves the existence of only one vertex on the sides of BMC. At this vertex can be at most one angle γ, so the other γ's lie at interior vertices. The same edge count shows that there is exactly one interior vertex, and since three triangles meet at it, all three angles must be γ. Thus $\gamma = 120°$, a contradiction.

Case 2. The triangle containing the vertex at α is as in Figure 1.4.15.

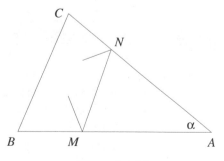

Figure 1.4.15

Without loss of generality, we can assume $\angle A = \alpha$, $\angle B = \beta$, and $\angle C = \gamma$. The observation made at the beginning of the proof shows that there is at least one more edge at M and one at N. Assume first that there are no other vertices on the sides of $BMNC$. Then, since any additional edge is counted twice as a side of a triangle, there exist exactly two other edges. But then there is exactly one vertex in the interior, say P, and the decomposition looks like in Figure 1.4.16.

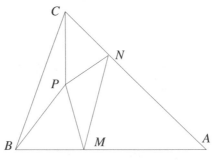

Figure 1.4.16

The angles at β can only be equal to α, so $\beta = 2\alpha$. Since $\gamma \neq \alpha + \beta$ (for the triangle is not right), the angles at C are both β. In this case $\angle A = \pi/7$, $\angle B = 2\pi/7$, $\angle C = 4\pi/7$. If $\angle MNA = \gamma$, the figure completes uniquely with $\angle CBP = \angle PBM = \angle PMN = \angle CNP = \alpha$, $\angle BCP = \angle PCN = \angle PNM = \angle BPM = \beta$, and $\angle CPB = \angle CPN = \angle NPM = \angle PMB = \gamma$. Since the triangles PCN and PCB are congruent, we have $PN = PB$. But then in triangles PBM and PMN, $PN < PM < PB$, which is impossible. A similar argument rules out the case $\angle NMA = \gamma$.

If there is one vertex on the sides of $BMNC$, then the parity of the number of the sides of the four triangles in the dissection implies the existence of yet another vertex. Since each of the two vertices on the sides can have at most one angle γ, there should be one more interior angle. But then, since there are at least 4 edges meeting at the vertices that lie on the sides of $BMNC$, counting the vertices of the four triangles from the dissection, we get 12 on the sides of $BMNC$, plus at least three more in the interior, which is impossible. This solves the problem.

(Proposed by R. Gelca for the USAMO, 2000)

1.5   Regular Polygons

1. *First solution:* Recall that if we rotate a point B around a point A by $60°$ to a point C, the triangle ABC is equilateral. This implies that an isosceles triangle with an angle of $60°$ is equilateral. This observation suggests that many problems involving equilateral triangles can be solved by finding a hidden $60°$ rotation. As we will see below, this is the case with our problem.

Let Q be the intersection of BM with the parallel through A to BC (see Figure 1.5.1). First we prove that the triangle AQN is equilateral. Since $\angle QAN = 60°$, it suffices to show that two sides of the triangle are equal. From the similar triangles MQA

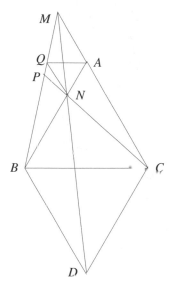

Figure 1.5.1

and MBC, we deduce that $AQ/BC = MA/MC = MA/(MA+BC)$. Also, from the similar triangles NMA and NDB, we deduce that $NA/NB = MA/BD$. Since $AB = BC = BD$, it follows that $NA/AB = MA/(MA+BC) = AQ/AB$. Therefore, $AQ = NA$.

Since the triangle AQN is equilateral, Q can be obtained from N by a $60°$ rotation around A. Also, since the triangle ABC is equilateral, B can be obtained from C by the same rotation. Hence the line BM can be obtained from CN by a $60°$ rotation, which shows that the two lines form a $60°$ angle.

Second solution: Let R be the intersection of CN with the circumcircle of ABC, and let M' be the intersection of AC with CR. The tangents BD and CD form with AB, BR, RC, and CA a degenerate cyclic hexagon $ABBRCC$. By Pascal's theorem, the pairs of opposite sides (AB, CR), (BB, CC), and (AR, AC) intersect at the collinear points N, D, M'. It follows that $M = M'$, and so BN and CM meet on the circumcircle of the triangle ABC. Consequently, these two lines form an angle of $60°$.

(Romanian high school textbook)

2. The center of a square can be obtained by rotating one of the vertices about an adjacent vertex by $45°$ and then contracting with respect, the center of rotation by a ratio of $1/\sqrt{2}$. As both of these operations can be easily expressed using complex numbers, it is natural to apply complex numbers to solve the problem.

Figure 1.5.2 should aid the intuition. To avoid working with fractions, we denote the complex coordinates of A, B, C, D respectively by $2a, 2b, 2c$, and $2d$. Then the coordinate m of M satisfies $m - b = \frac{1}{\sqrt{2}}\left(\frac{\sqrt{2}}{2} + i\frac{\sqrt{2}}{2}\right)(2a - 2b)$. Hence $m = (1+i)a + (1-i)b$. Similarly the coordinates of N, P, and Q are $n = (1+i)b + (1-i)c$, $p = (1+i)c + (1-i)d$, and $q = (1+i)d + (1-i)a$. Note that $q - n = (1+i)(d-b) + (1-i)(a-c) = i(p-m)$, from which it follows that NQ is obtained by rotating MP by an angle of $90°$.

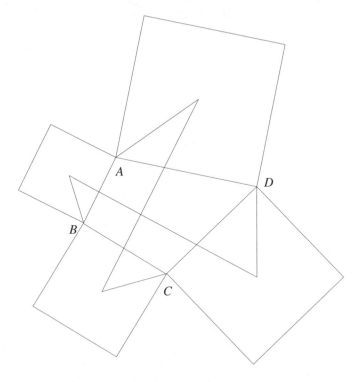

Figure 1.5.2

3. We will prove that the triangle MCD is equilateral. Since an equilateral triangle is a more symmetric object than is an isosceles one, it is easier to do the problem backwards and use the uniqueness of the geometric picture.

For this let M' be the point inside the regular pentagon such that $M'CD$ is equilateral (Figure 1.5.3). Then the triangles $CM'B$ and $DM'E$ are both isosceles, having two equal sides. We have $\angle M'CB = \angle DCB - \angle DCM' = 108° - 60° = 48°$, and, by symmetry,

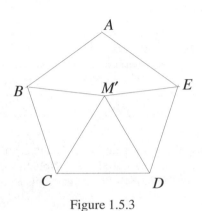

Figure 1.5.3

$\angle M'DE = 48°$. Therefore, $\angle M'BC = \angle M'ED = (180° - 48°)/2 = 66°$, It follows that $\angle M'BA = \angle M'EA = 42°$, so $M = M'$, and the assertion is proved.

4. The solution uses trigonometry, more precisely, complex numbers written in trigonometric form. By placing the geometric figure in the complex plane, we associate to each vertex a complex number coordinate, which we denote by the same letter as the point. Addition formulas for sine and cosine imply that multiplication by $e^{i\alpha}$ yields a counterclockwise rotation of angle α around the origin. Choose the system of coordinates such that the origin is at the center of symmetry of the hexagon.

Let $A_1A_2A_3A_4A_5A_6$ be the hexagon, oriented clockwise, $A_1A_2B_1$, $A_2A_3B_2$, $A_3A_4B_3$, $A_4A_5B_4$, $A_5A_6B_5$, $A_6A_1B_6$ the equilateral triangles, and C_1, C_2, C_3, C_4, C_5, C_6 the midpoints of the segments B_1B_2, B_2B_3, ..., B_6B_1, respectively.

Since B_1 is obtained by rotating A_2 around A_1 counterclockwise by $\frac{\pi}{3}$,

$$B_1 = A_1 + e^{i\frac{\pi}{3}}(A_2 - A_1) = (1 - e^{i\frac{\pi}{3}})A_1 + e^{i\frac{\pi}{3}}A_2 = e^{-i\frac{\pi}{3}}A_1 + e^{i\frac{\pi}{3}}A_2.$$

Similarly,

$$B_2 = e^{-i\frac{\pi}{3}}A_2 + e^{i\frac{\pi}{3}}A_3.$$

It follows that

$$C_1 = \frac{1}{2}\left(e^{-i\frac{\pi}{3}}A_1 + A_2 + e^{i\frac{\pi}{3}}A_3\right),$$
$$C_2 = \frac{1}{2}\left(e^{-i\frac{\pi}{3}}A_2 + A_3 + e^{i\frac{\pi}{3}}A_4\right).$$

On the other hand, if we rotate C_1 around the origin clockwise by $\frac{\pi}{3}$, we get a point of coordinate

$$\frac{1}{2}\left(e^{-i\frac{2\pi}{3}}A_1 + e^{-i\frac{\pi}{3}}A_2 + A_3\right).$$

It is the same as C_2, since $A_4 = -A_1$ by symmetry, and $e^{-2\pi i/3} = -e^{\pi i/3}$. The same argument works to show that C_i is obtained by rotating C_{i+1} clockwise around the origin by $\frac{\pi}{3}$. Hence $C_1C_2C_3C_4C_5C_6$ is a regular hexagon centered at the origin.

5. This problem is very similar to the one we solved in the introductory part of the section, using the same kind of trigonometric manipulations.

Inscribe the heptagon in a circle of radius R. The sides AB, AC, and AD subtend arcs of measure $2\pi/7, 4\pi/7$, and $6\pi/7$; hence $AB = 2R\sin\pi/7, AC = 2R\sin 2\pi/7$, and $AD = 2R\sin 3\pi/7$. The identity we want to prove is equivalent to the trigonometric identity

$$\frac{1}{\sin\frac{\pi}{7}} = \frac{1}{\sin\frac{2\pi}{7}} + \frac{1}{\sin\frac{3\pi}{7}}.$$

By eliminating the denominators, we get

$$\sin\frac{2\pi}{7}\sin\frac{3\pi}{7} = \sin\frac{\pi}{7}\sin\frac{3\pi}{7} + \sin\frac{\pi}{7}\sin\frac{2\pi}{7}.$$

Let us prove this equality. We can use the product-to-sum formula for each of the terms to get

$$-\cos\frac{5\pi}{7}+\cos\frac{\pi}{7}=-\cos\frac{\pi}{7}+\cos\frac{2\pi}{7}-\cos\frac{3\pi}{7}+\cos\frac{\pi}{7}.$$

Since $2\pi/7+5\pi/7=3\pi/7+4\pi/7=\pi$, it follows that $\cos 2\pi/7=-\cos 5\pi/7$ and $\cos 3\pi/7=-\cos 4\pi/7$, which proves the equality.

Remark. This identity also follows from Ptolemy's theorem applied to the cyclic quadrilateral $ABCD$.

(Romanian high school textbook)

6. *First solution:* Let $A_1A_2=a$, $A_1A_3=b$, $A_1A_4=c$. The previous problem shows that $a/b+a/c=1$. Since the triangles $A_1A_2A_3$ and $B_1B_2B_3$ are similar, $B_1B_2/B_1B_3=a/b$; hence $B_1B_2=a^2/b$. Analogously, $C_1C_2=a^2/c$; therefore,

$$\frac{S_B+S_C}{S_A}=\frac{a^2}{b^2}+\frac{a^2}{c^2}.$$

Then

$$\frac{a^2}{b^2}+\frac{a^2}{c^2}>\frac{1}{2}\left(\frac{a}{b}+\frac{a}{c}\right)^2=\frac{1}{2}.$$

Note that equality is not possible because $a/b\neq a/c$. This proves half of the inequality.

On the other hand,

$$\frac{a^2}{b^2}+\frac{a^2}{c^2}=\left(\frac{a}{b}+\frac{a}{c}\right)^2-\frac{2a^2}{bc}=1-\frac{2a^2}{bc}.$$

In the triangle $A_1A_3A_4$, $A_3A_4=a$, and hence by using the law of sines, we obtain

$$\frac{a^2}{bc}=\frac{\sin^2\frac{\pi}{7}}{\sin\frac{2\pi}{7}\sin\frac{4\pi}{7}}=\frac{\sin^2\frac{\pi}{7}}{8\sin^2\frac{\pi}{7}\cos^2\frac{\pi}{7}\cos\frac{2\pi}{7}}$$

$$=\frac{1}{8\cos^2\frac{\pi}{7}\cos\frac{2\pi}{7}}=\frac{1}{4\left(\cos\frac{2\pi}{7}+1\right)\cos\frac{2\pi}{7}},$$

where the denominator was transformed by applying double-angle formulas.

Since $2\pi/7>\pi/4$, $\cos 2\pi/7<\cos\pi/4=\sqrt{2}/2$, this gives

$$\frac{a^2}{bc}>\frac{1}{4\frac{\sqrt{2}}{2}\left(1+\frac{\sqrt{2}}{2}\right)}=\frac{\sqrt{2}-1}{2}.$$

It follows that $a^2/b^2+a^2/c^2<1-(\sqrt{2}-1)=2-\sqrt{2}$, which proves the right side of the inequality of the problem.

Second solution: Here is a proof that does not use trigonometry. By Ptolemy's theorem applied to the quadrilateral $A_1A_4A_5A_6$, we obtain

$$ab+ac=bc.$$

Dividing the last of these four equalities by bc, we obtain

$$\frac{a}{c} + \frac{a}{b} = 1.$$

Hence

$$\frac{x^2}{y^2} + \frac{x^2}{z^2} \geq \frac{1}{2}\left(\frac{x}{y} + \frac{x}{z}\right) = \frac{1}{2},$$

from which we obtain the inequality on the left.

On the other hand, from Ptolemy's theorem applied to the quadrilaterals $A_1A_2A_3A_4$ and $A_2A_2A_4A_5$, we obtain

$$a^2 + ac = b^2 \text{ and } a^2 + bc = c^2.$$

Hence

$$\frac{a^2}{b^2} = 1 - \frac{ac}{b^2} \text{ and } \frac{a^2}{c^2} = 1 - \frac{b}{c}.$$

Proving the inequality on the right is equivalent to

$$\frac{a^2}{b^2} + \frac{a^2}{c^2} < 2 - \sqrt{2}$$

which reduces to

$$\frac{b}{c} + \frac{ac}{b^2} > \sqrt{2}.$$

Note that by the triangle inequality in the triangle $A_1A_2A_3$, $2a > b$, hence $a/b > 1/2$. Then by the AM–GM inequality

$$\frac{b}{c} + \frac{ac^2}{b} \geq 2\sqrt{\frac{a}{b}} > \sqrt{2}$$

and we are done.

(Bulgarian Mathematical Olympiad, 1995, second solution by A. Hesterberg)

7. The solution is based on the property of the regular dodecagon that it can be decomposed as shown in Figure 1.5.4 into six equilateral triangles $P_1P_2Q_1$, $P_3P_4Q_2$, $P_5P_6Q_3$, $P_7P_8Q_4$, $P_9P_{10}Q_5$, $P_{11}P_{12}Q_6$; six squares $P_2P_3Q_2Q_1$, $P_4P_5Q_3Q_2$, $P_6P_7Q_4Q_3$, $P_8P_9Q_5Q_4$, $P_{10}P_{11}Q_6Q_5$, $P_{12}P_1Q_1Q_6$; and a regular hexagon $Q_1Q_2Q_3Q_4Q_5Q_6$. In the isosceles triangle $Q_1P_1Q_2$, $\angle P_1Q_1Q_2 = 150°$; hence $\angle Q_1Q_2P_1 = 15°$. Since

$$\angle Q_1Q_2P_1 + \angle Q_1Q_2Q_3 + \angle Q_3Q_2P_5 = 15° + 120° + 45° = 180°,$$

the points P_1, Q_2, and P_5 are collinear. Similarly, P_4, Q_3, and P_8 are collinear. All that remains to show is that the line P_3P_6 passes through the center of the square $P_4P_5Q_3Q_2$. This follows from the fact that the hexagon $P_3P_4P_5P_6Q_3Q_2$ is symmetric with respect to the center of this square.

(23rd W.L. Putnam Mathematical Competition, 1963)

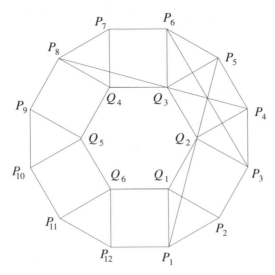

Figure 1.5.4

8. A purely geometric solution is possible. However, complex numbers enable us to organize the information better. We associate to the vertices of the square the following coordinates: $A(-1-i)$, $B(1-i)$, $C(1+i)$, $D(-1+i)$. Then the coordinates of K, L, M, and N are respectively $(\sqrt{3}-1)i$, $-(\sqrt{3}-1)$, $-(\sqrt{3}-1)i$, and $(\sqrt{3}-1)$. Consequently, the midpoints of the segments KL, LM, MN, NK have the coordinates $\pm(\sqrt{3}-1)\pm(\sqrt{3}-1)i$, and those of the segments AK, BK, BL, CL, CM, DM, DN, AN have the coordinates $\pm(2-\sqrt{3})\pm i$ and $\pm 1 \pm (2-\sqrt{3})i$.

If we rescale everything by a factor of $\sqrt{2}/(2(\sqrt{3}-1))$, we see that the 12 vertices of the hexagon are

$$\pm\frac{\sqrt{2}}{2}\pm\frac{\sqrt{2}}{2}i, \pm\frac{\sqrt{6}-\sqrt{2}}{4}\pm\frac{\sqrt{6}+\sqrt{2}}{4}i, \pm\frac{\sqrt{6}+\sqrt{2}}{4}\pm\frac{\sqrt{6}-\sqrt{2}}{4}i,$$

with all possible choices of the plus and minus signs. Writing these numbers in trigonometric form, we see that they are $\cos 2k\pi/12 + i\sin 2k\pi/12$, $k = 0, 1, 2, \ldots, 11$. Hence the complex coordinates of the vertices of the hexagon are the twelfth roots of unity, which proves that the dodecagon is regular.

(19th IMO, 1977)

9. We hunt for an isosceles triangle with an angle of $60°$. Note that A, F, B, E, D, and C are the vertices A_1, A_7, A_{10}, A_{11}, A_{17} and A_{18} of a regular polygon with 18 sides, $A_1A_2A_3\ldots A_{18}$ (see Figure 1.5.5). This polygon has a special property, namely that the diagonals A_3A_{13}, A_7A_{15}, $A_{10}A_{17}$, and $A_{11}A_{18}$ are concurrent.

Indeed, if we let M' be the intersection of A_3A_{13} and A_7A_{15}, then $\angle A_3A_7A_{15} = 60° = \angle A_7A_3A_{13}$. Hence $\angle A_3M'A_7$ has $60°$. We get $M'A_3 = A_3A_7 = A_3A_{17} = A_7A_{11}$, and hence the triangle $A_7M'A_{11}$ is isosceles. It follows that $\angle M'A_{11}A_7 = 180° - \angle M'A_{11}A_7/2 = 70°$, so $A_{11}M'$ coincides with the diagonal $A_{11}A_{18}$. Similarly, $A_{17}M'$ coincides with the diagonal $A_{17}A_{10}$. It follows that $M' = M$; hence $EF = A_7A_{11} = FM' = FM$.

Here is another argument that M' is on $A_{11}A_{18}$ (and also on $A_{17}A_{10}$) noted by R. Stong: A_3A_{13} bisects $\angle A_{11}A_3A_{15}$ and A_7A_{15} bisects $A_3A_{15}A_{11}$ hence M' is the incenter to $A_3A_{11}A_{15}$ (and similarly it is the incenter of $A_7A_{13}A_{17}$). Hence M' lies on the third bisector $A_{11}A_{18}$.

(Proposed by R. Gelca for the USAMO, 1998)

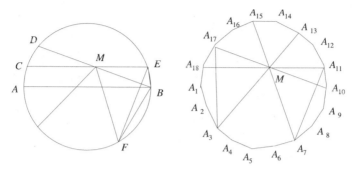

Figure 1.5.5

10. We reduce the problem to trigonometry. In the triangle $A_1A_7A_9$, $\angle A_1 = \pi/13$, $\angle A_7 = 9\pi/13$, and $\angle A_9 = 3\pi/13$, which can be shown by inscribing the regular polygon in a circle.

Let R be the circumradius of triangle $A_1A_7A_9$. Using some simple trigonometric relations in this triangle (see Figure 1.5.6), we get

$$HA_1' = A_1'A_7 \cot A_9 = -A_1A_7 \cos A_7 \cot A_9 = -2R \cos A_7 \cos A_9.$$

Similarly, $HA_7' = 2R \cos A_1 \cos A_9$ and $HA_9' = -2R \cos A_1 \cos A_7$.

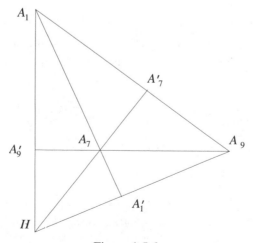

Figure 1.5.6

The identity to be proved reduces to

$$\cos\frac{9\pi}{13}\cos\frac{3\pi}{13}+\cos\frac{3\pi}{13}\cos\frac{\pi}{13}+\cos\frac{9\pi}{13}\cos\frac{\pi}{13}=-\frac{1}{4}.$$

By applying product-to-sum formulas, we transform this equality into

$$\sum_{k=1}^{6}\cos\frac{2k\pi}{13}=-\frac{1}{2}.$$

It is not hard to see that the latter equality is true. Indeed, $\cos 2k\pi/13 = \cos(26-2k)\pi/13$, $k = 1,2,\dots 6$, and $\sum_{k=0}^{12}\cos 2k\pi/13 = 0$, being the real part of the sum of the 13th roots of unity. Hence

$$\sum_{k=1}^{6}\cos\frac{2k\pi}{13}=\frac{1}{2}\sum_{k=1}^{12}\cos\frac{2k\pi}{13}=-\frac{1}{2}\cos 0=-\frac{1}{2},$$

and we are done.

(P.S. Modenov, *Zadači po geometrii (Problems in geometry)*, Nauka, Moscow, 1979)

11. Again, complex numbers and roots of unity. The polygon can be placed in the complex plane such that the vertex A_i has the coordinate $R\varepsilon_i$, $i = 1,2,\dots,n$ where ε_i are the nth roots of unity with $\varepsilon_1 = 1$. The coordinate of P is a real number. Let $x > 1$ be such that the coordinate of P is Rx. We have

$$\prod_{i=1}^{n}PA_i=\prod_{i=1}^{n}|Rx-R\varepsilon_i|=R^n\prod_{i=1}^{n}|x-\varepsilon_i|$$

$$=R^n\left|\prod_{i=1}^{n}(x-\varepsilon_i)\right|=R^n(x^n-1)=(Rx)^n-R^n=PO^n-R^n,$$

which proves the identity.

(15th W.L. Putnam Mathematical Competition, 1955)

12. *First solution:* Let l be the length of the side of the polygon and let $d = A_1A_3$. By writing Ptolemy's theorem in the quadrilateral $A_{2n}A_{2n+1}A_1A'$, we get

$$A_{2n+1}A'\cdot A_1A_{2n}=A_{2n+1}A_1\cdot A_{2n}A'+A_{2n}A_{2n+1}\cdot A_1A'.$$

Hence $dR = la_1$. Similarly Ptolemy's theorem in the quadrilaterals

$$A_{2n+1}A_1A_2A',\ A_1A_2A_3A',\dots,A_{n-1}A_nA'A_{n+1}$$

yields

$$da_1 = l(2R+a_2),\ da_2 = l(a_1+a_3),$$
$$da_3 = l(a_2+a_4),\dots,da_{n-1}=l(a_{n-2}+a_n),\ da_n=l(a_{n-1}-a_n).$$

Note that in the last relation, we used the fact that $A'A_n = A'A_{n+1} = a_n$.

By adding and subtracting these relations alternately, we get

$$d(R - a_1 + a_2 - a_3 + \cdots \mp a_{n-1} \pm a_n)$$
$$= l(a_1 - 2R - a_2 + a_1 + a_3 - a_2 - \cdots$$
$$\mp a_{n-2} \mp a_n \pm a_{n-1} \mp a_n).$$

Move everything to the left to obtain

$$(d + 2l)(R - a_1 + a_2 - a_3 + \cdots \pm a_n) = 0.$$

Since the first factor is nonzero, the second factor must be zero, which proves the identity.

Second solution: Scaling such that $R = 1$, we can assume that the vertices are placed on the unit circle such that $A' = 1$. Then the identity from the statement reduces to the trigonometric identity

$$\sum_{k=1}^{n} (-1)^{k-1} \cos \frac{k\pi}{2n+1} = \frac{1}{2}.$$

This can be proved using the telescopic method (which will be explained in more detail later in the book) as follows. Set $\alpha = \frac{\pi}{2n+1}$, and multiply both sides of the desired identity by $2\cos \frac{\alpha}{2}$. Since

$$2\cos k\alpha \cos \frac{\alpha}{2} = \cos \frac{(2k-1)\alpha}{2} + \cos \frac{(2k+1)}{2}$$

the sum telescopes to $\cos \frac{\alpha}{2} \pm \cos \frac{(2n+1)\alpha}{2}$. But $\cos \frac{(2n+1)\alpha}{2} = \cos \frac{\pi}{2}$. Hence the sum is equal to $\cos \alpha$. Dividing back by $2\cos \alpha$, we obtain the desired identity.

(Gh. Țițeica, *Probleme de geometrie (Problems in geometry)*, Ed. Tehnică, Bucharest, 1929)

13. First, recall that the area of the regular n-gon of side a is equal to $(na^2)/(4\tan \pi/n)$. This area can also be written as the sum of the areas of the triangles formed by two consecutive vertices and the point M. Therefore,

$$a(x_1 + x_2 + \cdots + x_n) = \frac{na^2}{2\tan \frac{\pi}{n}}.$$

We use two well-known inequalities: $\tan x > x$ for $x \in (0, \pi/2)$ and the arithmetic-harmonic mean inequality to obtain

$$(x_1 + x_2 + \cdots + x_n)\left(\frac{1}{x_1} + \frac{1}{x_2} + \cdots + \frac{1}{x_n}\right) \geq n^2.$$

We have

$$\frac{1}{x_1} + \frac{1}{x_2} + \cdots + \frac{1}{x_n} = \left(\frac{1}{x_1} + \frac{1}{x_2} + \cdots + \frac{1}{x_n}\right)(x_1 + x_2 + \cdots + x_n)\frac{2\tan \frac{\pi}{n}}{na^2}$$

$$> \frac{2\pi}{na^2}\left(\frac{1}{x_1} + \frac{1}{x_2} + \cdots + \frac{1}{x_n}\right)(x_1 + x_2 + \cdots + x_n) \geq \frac{2\pi}{a}$$

and the problem is solved.

(M. Pimsner and S. Popa, *Probleme de geometrie elementară (Problems in elementary geometry)*, Ed. Didactică şi Pedagogică, Bucharest, 1979)

14. The solution reduces to algebraic manipulations with complex numbers. The vertices of an arbitrary n-gon in the complex plane are of the form $z + w\zeta^j$, $j = 1, \ldots, n$, where $\zeta = e^{2\pi i/n}$ is a primitive nth root of the unity. Hence for any two complex numbers z and w,

$$\sum_{j=1}^{n} f(z + w\zeta^j) = 0.$$

In particular, replacing z by $z - \zeta^k$ for each $k = 1, 2, \ldots, n$, we have

$$\sum_{j=1}^{n} f(z - \zeta^k + \zeta^j) = 0.$$

Summing over k, and changing the order of summation, we get

$$\sum_{m=1}^{n} \sum_{k=1}^{n} f(z - (1 - \zeta^m)\zeta^k) = 0.$$

For $m = n$, the inner sum is $nf(z)$; for other m, the inner sum runs over a regular polygon, hence is zero. Thus $f(z) = 0$ for all $z \in \mathbf{C}$.

(Romanian IMO Team Selection Test, 1996, proposed by G. Barad)

15. Represent the given points by n complex numbers $z_1, z_2, \ldots, z_n$ on the unit circle. By rotating the plane, we may assume that $z_1 z_2 \cdots z_n = (-1)^n$. Consider the polynomial $P(z) = (z - z_1)(z - z_2) \cdots (z - z_n)$. Then $P(0) = (-z_1)(-z_2) \cdots (-z_n) = 1$, and the given condition is equivalent to the fact that for every $z \in \mathbf{C}$ with $|z| = 1$,

$$|P(z)| = |z - z_1| \cdot |z - z_2| \cdots |z - z_n| \leq 2.$$

Writing $P(z) = \sum_k a_k z^k$ and setting $\zeta = e^{2\pi i/n}$, we get

$$\sum_{j=0}^{n-1} P(\zeta^j) = n(z^0 P(z) + z^n P(z)) = 2n.$$

Since $|P(\zeta^j)| \leq 2$ for each j, the above equality implies that $P(\zeta^j) = 2$ for every j. We also know that $P(0) = 1$. Since a polynomial of degree n is uniquely determined by its values at $n + 1$ points, we must have $P(z) = z^n + 1$. Therefore, the points $z_1, z_2, \ldots, z_n$ are the vertices of a regular polygon.

16. Suppose that for some $n > 6$ there is a regular n-gon with vertices having integer coordinates. Let $A_1 A_2 \ldots A_n$ be the smallest such n-gon, and let a be its side-length. Let O be the origin and consider the vectors $\overrightarrow{OB_i} = \overrightarrow{A_{i-1} A_i}$, $i = 1, 2, \ldots, n$ (where $A_0 = A_n$). Then B_i has also integer coordinates for each i, and $B_1 B_2 \ldots B_n$ is a regular n-gon with side-length $2a \sin(\pi/n) < a$, which contradicts the minimality of the original polygon. Hence this is impossible.

We are left with analyzing the cases $n = 3, 4, 5$, and 6. If $n = 3, 5, 6$, consider the center of the polygon. Being the centroid, it has rational coordinates, and by changing the scale, we may assume that both the center and the vertices have integer coordinates. Rotating the polygon three times by $90°$ around its center, we obtain a regular 12-gon or a regular 20-gon whose vertices have integer coordinates. But we saw above that this is impossible. Hence we must have $n = 4$, which is indeed a solution.

(Proposed by Israel for the IMO, 1985)

1.6 Geometric Constructions and Transformations

1. Construct the symmetric image of the polygonal line with respect to P (see Figure 1.6.1).

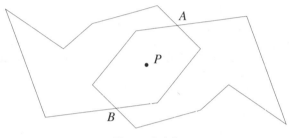

Figure 1.6.1

Both polygonal lines contain P in their interior, and since one cannot contain the other in its interior, for reasons of symmetry, they must intersect. Let A be an intersection point. Then B, the symmetric of A with respect to P, is also an intersection point, since the whole figure is symmetric with respect to P. Thus A and B lie on the first polygonal line, and P is the midpoint of AB. This solves the problem.

(W.L. Putnam Mathematical Competition)

2. This problem is similar to the previous one and relies on a $60°$ rotation. Choose a point A that lies on the polygonal line and is not a vertex. Rotate the polygonal line around A by $60°$. The image through the rotation intersects the original line once in A, so the two polygonal lines should intersect at least one more time (here we use the fact that A is not a vertex). Choose B an intersection point different from A. Note that B is on both polygonal lines, and its preimage C through rotation is on the initial polygonal line (see Figure 1.6.2). Then, the triangle ABC is equilateral, and we are done.

(M. Pimsner and S. Popa, *Probleme de geometrie elementară (Problems in elementary geometry)*, Ed. Didactică şi Pedagogică, Bucharest, 1979)

3. Let the given segment be AB. Using the straightedge, extend the segment in both directions, then construct equilateral triangles AA_1A_2 and BB_1B_2 on both sides of the line AB, as shown on the left in Figure 1.6.3. The two equilateral triangles are symmetric with respect to the midpoint of AB, hence the line A_2B_2 passes through the midpoint of the segment AB. This solves part (a). For part (b), we use homothety instead of

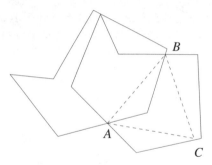

Figure 1.6.2

symmetry. Drawing four equilateral triangles as shown on the right in Figure 1.6.4, we can obtain an equilateral triangle twice the size of the template. Do this at the endpoint B of the segment, while drawing at A just one triangle as shown in the figure, to obtain the equilateral triangles AA_1A_2 and BB_1B_2. These triangles are homothetic, with center of homothety the intersection M of the line AB with the line A_2B_2. The ratio of homothety is $1/2$, hence $AM/AB = 1/3$. Doing this construction on the other side yields a point N such that $AM = MN = NB$, and we are done.

(Russian Mathematical Olympiad, 1988–1989)

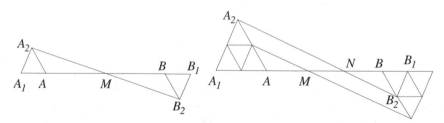

Figure 1.6.3

4. Let $ABCD$ be the trapezoid to be constructed, with AB and CD the bases. First construct the side AB, then the circles $\mathcal{C}_1$ of center A and radius AD and $\mathcal{C}_2$ of center B and radius BC (see Figure 1.6.4). Since the point C is obtained by translating D by a vector $\vec{v}$ parallel to AB and of length DC, it follows that C can be constructed as the

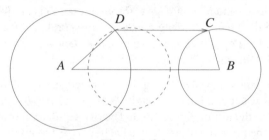

Figure 1.6.4

intersection of the circle $\mathscr{C}_2$ with the translation by $\vec{v}$ of the circle $\mathscr{C}_1$. Finally, D can be obtained by intersecting $\mathscr{C}_1$ and the circle of center C and radius CD.

(L. Duican and I. Duican, *Trasformări geometrice (Geometric transformations)*, Ed. Științifică și Enciclopedică, Bucharest, 1987)

5. Since the segment connecting the midpoints of the nonparallel sides of the trapezoid has its length equal to half the sum of lengths of the two bases, the problem reduces to constructing the trapezoid $ABCD$ knowing the lengths of AC, BD, and $AB+CD$, and the measure of the angle $\angle DAB$ (see Figure 1.6.5). Note that by translating the diagonal BD by the vector $\overrightarrow{DC}$, one obtains a triangle $AB'C$ whose lengths of sides are known (since $AB' = AB + CD$). Thus we can start by constructing this triangle first, then the angle $\angle D'AB'$ (equal to $\angle DAB$). The point D is obtained by intersecting AD' with the parallel through C to AB'. Finally, B is obtained by intersecting AB' with the parallel to $B'C$ through D.

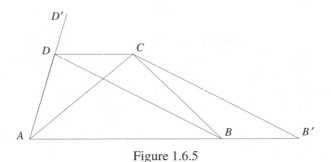

Figure 1.6.5

(L. Duican and I. Duican, *Trasformări geometrice (Geometric transformations)*, Ed. Științifică și Enciclopedică, Bucharest, 1987)

6. It is important to note that B is the image of C through the dilation ρ of center A and ratio $\frac{1}{2}$. Choose A on the exterior circle $\mathscr{C}_1$ and construct $\rho(\mathscr{C}_2)$ (Figure 1.6.6). Choose B one of the points of intersection of $\mathscr{C}_2$ and $\rho(\mathscr{C}_2)$. Then the line AB satisfies the required condition, since if we denote by C and D the other points of intersection of

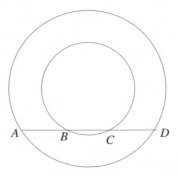

Figure 1.6.6

AB with $\mathscr{C}_2$ and $\mathscr{C}_1$, respectively, then $AB = BC$ from the construction, and $AB = CD$ from the symmetry of the figure.

Note that the construction is impossible if $\mathscr{C}_2$ and $\rho(\mathscr{C}_2)$ are disjoint.

7. Let d_1 and d_2 be the two parallel lines, and let N and P be the intersections of m and d_1 and n and d_2, respectively (Figure 1.6.7).

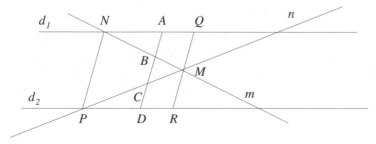

Figure 1.6.7

Let Q and R be the intersections with d_1 and d_2 of the parallel to NP passing through M. Then the line with the required property is the translation of the line NP by the vector $2\overrightarrow{NQ}/3$. Indeed, by similarity of triangles, the segment BC determined on this line by m and n has length $NP/3$. Also, if A and D are the intersections of the line BC with d_1 and d_2, then $AB = 2QM/3 = NP/3$, and of course, $CD = NP/3$ as well. Thus the three segments are equal.

(D. Smaranda and N. Soare, *Trasformări geometrice (Geometric transformations),* Ed. Academiei, Bucharest, 1988)

8. The solution is based on the following observation. If N and P are two points in the plane and N' and P' are their respective images through a rotation of center M and angle α, then P' can also be obtained by translating P by the vector $\overrightarrow{NN'}$ to P'' and then rotating P'' by the angle α around N'. This follows from the fact that the segments NP and $N'P'$ are equal and the angle between them is α.

The problem asks us to rotate B around A through $60°$. Choose the points $P_0 = A$, $P_1, \ldots, P_n = B$ such that $P_1P_2 = P_2P_3 = \cdots = P_{n-1}P_n = 1$, assuming that the opening of the compass is also equal to 1. One can easily construct the equilateral triangle $P_0P_1Q_1$, in which case Q_1 is the rotation of P_1 around A by $60°$. One then translates inductively P_k to R_k by the vector $\overrightarrow{P_{k-1}Q_{k-1}}$ (as explained in the first part of the book), and then rotates R_k around R_{k-1} by $60°$ to get Q_k. It follows by induction that Q_k is the rotation of P_k around A through $60°$. In particular, this is true for $C = Q_n$, which is the rotation of B, and the problem is solved.

(R. Gelca)

9. The problem is analogous to the previous one, but the rotation is combined with a dilation. As before, let the opening of the compass be equal to 1. If $AB = 1$, the problem can be solved using three equilateral triangles, as seen in Figure 1.6.8. In this case, $AC/BC = \sqrt{3}$.

We want to perform in general the composition of a $90°$ rotation with a dilation of ratio $\sqrt{3}$. For this, we use the following property. If N' and P' are the images of N

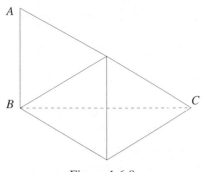

Figure 1.6.8

and P through the rotation of center M and angle α followed by the dilation of center M and ratio r, then P' can be obtained by translating P to P'' with the vector $\overrightarrow{NN'}$, and then rotating P'' around N' through the angle α and dilating with ratio r and center N' (see Figure 1.6.9). Indeed, the segment $P''N'$ is parallel to the segment PN, and the segments $P'N'$ and PN are each the image of the other through the rotation–dilation of center M, thus their lengths have ratio r, and the angle between them is α.

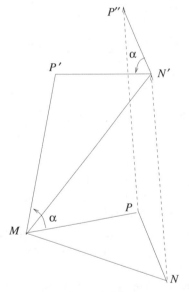

Figure 1.6.9

From here the argument follows *mutatis mutandis* the steps of the solution to problem 7.   Choose $P_0 = A, P_2, \ldots, P_n = B$ such that $P_0P_1 = P_1P_2 = \cdots = P_{n-1}P_n = 1$. Do the rotation–dilation of P_1 around $P_0 = A$ of angle $90°$ and ratio $\sqrt{3}$; then successively let R_k be the translation of P_k of vector $\overrightarrow{P_{k-1}Q_{k-1}}$ and Q_k the rotation–dilation of R_k around Q_{k-1} by an angle of $90°$ and ratio $\sqrt{3}$. Inductively we get that Q_k is the image of P_k through the rotation–dilation of center A. Thus if we choose $C = Q_n$, then $\angle CAB = 90°$ and we are done.

(R. Gelca)

10. There are three important facts necessary for finding the center of a circle with a compass:

(a) Using only a compass, a point may be inverted with respect to a circle of known center.

(b) Using a compass, a point may be reflected across a line when only two points of the line are given.

(c) Given a circle centered at O passing through the center of inversion Q and its image line AB, the image of O through the inversion is the reflection of Q across AB.

For (a), note that the inverse of a point P with respect to a circle of center Q can be constructed as follows. First intersect the given circle with a circle centered at P. Let M and N be the two intersection points (Figure 1.6.10). The circles of centers M and N and radii equal to QM intersect a second time at P'. From the similarity of triangles PMQ and $MP'Q$, it follows that $PQ \cdot P'Q = MQ^2$; hence P' is the inverse of P.

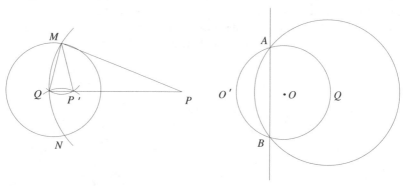

Figure 1.6.10   Figure 1.6.11

The second claim follows from the fact that the reflection of P with respect to the line AB can be constructed as the second intersection of the circle of center A and radius AP with the circle of center B and radius BP.

Finally, to prove the third claim, note that if M and M' are the intersections of the line OQ with the circle, respectively the line, and if O' is the image of O through the inversion, then $QM \cdot QM' = QO \cdot QO'$, while $QM = 2QO$. Thus $QO' = QM'$, which shows that O' is the symmetric image of O with respect to the line.

To solve the problem, we will first construct the image through an inversion of the center O we are trying to find. To this end, choose an arbitrary point Q on the circle and draw a circle larger than the given one, centered at Q (Figure 1.6.11). Denote by A and B the intersections of the two circles. Reflect Q across AB to O'. By property (c) mentioned above, O' is the image of O through the inversion with respect to the circle centered at Q. Inverting O' with respect to the circle of center Q to O solves the problem.

1.7 Problems with Physical Flavor

1. Let $A_1 A_2 \ldots A_n$ be the polygon and $\overrightarrow{v}_i$ the vector orthogonal to $A_i A_{i+1}$, $i = 1, 2, \ldots, n$ ($A_{n+1} = A_1$). The sum $\overrightarrow{v}_1 + \overrightarrow{v}_2 + \cdots + \overrightarrow{v}_n$ is proportional to the rotation of $\overrightarrow{A_1 A_2} + \overrightarrow{A_2 A_3} + \cdots + \overrightarrow{A_n A_1}$ by $90°$. Since the latter sum is zero (the vectors form a closed polygonal line), the sum of the vectors $\overrightarrow{v}_i$ is zero as well.

2. This is the spatial version of the previous problem. It has the following physical interpretation. If the polyhedron is filled with gas at pressure numerically equal to one, then by Pascal's law the vectors are the forces that act on the faces. The property states that the sum of these forces is zero, namely that the polyhedron is in equilibrium. This is obvious from the physical point of view, since no exterior force acts on it

Note that it is sufficient to prove the property for a tetrahedron. The case of the general polyhedron is then easily obtained by dividing the polyhedron into tetrahedra and using the fact that the forces on the interior walls cancel each other.

First solution· Let $ABCD$ be the tetrahedron and $\overrightarrow{v}_1$, $\overrightarrow{v}_2$, $\overrightarrow{v}_3$, and $\overrightarrow{v}_4$ be the four vectors perpendicular to the faces, as shown in Figure 1.7.1.

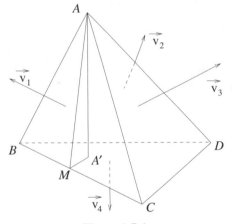

Figure 1.7.1

Let α_1, α_2, and α_3 be the angles that the planes (ABC), (ABD), and (ACD) form with the plane (BCD). For simplicity, we will carry out the proof in the case where all three angles are acute, the other cases being analogous.

The vertical component of $\overrightarrow{v}_1 + \overrightarrow{v}_2 + \overrightarrow{v}_3$ points upward and has length

$$\| \overrightarrow{v}_1 \| \cos \alpha_1 + \| \overrightarrow{v}_2 \| \cos \alpha_2 + \| \overrightarrow{v}_3 \| \cos \alpha_3.$$

This is the same as

$$\sigma_{ABC} \cos \alpha_1 + \sigma_{ABD} \cos \alpha_2 + \sigma_{ACD} \cos \alpha_3,$$

where we denote by σ_{XYZ} the area of the triangle XYZ. If we let A' be the projection of A onto the plane (BCD), then the three terms of the above sum are the areas of the triangles $A'BC$, $A'BD$, and $A'CD$, so they add up to the area of the triangle BCD. This shows that the vertical component of the sum of the four vectors is zero.

On the other hand, the horizontal component of $\overrightarrow{v}_1$ is orthogonal to BC and has length $\sigma_{ABC} \sin \alpha_1$. If we let $M \in BC$ such that $AM \perp BC$ (Figure 1.7.1), then $\alpha_1 = \angle AMA'$; hence $\sin \alpha_1 = AA'/AM$. This implies $\sigma_{ABC} \sin \alpha_1 = AA' \cdot BC/2$. Similarly, $\sigma_{ABD} \sin \alpha_2 = AA' \cdot BD/2$ and $\sigma_{ACD} \sin \alpha_3 = AA' \cdot CD/2$. It follows from the previous problem that the three horizontal components add up to zero, and since the fourth vector has no horizontal component, the problem is solved.

Second solution: Here is another solution suggested by R. Stong. Let $P_{BCD}(x)$ be the volume of the tetrahdron with base BCD and vertex $\vec{x}$. Then

$$P_{BCD}(x) - P_{BCD}(y) = \frac{1}{3}(\vec{y} - \vec{x}) \cdot \vec{v}_4$$

(for x and y exterior this still works with signed volumes). Since the sum of the left-hand side over all four faces of the tetrahedron is just *volume(ABCD)* − *volume(ABCD)* $= 0$, we see that

$$(\vec{y} - \vec{x}) \cdot (\vec{v}_1 + \vec{v}_2 + \vec{v}_3 + \vec{v}_4) = 0,$$

for all interior $\vec{x}$ and $\vec{y}$. Hence the sum is zero.

3. Consider the unit vectors $\overrightarrow{v}_1$, $\overrightarrow{v}_2$, $\overrightarrow{v}_3$, and $\overrightarrow{v}_4$, perpendicular to the faces and pointing outwards. The sum S of cosines of the dihedral angles of tetrahedron is the negative of the sum of the dot products $\overrightarrow{v}_i \cdot \overrightarrow{v}_j$, for all $i \neq j$. Thus,

$$-2S + (\overrightarrow{v}_1^2 + \overrightarrow{v}_2^2 + \overrightarrow{v}_3^2 + \overrightarrow{v}_4^2) = (\overrightarrow{v}_1 + \overrightarrow{v}_2 + \overrightarrow{v}_3 + \overrightarrow{v}_4)^2.$$

From this it follows that

$$S = 2 - \frac{1}{2}(\overrightarrow{v}_1 + \overrightarrow{v}_2 + \overrightarrow{v}_3 + \overrightarrow{v}_4)^2,$$

and the inequality is established.

For the equality, $S = 2$ if and only if $\overrightarrow{v}_1 + \overrightarrow{v}_2 + \overrightarrow{v}_3 + \overrightarrow{v}_4 = 0$. On the other hand, by the previous problem, the vectors that are perpendicular to the faces and have lengths equal to the areas of the faces add up to zero as well. Since these vectors are not coplanar, it follows that they are proportional to the $\overrightarrow{v}_i$'s, hence the areas of the faces are equal, and the problem is solved.

(I. Sharigyn)

4. Let $ABCD$ be the tetrahedron. Place at each vertex of the tetrahedron a planet of mass 1, and assume that the gravitational potential is linear (which is true in a first-order approximation). Then the point P has minimal potential with respect to the gravitational fields of the four planets; hence an object placed at P must be in equilibrium. It follows that the sum of the attractive forces of the planets A and B has the same magnitude and opposite direction as the sum of the attractive forces of planets C and D. The direction of the first resultant is given by the bisector of $\angle APB$ and the direction of the second by the bisector of the angle $\angle CPD$.

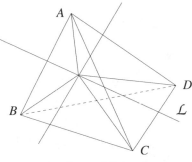

Figure 1.7.2

For a rigorous proof, let $\mathcal{E}_1$ be the ellipsoid of revolution defined by the equation

$$AX + BX = AP + PB$$

and let $\mathcal{E}_2$ be the ellipsoid of revolution defined by

$$CY + DY = CP + DP.$$

Because of the minimality of the sum of distances, the interiors of $\mathcal{E}_1$ and $\mathcal{E}_2$ have no common points; hence the two ellipsoids are tangent. The bisectors of $\angle APB$ and $\angle CPD$ are normal to the common tangent plane, so they have the same supporting line. The same argument works for the other pairs of angles.

Let $\mathcal{L}$ be the supporting line of the bisectors of $\angle APB$ and $\angle CPD$. If we rotate the whole figure around $\mathcal{L}$ by $180°$, then the line AP becomes the line BP, and DP becomes CP (see Figure 1.7.2). Consequently, the common bisector of $\angle APD$ and $\angle BPC$ is invariant under the rotation; hence is orthogonal to $\mathcal{L}$.

(9th W.L. Putnam Mathematical Competition, 1949)

5. Assume that this is not true. Construct the polyhedron out of some inhomogeneous material in such a way that the given point is its center of mass. Then, since the point always projects outside any face, if placed on a plane, the polyhedron will roll forever. Thus we have constructed a perpetuum mobile. This is physically impossible. The move clearly stops when the point reaches its lowest potential.

This suggests that the point projects inside the face that is closest to it. Suppose this is not so, and let P be the point, F the face closest to it, and P' the projection onto the plane of F (see Figure 1.7.3). Let F' be a face that is intersected by PP', and let M be the intersection point. Then PM is not orthogonal to F', and hence the distance from P to F' is strictly less than PM, which in turn is less than the distance from P to F. This contradicts the minimality of the distance from P to F, and the problem is solved.

6. One can think about the road as the path traced by a beam of light that passes through a very dense medium (the river). Since the beam will follow the quickest path, the dense medium will be crossed in a direction perpendicular to its sides. Consequently, the beam will trace the road $AMNB$. The reader with some knowledge of physics will know that the beam enters the medium at the same angle that it exits. Thus

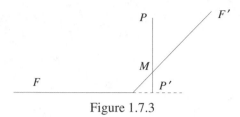

Figure 1.7.3

we can actually forget the existence of the river and shift the cities by the width of the river toward each other, take the shortest path, then insert the river (see Figure 1.7.4). Rigorously, since the length of the bridge is always the same, consider the translation A' of A toward B with vector perpendicular to the shore of the river and of length equal to the length of the bridge. Minimizing the length of $AMNB$ is the same as minimizing the length of $A'NB$. The latter is minimized when A',N, and B are collinear. This locates the position of N, and we are done.

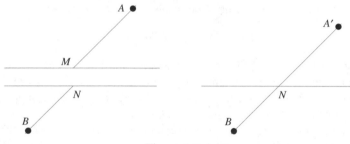

Figure 1.7.4

7. We will prove the following general statement.

Given n points on a circle, a perpendicular is drawn from the centroid of any $n-2$ of them to the chord connecting the remaining two. Prove that all lines obtained in this way have a common point.

First solution: We base the proof on the following property of the centroid:

The centroid of the system of $k+j$ points $A_1,A_2,\ldots,A_k,B_1,B_2,\ldots,B_j$ divides the line segment G_1G_2 in the ratio $j{:}k$ where G_1 and G_2 are the centroids of the systems $A_1,A_2,\ldots,A_k$ and $B_1,B_2,\ldots,B_j$.

Returning to the problem, let O be the center of the circle, G the centroid of the given n points, G_1 the centroid of some $n-2$ of them, and M the midpoint of the chord joining the remaining two (which is also the centroid of the system formed by these two points). Let l be the perpendicular from G_1 onto the chord formed by the remaining two points. By the above-mentioned property, the points G, G_1, and M are collinear, and $G_1G{:}GM = 2{:}(n-2)$. Denote by P the point of intersection of the lines l and OG. The triangles GG_1P and GMO are similar, since OM is parallel to l. Hence $GP{:}OG = 2{:}(n-2)$, and thus the point P is uniquely determined by O and G and so is common to all lines under consideration.

Second solution: An alternate solution is to use complex coordinates. First note that for points $\omega_i \in \mathbf{C}$ with $|\omega_i| = 1$, the ray from the origin through $\omega_i + \omega_j$ is always perpendicular to the line through ω_i and ω_j. It follows that $\sum_{i=1}^{n} \omega_i/(n-2)$ lies on all the perpendiculars, and we are done.

(Proposed by T. Andreescu for the USAMO, 1995, second solution by R. Stong)

8. Focus on the centroid of S. We know that the centroid lies on the perpendicular bisector of the segment determined by any two points in S. Thus all points in S lie on a circle centered at the centroid. From here the problem is simple. Just pick three consecutive points A, B, C. Since S is symmetric with respect to the perpendicular bisector of AC, B must be on this perpendicular bisector. Hence $AB = BC$. Repeating the argument for all triples of consecutive points, we conclude that S is a regular polygon. Clearly, all regular polygons satisfy the given condition.

(40th IMO, 1999; proposed by Estonia)

9. *First solution.* Say that the $\binom{n}{k}$ vertices with k ones and $n-k$ zeros are at level k. Suppose that a current enters the network at $A = (0,0,\ldots,0)$ and leaves from $B = (1,1,\ldots,1)$. By symmetry, all vertices at level k have the same potential, so they might be collapsed into one vertex without changing the way the current flows. Thus the resistance between A and B in the network is the same as that of a series combination of n resistors, the kth being the parallel combination of the $\binom{n}{k}(n-k)$ edges (1-ohm resistors) joining level k to level $k+1$. Thus the desired resistance is

$$R_n = \sum_{k=0}^{n-1} \frac{1}{\binom{n}{k}(n-k)} = \frac{1}{n}\sum_{k=0}^{n-1}\binom{n-1}{k}^{-1}.$$

It remains to show that

$$\frac{2^n}{n}\sum_{k=0}^{n-1}\binom{n-1}{k}^{-1} = \sum_{k=1}^{n}\frac{2^k}{k}.$$

We will prove this identity by induction. It is clearly true if $n = 1$. For the induction step note that

$$\sum_{k=1}^{n+1}\frac{2^k}{k} = \sum_{k=1}^{n}\frac{2^k}{k} + \frac{2^{n+1}}{n+1}.$$

Thus it remains to show that

$$\frac{2^n}{n}\sum_{k=0}^{n-1}\binom{n-1}{k}^{-1} = 2\frac{n}{n+1}\sum_{k=0}^{n-1}\binom{n}{k}^{-1}.$$

We have

$$2\frac{n}{n+1}\sum_{k=0}^{n-1}\binom{n}{k}^{-1} = \frac{n}{n+1}\sum_{k=1}^{n-1}\left[\binom{n}{k}^{-1} + \binom{n}{k-1}^{-1}\right]$$
$$= \frac{n}{n+1}\sum_{k=1}^{n-1}\frac{(k-1)!(n-k-1)!(n+1)}{n(n-1)!} = \sum_{k=1}^{n-1}\binom{n-1}{k}^{-1},$$

and the identity is proved.

Second solution. The n-dimensional cube consists of two copies of the $(n-1)$-dimensional cube, namely $Q_{n-1}(0) = \{(x_1,x_2,\ldots,x_{n-1},0)|x_i=0 \text{ or } 1\}$ and $Q_{n-1}(1) = \{(x_1,x_2,\ldots,x_{n-1},1)|x_i=0 \text{ or } 1\}$. Consider the response of the network to a current of 1 ampere entering at $A = (0,0,\ldots,0)$ and leaving from $B = (1,1,\ldots,1)$. With B taken as the ground, by Ohm's law the potential (voltage) at A is R_n. By symmetry, the current is $1/n$ in each of the edges leaving A or entering B, so this solution yields potentials of $R_n - 1/n$ at $C = (0,0,\ldots,0,1)$ and $1/n$ at $D = (1,1,\ldots,1,0)$. Similarly, if a current of 1 ampere enters at $C = (0,0,\ldots,0,1)$ and leaves from D, the potentials at A, B, C, D are respectively $R_n - 1/n, 1/n, R_n, 0$.

Now suppose currents of 1 ampere enter at A and C and leave at B and D. Superposition of the two results above yields potentials of $2R_n - 1/n, 1/n, 2R_n - 1/n, 1/n$ at A, B, C, D, respectively. Thus, the potential of A above D is $2R_n - 2/n$.

On the other hand, symmetry shows that in this case, the net current between each vertex in $Q_{n-1}(0)$ and its corresponding vertex in $Q_{n-1}(1)$ is zero, so it is as if there were no connection between the two $(n-1)$-dimensional cubes. Thus the potential of A above D must be R_{n-1}. Consequently,

$$2R_n - \frac{2}{n} = R_{n-1},$$

and hence

$$R_n = \sum_{k=1}^{n} \frac{1}{k2^{n-k}}.$$

Note that by putting the two solutions together, one gets a physical proof of a combinatorial identity.

(Proposed by C. Rousseau for USAMO, 1996)

1.8   Tetrahedra Inscribed in Parallelepipeds

1. Inscribe the tetrahedron in a parallelepiped as in Figure 1.8.1. If x, y, z are the edges of the parallelepiped, and a, b, c are the edges of the tetrahedron, then since the parallelepiped is right, the Pythagorean theorem gives

$$x^2 + y^2 = a^2,$$
$$y^2 + z^2 = c^2,$$
$$x^2 + z^2 = b^2.$$

This implies

$$x = \sqrt{\frac{a^2 + b^2 - c^2}{2}}; \quad y = \sqrt{\frac{a^2 + c^2 - b^2}{2}}; \quad z = \sqrt{\frac{b^2 + c^2 - a^2}{2}}.$$

The volume of the parallelepiped is xyz, and the volume of the tetrahedron is one third of that; this is equal to

$$\frac{1}{6\sqrt{2}} \sqrt{(a^2 + b^2 - c^2)(a^2 + c^2 - b^2)(b^2 + c^2 - a^2)}.$$

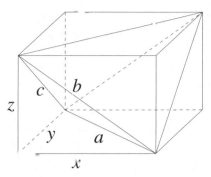

Figure 1.8.1

(H. Steinhaus, *One hundred problems in elementary mathematics*, Dover Publ. Inc., New York, 1979)

2.   The answer is immediate once we notice that the circumsphere of the tetrahedron is also the circumsphere of the right parallelepiped in which it is inscribed. The circumradius of a right parallelepiped is half its main diagonal. If a, b, c are the lengths of the edges of the tetrahedron, then the edges x, y, and z of the parallelepiped satisfy $x^2 + y^2 = a^2$, $y^2 + z^2 = b^2$, and $x^2 + z^2 = c^2$. The length of the diagonal of the parallelepiped is

$$\sqrt{x^2 + y^2 + z^2} = \sqrt{\frac{a^2 + b^2 + c^2}{2}},$$

so the circumradius is $\sqrt{(a^2 + b^2 + c^2)/8}$.

3.   Inscribe the tetrahedron in the parallelepiped $AECFGBHD$ as in Figure 1.8.2. Then M and N are the centers of the faces $AEBG$ and $CHDF$, so the centroid O of the parallelepiped is on MN. Repeating the argument, we get that O is the intersection of MN, PQ, and RS. Note that O is also the centroid of $ABCD$.

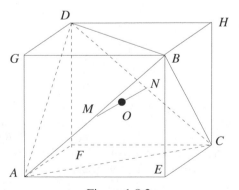

Figure 1.8.2

4. Inscribe the tetrahedron in a parallelepiped. Since one of the diagonals of the face $AEBG$ is AB and the other one is parallel to CD, it follows that $AEBG$ is a parallelogram with orthogonal diagonals, hence a rhombus. A similar argument shows that $AECF$ is a rhombus. Thus $BH = BG = BE$, which shows that the face $BHCE$ is a rhombus, and consequently $AD \perp BC$.

5. This problem is an application of the following theorem.

Given a point A and a plane not containing A, let l be a line in the plane, B the projection of A onto the plane, and C a point in the plane. Then BC is orthogonal to l if and only if AC is orthogonal to l.

For the proof of this result, note that AB is orthogonal to the plane, hence to l. Then any of the two orthogonalities implies that l is orthogonal to the plane ABC, hence implies the other orthogonality.

Let $ABCD$ be the tetrahedron. If $AH_1 \perp (BCD)$, where $H_1 \in (BCD)$ (see Figure 1.8.3), then H_1 is the orthocenter of the triangle BCD. Indeed, since $AH_1 \perp (BCD)$ and $AB \perp CD$, then $BH_1 \perp CD$ by the above theorem. Similarly, $CH_1 \perp BD$ and $DH_1 \perp BC$. Let $BH_1 \cap CD = \{M\}$ and $H_2 \in AM$ with $BH_2 \perp AM$. Since $CD \perp AH_1$ and $CD \perp AB$, we have $CD \perp (ABM)$; hence $CD \perp BH_2$. This shows that $BH_2 \perp (ACD)$, and hence H_2 is the orthocenter of the triangle ACD. As a consequence of this construction, AH_1 and BH_2 intersect, and by symmetry, if CH_3 and DH_4 are the other altitudes, then any two of the lines AH_1, BH_2, CH_3, and DH_4 intersect. Since no three of these lines lie in the same plane, it follows that they all intersect at the same point. Their intersection point is usually called the orthocenter of the tetrahedron.

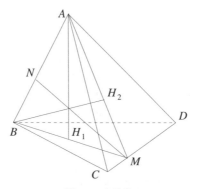

Figure 1.8.3

6. Recall the solution to the previous problem. If we let N be on AB such that $MN \perp AB$, then MN contains the orthocenter of the tetrahedron, and MN is the common perpendicular of the lines AB and CD. The tetrahedron $A_1B_2C_1D_2$ is orthogonal; hence the common perpendiculars of the pairs of lines from the statement contain the orthocenter of the tetrahedron; hence they intersect.

7. Recall the parallelogram identity, which states that in a parallelogram $MNPQ$, $MP^2 + NQ^2 = MN^2 + NP^2 + PQ^2 + QM^2$ holds. Inscribe the tetrahedron in a rhomboidal parallelepiped with edge of length equal to a. By writing the parallelogram

identity for each face, we get $AB^2 + CD^2 = 4a^2 = BC^2 + AD^2 = AD^2 + BC^2$, and the problem is solved.

8. Let $ABCD$ be the given tetrahedron, inscribed in the rhomboidal parallelepiped $AHDGECFB$. The volume of the tetrahedron is one third of the volume of the parallelepiped. On the other hand, the volume of the parallelepiped is six times the volume of the tetrahedron $EABC$. Thus let us compute the latter.

Set $AB = a$, $BC = b$, $AC = c$, and $EH = d$ (note that $BD = d$). As seen in the previous problem, these four edges completely determine the remaining two. The area of the face ABC is, by Hero's formula, equal to $s = \sqrt{p(p-a)(p-b)(p-c)}$, where $p = (a+b+c)/2$. On the other hand, since the edges EA, EB, and EC are equal, the vertex E projects in the circumcenter O of face ABC. To compute the altitude EO, note that since the face $AECH$ is a rhombus, $AE = 1/2\sqrt{c^2 + d^2}$, and in triangle ABC, $AO = abc/(4s) = abc/(4\sqrt{p(p-a)(p-c)(p-c)})$. The Pythagorean theorem implies

$$EO = \sqrt{EA^2 - AO^2} = \frac{1}{2}\sqrt{c^2 + d^2 - \frac{a^2b^2c^2}{4p(p-a)(p-b)(p-c)}}.$$

Hence the volume of the tetrahedron $ABCD$ is

$$\frac{2}{3}EO \cdot s = \frac{1}{6}\sqrt{4(c^2 + d^2)p(p-a)(p-b)(p-c) - a^2b^2c^2}.$$

9. Let $ABCD$ be the tetrahedron, inscribed as before, in the parallelepiped $AHDGECFB$. Let M and N be the projections of A and D on the plane $(ECFB)$. Let $MP \perp BC$ and $NQ \perp BC$, $P, Q \in BC$ (Figure 1.8.4). By the result mentioned in the solution to Problem 5, $AP \perp BC$ and $DQ \perp BC$, and since the triangles ABC and DBC have the same area, we get $AP = DQ$. This implies that the triangles AMP and DNQ are congruent, so $MP = NQ$. Thus M and N are at the same distance from BC. Since A and D are also at equal distance from GH, it follows that BC is the projection of GH onto the plane $(ECFB)$; hence the planes $(GBCH)$ and $(ECFB)$ are orthogonal. Here we used the fact that the planes $(AHDG)$ and $(EDFB)$ are parallel, and that the lines BC and GH are parallel as well. This shows that CH is orthogonal to both EC and CF.

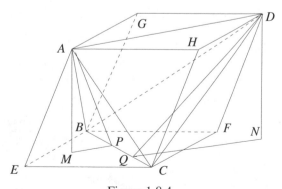

Figure 1.8.4

A similar argument gives $CE \perp CF$, thus the parallelepiped $AHDGECFB$ is right, so the tetrahedron is isosceles.

10. Let $ABCD$ be the tetrahedron and suppose that A projects in the orthocenter H of the triangle BCD. Let CM be a an altitude of this triangle. Since AH is orthogonal to the plane (BCD) and $CM \perp BD$, by the theorem mentioned in the solution to Problem 5, it follows that $AC \perp BD$. Similarly, $AD \perp BC$ and $AB \perp CD$. Thus the tetrahedron is orthogonal, which implies that the associated parallelepiped is rhomboidal.

On the other hand, the volume formula and the equality of altitudes implies the equality of the areas of the four faces. Problem 9 implies that the tetrahedron is isosceles; thus the associated parallelepiped is right. But a rhomboidal right parallelepiped is a cube, and its associated tetrahedron is regular.

(Romanian Mathematical Olympiad, 1975; proposed by N. Popescu)

1.9   Telescopic Sums and Products in Trigonometry

1. The addition formula for cosine implies

$$\sin kx \sin x = \cos kx \cos x - \cos(k+1)x,$$

where k is an arbitrary positive integer. Dividing both sides of the equality by $\sin x \cos^k x$ yields

$$\frac{\sin kx}{\cos^k x} = \frac{\cos kx}{\sin x \cos^{k-1} x} - \frac{\cos(k+1)x}{\sin x \cos^k x}.$$

We have

$$\frac{\sin x}{\cos x} + \frac{\sin 2x}{\cos^2 x} + \frac{\sin 3x}{\cos^3 x} + \cdots + \frac{\sin nx}{\cos^n x}$$
$$= \frac{\cos x}{\sin x} - \frac{\cos 2x}{\sin x \cos x} + \frac{\cos 2x}{\sin x \cos x} - \frac{\cos 3x}{\sin x \cos^2 x}$$
$$+ \cdots + \frac{\cos nx}{\sin x \cos^{n-1} x} - \frac{\cos(n+1)x}{\sin x \cos^n x}$$
$$= \cot x - \frac{\cos(n+1)x}{\sin x \cos^n x},$$

and the identity is proved.

(C. Ionescu-Ţiu and M. Vidraşcu, *Exerciţii şi probleme de trigonometrie (Exercises and problems in trigonometry)*, Ed. Didactică şi Pedagogică, Bucharest, 1969)

2. Multiplying the relation by $\sin 1°$, we obtain

$$\frac{\sin 1°}{\cos 0° \cos 1°} + \frac{\sin 1°}{\cos 1° \cos 2°} + \cdots + \frac{\sin 1°}{\cos 88° \cos 89°} = \frac{\cos 1°}{\sin 1°}.$$

This can be rewritten as

$$\frac{\sin(1° - 0°)}{\cos 1° \cos 0°} + \frac{\sin(2° - 1°)}{\cos 2° \cos 1°} + \cdots + \frac{\sin(89° - 88°)}{\cos 89° \cos 88°} = \cot 1°.$$

From the identity

$$\frac{\sin(a-b)}{\cos a \cos b} = \tan a - \tan b,$$

it follows that the left side equals

$$\sum_{k=1}^{89} [\tan k° - \tan(k-1)°] = \tan 89° - \tan 0° = \cot 1°,$$

and the identity is proved.
   (USAMO, 1992)

3. Transform the sum as

$$\sum_{k=1}^{n} \frac{1}{\cos a - \cos(2k+1)a} = \frac{1}{2} \sum_{k=1}^{n} \frac{1}{\sin ka \sin(k+1)a}.$$

As in the solution to the previous problem, after multiplication by $\sin a$, the sum becomes

$$\frac{1}{2} \sum_{k=1}^{n} \frac{\sin((k+1)a - ka)}{\sin ka \sin(k+1)a} = \frac{1}{2} \sum_{k=1}^{n} (\cot ka - \cot(k+1)a)$$

$$= \frac{1}{2}(\cot a - \cot(n+1)a).$$

Hence the answer to the problem is $(\cot a - \cot(n+1)a)/(2\sin a)$.

4. We have

$$\sum_{k=1}^{n} \tan^{-1} \frac{1}{2k^2} = \sum_{k=1}^{n} \tan^{-1} \frac{(2k+1) - (2k-1)}{1 + (2k-1)(2k+1)}.$$

Using the subtraction formula for the arctangent (see the introduction to Section 1.9), the latter sum becomes

$$\sum_{k=1}^{n} (\tan^{-1}(2k+1) - \tan^{-1}(2k-1)).$$

This is equal to

$$\tan^{-1}(2n+1) - \tan^{-1}(1) = \tan^{-1} \frac{(2n+1) - 1}{1 + (2n+1)} = \tan^{-1} \frac{n}{n+1}.$$

5. Note that

$$\tan x = \frac{1}{\tan x} - \frac{1 - \tan^2 x}{\tan x} = \frac{1}{\tan x} - 2\frac{1}{\tan 2x} = \cot x - 2\cot 2x.$$

This implies that

$$\frac{1}{2^n} \tan \frac{a}{2^n} = \frac{1}{2^n} \cot \frac{a}{2^n} - \frac{1}{2^{n-1}} \cot \frac{a}{2^{n-1}}.$$

The sum telescopes and yields the answer

$$\lim_{n\to\infty} \frac{1}{2^n}\cot\frac{a}{2^n} - \cot a.$$

The latter limit is equal to $1/a$, since

$$\lim_{x\to 0} x\cot ax = \lim_{x\to 0}\cos ax\frac{x}{\sin ax} = \frac{1}{a}.$$

Thus the answer to the problem is $1/a - \cot a$.
(T. Andreescu)

6. Using the identity $\sin 3x = 3\sin x - 4\sin^3 x$, we see that

$$3^{n-1}\sin^3\frac{a}{3^n} = \frac{1}{4}\left(3^n\sin\frac{a}{3^n} - 3^{n-1}\sin\frac{a}{3^{n-1}}\right).$$

Hence the sum telescopes and is equal to

$$\frac{1}{4}\lim_{n\to\infty}\frac{\sin\frac{a}{3^n}}{\frac{1}{3^n}} - \frac{1}{4}\sin a.$$

Using the fact that

$$\lim_{n\to\infty}\frac{\sin\frac{a}{3^n}}{\frac{1}{3^n}} = \lim_{x\to 0}\frac{\sin ax}{x} = a,$$

we conclude that the value of the sum is equal to $(a - \sin a)/4$.

7. Let us compute

$$\sum_{k=1}^{90} 2k\sin(2k)^\circ \sin 1^\circ.$$

Transforming the products into sums gives

$$\sum_{k=1}^{90} k\cos(2k-1)^\circ - k\cos(2k+1)^\circ$$

$$= \sum_{k=1}^{90}(k - (k-1))\cos(2k-1)^\circ - 90\cos 181^\circ$$

$$= \sum_{k=1}^{90}\cos(2k-1)^\circ + 90\cos 1^\circ.$$

Since $\cos(180^\circ - x) = -\cos x$, the terms in the last sum cancel pairwise; hence the expression computed is equal to $90\cos 1^\circ$. Dividing by $\sin 1^\circ$ and taking the average, we get $\cot 1^\circ$.

An alternative solution is possible using complex numbers. One expresses $\sin n°$ as $(e^{i\pi n/180} - e^{-i\pi n/180})/(2i)$ and uses the fact that

$$x + 2x^2 + \cdots + nx^n = \frac{nx^{n+1}}{x-1} - \frac{x^{n+1} - x}{(x-1)^2}.$$

This formula can be proved either by induction or by differentiating $1 + x + x^2 + \cdots + x^n = (x^{n+1} - 1)/(x - 1)$.

(USAMO, 1996; proposed by T. Andreescu)

8. We have

$$\frac{1}{\sin 2x} = \frac{2\cos^2 x - (2\cos^2 x - 1)}{\sin 2x} = \frac{2\cos^2 x}{2\sin x \cos x} - \frac{2\cos^2 x - 1}{\sin 2x}$$

$$= \frac{\cos x}{\sin x} - \frac{\cos 2x}{\sin 2x} = \cot x - \cot 2x.$$

This shows that the left side telescopes, and the identity follows.
(8th IMO, 1966)

9. We have

$$\frac{\tan a}{\cos 2a} = \frac{\tan a(1 + \tan^2 a)}{1 - \tan^2 a} = \frac{2\tan a - \tan a(1 - \tan^2 a)}{1 - \tan^2 a}$$

$$= \frac{2\tan a}{1 - \tan^2 a} - \tan a = \tan 2a - \tan a.$$

Thus the sum in the problem becomes

$$\tan 2 - \tan 1 + \tan 4 - \tan 2 + \cdots + \tan 2^{n+1} - \tan 2^n = \tan 2^{n+1} - \tan 1.$$

10. From the double-angle formula for sine, we get $\cos u = \sin 2u/(2\sin u)$. Hence we can write

$$\prod_{n=1}^{\infty} \cos \frac{x}{2^n} = \lim_{k \to \infty} \prod_{n=1}^{k} \cos \frac{x}{2^n} = \lim_{k \to \infty} \prod_{n=1}^{k} \frac{1}{2} \cdot \frac{\sin \frac{x}{2^{n-1}}}{\sin \frac{x}{2^n}}$$

$$= \lim_{k \to \infty} \frac{1}{2^k} \frac{\sin x}{\sin \frac{x}{2^k}} = \frac{\sin x}{x} \lim_{k \to \infty} \frac{\frac{x}{2^k}}{\sin \frac{x}{2^k}} = \frac{\sin x}{x}.$$

For $x = \pi$, this limit was used by Archimedes to find an approximate value for π. In fact, what Archimedes did was to approximate the length of the circle by the perimeter of the inscribed regular polygon with 2^k sides, and his computation reduces to the above formula.

11. If we multiply the left side by $\sin \frac{2\pi}{2^n - 1}$ and apply successively the double-angle formula for sine, we obtain

$$2^n \sin \frac{2\pi}{2^n - 1} \cos \frac{2\pi}{2^n - 1} \cos \frac{4\pi}{2^n - 1} \cdots \cos \frac{2^n \pi}{2^n - 1}$$

$$= 2^{n-1} \sin \frac{4\pi}{2^n - 1} \cos \frac{4\pi}{2^n - 1} \cdots \cos \frac{2^n \pi}{2^n - 1} = \cdots$$

$$= \sin \frac{2^{n+1} \pi}{2^n - 1} = \sin \left(2\pi + \frac{2\pi}{2^n - 1} \right) = \sin \frac{2\pi}{2^n - 1}.$$

Since the product we intended to calculate is obtained from this by dividing by $2^n \sin((2\pi)/(2^n - 1))$, the conclusion follows.

(Romanian high school textbook)

12. The identity

$$1 - \tan^2 x = \frac{2\tan x}{\tan 2x}$$

yields

$$1 - \tan^2 \frac{2^k \pi}{2^n + 1} = 2\frac{\tan \frac{2^k \pi}{2^n + 1}}{\tan \frac{2^{k+1}\pi}{2^n + 1}}.$$

It follows that the product telescopes to

$$2^n \frac{\tan \frac{2\pi}{2^n + 1}}{\tan \frac{2^{n+1}\pi}{2^n + 1}} = 2^n \frac{\tan \frac{2\pi}{2^n + 1}}{\tan\left(2\pi - \frac{2\pi}{2^n + 1}\right)} = -2^n.$$

(T. Andreescu)

13. Note that

$$1 - \cot x = \sqrt{2}\left(\cos \frac{\pi}{4} - \cos \frac{\pi}{4}\cot x\right) = \sqrt{2}\frac{\sin(x - \frac{\pi}{4})}{\sin x}.$$

Then

$$(1 - \cot 1°)(1 - \cot 2°)\cdots(1 - \cot 44°) = 2^{22}\frac{\sin(-44°)}{\sin 1°}\cdot\frac{\sin(-43°)}{\sin 2°}\cdots\frac{\sin(-1°)}{\sin 44°}$$

$$= 2^{22}.$$

14. Note that

$$\frac{\cos 3x}{\cos x} = 4\cos^2 x - 3 = 2(1 + \cos 2x) - 3 = 2\cos 2x - 1.$$

Therefore,

$$\frac{1}{2} + \cos x = -\frac{1}{2}\left(\frac{\cos(\frac{3}{2}x + \frac{3\pi}{2})}{\cos(\frac{x}{2} + \frac{\pi}{2})}\right)$$

$$= \frac{1}{2}\frac{\sin \frac{3}{2}x}{\sin \frac{x}{2}}.$$

This means that

$$\left(\frac{1}{2} + \cos \frac{\pi}{20}\right)\left(\frac{1}{2} + \cos \frac{3\pi}{20}\right)\left(\frac{1}{2} + \cos \frac{9\pi}{20}\right)\left(\frac{1}{2} + \cos \frac{27\pi}{20}\right)$$

$$= \frac{1}{16}\frac{\sin \frac{3\pi}{40}}{\sin \frac{\pi}{40}}\cdot\frac{\sin \frac{9\pi}{40}}{\sin \frac{3\pi}{40}}\cdot\frac{\sin \frac{27\pi}{40}}{\sin \frac{9\pi}{40}}\cdot\frac{\sin \frac{81\pi}{40}}{\sin \frac{27\pi}{40}}$$

$$= \frac{1}{16}\cdot\frac{\sin \frac{81\pi}{40}}{\sin \frac{\pi}{40}} = \frac{1}{16}.$$

15. We can write

$$1 - 2\cos a = \frac{1 - 4\cos^2 a}{1 + 2\cos a} = \frac{1 - 2(1 + \cos 2a)}{1 + 2\cos a}$$
$$= -\frac{1 + 2\cos 2a}{1 + 2\cos a}.$$

It follows that the product in the problem telescopes to

$$(-1)^n \frac{1 + 2\cos x}{1 + 2\cos \frac{x}{2^n}},$$

and we are done.

16. We have

$$1 + 2\cos 2a_k = 1 + 2(1 - 2\sin^2 a_k) = 3 - 4\sin^2 a_k$$
$$= \frac{\sin 3a_k}{\sin a_k},$$

where $a_k = \frac{3^k \pi}{3^n + 1}$. Hence our product telescopes to

$$\frac{\sin 3a_n}{\sin a_1} = \frac{\sin \frac{3^{n+1} \pi}{3^n + 1}}{\sin \frac{3\pi}{3^n + 1}}.$$

The numerator of the last fraction is equal to

$$\sin \left(3\pi - \frac{3\pi}{3^n + 1} \right) = \sin \frac{3\pi}{3^n + 1},$$

and hence equals the denominator of the same fraction. This completes the solution.
  (T. Andreescu)

1.10 Trigonometric Substitutions

1. Choose $t \in [0, \pi]$ such that $\cos t = x$. This is possible because the value of $|x|$ cannot exceed 1. We have $\sqrt{1 - x^2} = \sin t$ because $\sin t$ is positive for $t \in [0, \pi]$. The inequality becomes $\sin t + \cos t \geq a$. The maximum of the function

$$f(t) = \sin t + \cos t = 2\sin \frac{\pi}{4} \cos \left(t - \frac{\pi}{4} \right) = \sqrt{2}\cos \left(t - \frac{\pi}{4} \right)$$

on the interval $[0, \pi]$ is $\sqrt{2}$; hence the range of a is the set of all real numbers not exceeding $\sqrt{2}$.

2. Let the numbers be a_1, a_2, a_3, a_4. Make the substitution $a_k = \sin t_k$, $t_k \in (0, \pi/2)$. The problem asks us to show that there exist two indices i and j with

$$0 < \sin t_i \cos t_j - \sin t_j \cos t_i < \frac{1}{2}.$$

But $\sin t_i \cos t_j - \sin t_j \cos t_i = \sin(t_i - t_j)$, and hence we must prove that there exist i and j with $t_i > t_j$ and $t_i - t_j < \pi/6$. This follows from the pigeonhole principle, since two of the four numbers must lie in one of the intervals $(0, \pi/6]$, $(\pi/6, \pi/3]$, $(\pi/3, \pi/2)$.

3. Any real number can be represented as $\tan x$ where $x \in (0, \pi)$. Note that

$$\frac{1 + \tan x \tan y}{\sqrt{1 + \tan^2 x} \cdot \sqrt{1 + \tan^2 y}} = \cos x \cos y + \sin x \sin y = \cos(x - y).$$

Now we can restate the problem: prove that among any four numbers from $[0, \pi)$ there are two x and y such that $0 < x - y < \frac{\pi}{3}$, which is obvious by the pigeonhole principle.

4. Substitute $x = \cos a$, where $0 \le a \le \pi$. Using triple equation formula, the given equation reduces to $\cos^2 a + \cos^2 3a = 1$. This is equivalent to

$$\frac{1 + \cos 2a}{2} + \frac{1 + \cos 6a}{2} = 1$$

or

$$\cos 2a + \cos 6a = 0.$$

It follows that $2\cos 2a \cos 4a = 0$, which gives either $2a = \frac{\pi}{2}, \frac{3\pi}{2}$ or $4a = \frac{\pi}{2}, \frac{3\pi}{2}, \frac{5\pi}{2}, \frac{7\pi}{2}$. We obtain the solutions $\pm\frac{\sqrt{2}}{2}$ and $\pm\sqrt{\frac{2 \pm \sqrt{2}}{2}}$, and they all satisfy the given equation.

5. With the substitution $x = 2\cos t$, we have

$$\sqrt{2 + \sqrt{2 + \cdots + \sqrt{2 + x}}} = \sqrt{2 + \sqrt{2 + \cdots + \sqrt{2 + 2\cos t}}}$$

$$= \sqrt{2 + \sqrt{2 + \cdots + 2\cos\frac{t}{2}}} = \cdots$$

$$= 2\cos\frac{t}{2^n}.$$

The integral becomes

$$I = -4\int \sin t \cos\frac{t}{2^n} dt = -2\int\left(\sin\frac{2^n + 1}{2^n}t - \sin\frac{2^n - 1}{2^n}t\right) dt$$

$$= \frac{2^{n+1}}{2^n + 1}\cos\left(\frac{2^n + 1}{2^n}\arccos\frac{x}{2}\right) - \frac{2^{n+1}}{2^n - 1}\cos\left(\frac{2^n - 1}{2^n}\arccos\frac{x}{2}\right) + C.$$

(C. Mortici, *Probleme Pregătitoare pentru Concursurile de Matematică (Training Problems for Mathematics Contests)*, GIL, 1999)

6. Because $0 \le x_n \le 2$ for all n, we can use the trigonometric substitution $x_n = 2\cos y_n$, where $0 \le y_n \le \pi/2$. From the inequality $\sqrt{x_{n+2} + 2} \le x_n$ and the double-angle formula $\cos 2\alpha + 1 = 2\cos^2 \alpha$, we get $\cos y_{n+2}/2 \le \cos y_n$. Since the cosine is a decreasing function on $[0, \pi/2]$, this implies $y_{n+2}/2 \ge y_n$ for all n. It follows by induction that for all n and k, $y_n \le y_{n+2k}/2^k$, and by letting k go to infinity, we obtain $y_n = 0$ for all n. Hence $x_n = 2$ for all n.

(Romanian IMO Team Selection Test, 1986; proposed by T. Andreescu)

7. If one of the unknowns, say x, is equal to ± 1, then $2x + x^2y = y$ leads to $0 = \pm 2$, which is impossible. Hence the system can be rewritten as

$$\frac{2x}{1-x^2} = y,$$
$$\frac{2y}{1-y^2} = z,$$
$$\frac{2z}{1-z^2} = x.$$

Because of the double-angle formula for the tangent,

$$\tan 2a = \frac{2\tan a}{1 - \tan^2 a},$$

it is natural to make the substitution $x = \tan\alpha$ for $\alpha \in (-\pi/2, \pi/2)$.

From the first two equations, $y = \tan 2\alpha$ and $z = \tan 4\alpha$, and the last equation implies $\tan 8\alpha = \tan\alpha$. Hence $8\alpha - \alpha = k\pi$, for some integer k, so $\alpha = k\pi/7$, and since $\alpha \in (-\pi/2, \pi/2)$, we must have $-3 \le k \le 3$. It follows that the solutions to the system are

$$\left(\tan\frac{k\pi}{7}, \tan\frac{2k\pi}{7}, \tan\frac{4k\pi}{7}\right), \quad k = -3, -2, -1, 0, 1, 2, 3.$$

8. The trigonometric formula hidden in the statement of the problem is the double-angle formula for the cotangent:

$$2\cot 2\alpha = \cot\alpha - \frac{1}{\cot\alpha}.$$

This can be proved from the double-angle formula for the tangent $\tan 2\alpha = 2\tan\alpha/(1 - \tan^2\alpha)$ by replacing $\tan\alpha$ by $1/\cot\alpha$.

If we set $x_1 = \cot\alpha$, $\alpha \in (0, \pi)$, then from the first equation $x_2 = \cot 2\alpha$, from the second $x_3 = \cot 4\alpha$, from the third $x_4 = \cot 8\alpha$, and from the last $x_1 = \cot 16\alpha$. Hence $\cot\alpha = \cot 16\alpha$, which implies $16\alpha - \alpha = k\pi$, for some integer k. We obtain the solutions $\alpha = k\pi/15$, $k = 1, 2, \ldots, 14$; hence the solutions to the given system are $x_1 = \cot k\pi/15$, $x_2 = \cot 2k\pi/15$, $x_3 = \cot 4k\pi/15$, $x_4 = \cot 8k\pi/15$, $k = 1, 2, 3, \ldots, 14$.

9. Let $x = \tan a$ and $y = \tan b$. Then

$$x + y = \tan a + \tan b = \frac{\sin(a+b)}{\cos a \cos b},$$
$$1 - xy = 1 - \tan a \tan b = \frac{\cos(a+b)}{\cos a \cos b},$$

$$\frac{1}{1+x^2} = \cos^2 a,$$
$$\frac{1}{1+y^2} = \cos^2 b.$$

The inequality we want to prove is equivalent to $-1 \le 2\sin(a+b)\cos(a+b) \le 1$, that is, $-1 \le \sin 2(a+b) \le 1$, and we are done.

10. First notice that
$$\frac{1}{1-x_n} - \frac{1}{1+x_n} = \frac{2x_n}{1-x_n^2}.$$
If we let $x_1 = \tan\beta$, with $\beta \in (-\pi/2, \pi/2)$, then
$$x_2 = \frac{2\tan\beta}{1-\tan^2\beta} = \tan 2\beta.$$

Inductively we get $x_n = \tan 2^{n-1}\beta$; thus $x_8 = \tan 2^7\beta = \tan 128\beta$. For the sequence to have length 8, we must have $\tan 128\beta = \pm 1$. This implies that $128\beta = \frac{(2k+1)\pi}{4}$, for some integer k. So
$$x = \tan\left(\pm\frac{(2k+1)\pi}{512}\right),$$
for some $k = -128, \ldots, 127$. It remains to check that none of the values produces a sequence of length less than eight. Indeed, for the sequence to terminate earlier, one should have
$$\pm\frac{(2k+1)\pi}{512} = \pm\frac{\pi}{2^{r+2}} + m\pi.$$
But this would then imply that $(2k+1)/512 = k'/128$ for some integer k', and this is impossible. Hence all 256 sequences have length equal to 8.

(Proposed for AIME, 1996)

11. If we define the sequence $b_1 = \tan^{-1} a_1$ and $b_{k+1} = b_k + \tan^{-1}(1/k)$, $k = 1, 2, 3, \ldots$, then the addition formula for the tangent,
$$\tan(x+y) = \frac{\tan x + \tan y}{1 - \tan x \tan y},$$
shows that $a_k = \tan b_k$, for all k. Since $\lim_{x\to 0}\frac{\tan x}{x} = 1$, it follows that
$$\lim_{k\to\infty} \frac{\tan^{-1} 1/k}{1/k} = 1.$$

This implies that, on the one hand, the series
$$b_0 + \sum_{k=1}^{\infty} \tan^{-1}\frac{1}{k}$$

is divergent, and on the other hand, the terms of the series tend to zero as k tends to infinity. Hence there are infinitely many partial sums of the series lying in intervals of the form $(2\pi n, 2\pi n + \pi/2)$ and infinitely many partial sums of the series lying in intervals of the form $(2\pi n + \pi/2, (2n+1)\pi)$. But a partial sum is a b_m for some m, and since $a_m = \tan b_m$, it follows that there are infinitely many positive a_m's and infinitely many negative a_m's.

(Leningrad Mathematical Olympiad, 1989)

12. It is natural to make the trigonometric substitution $a_i = \cos x_i$ for some $x_i \in [0, \pi]$, $i = 1, 2, \ldots, n$. Note that the monotonicity of the cosine function combined with the given inequalities shows that the x_i's form a decreasing sequence. The expression on the left becomes

$$\sum_{i=1}^{n-1} \sqrt{1 - \cos x_i \cos x_{i+1} - \sin x_i \sin x_{i+1}} = \sum_{i=1}^{n-1} \sqrt{1 - \cos(x_{i+1} - x_i)}$$

$$= \sqrt{2} \sum_{i=1}^{n-1} \sin \frac{x_{i+1} - x_i}{2}.$$

Here we used a subtraction and a double-angle formula. The sine function is concave down on $[0, \pi]$; hence we can use Jensen's inequality to obtain

$$\frac{1}{n-1} \sum_{i=1}^{n-1} \sin \frac{x_{i+1} - x_i}{2} \le \sin \left(\frac{1}{n-1} \sum_{i=1}^{n-1} \frac{x_{i+1} - x_i}{2} \right).$$

Hence,

$$\sqrt{2} \sum_{i=1}^{n-1} \sin \frac{x_{i+1} - x_i}{2} \le (n-1)\sqrt{2} \sin \frac{x_n - x_1}{2(n-1)} \le \sqrt{2}(n-1) \sin \frac{\pi}{2(n-1)},$$

since $x_n - x_1 \in (0, \pi)$. Using the fact that $\sin x < x$ for $x > 0$ yields $\sqrt{2}(n-1)$ $\sin \pi/(2(n-1)) \le \sqrt{2}\pi/2$.
(Mathematical Olympiad Summer Program, 1996)

13. Since the x_i's are positive and add up to 1, we can make the substitutions $x_0 + x_1 + \cdots + x_k = \sin a_k$, with $a_0 = 0 < a_1 < \ldots < a_n = \pi/2$, $k = 0, 1, \ldots, n$. The inequality becomes

$$\sum_{k=1}^{n} \frac{\sin a_k - \sin a_{k-1}}{\sqrt{1 + \sin a_{k-1}} \sqrt{1 - \sin a_{k-1}}} < \frac{\pi}{2},$$

which can be rewritten as

$$\sum_{k=1}^{n} \frac{2 \sin \frac{a_k - a_{k-1}}{2} \cos \frac{a_k + a_{k-1}}{2}}{\cos a_{k-1}}.$$

For $0 < x \le \pi/2$, $\cos x$ is a decreasing function and $\sin x < x$. Hence the left side of the inequality is strictly less than

$$\sum_{k=1}^{n} \frac{2 \frac{a_k - a_{k-1}}{2} \cos a_{k-1}}{\cos a_{k-1}} = \sum_{k=1}^{n} (a_k - a_{k-1}) = \frac{\pi}{2},$$

and the problem is solved.
(Chinese Mathematical Olympiad, 1996)

14. The triples satisfying the equation are the cosines of the angles of an acute triangle. Let us first show that if A, B, C are the angles of a triangle, then

$$\cos^2 A + \cos^2 B + \cos^2 C + 2 \cos A \cos B \cos C = 1.$$

Indeed, $\cos A = -\cos(B+C) = \sin B \sin C - \cos B \cos C$, so

$$\cos^2 A + \cos^2 B + \cos^2 C + 2\cos A \cos B \cos C$$
$$= (\cos A + \cos B \cos C)^2 + 1 - (1 - \cos^2 B)(1 - \cos^2 C)$$
$$= (\sin B \sin C)^2 + 1 - \sin^2 B \sin^2 C = 1.$$

To prove that these are the only solutions, note that from the equation in the statement it follows that a given solution satisfies $x^2 + y^2 < 1$, and so the equation in z

$$z^2 + 2xyz - 1 + x^2 + y^2 = 0$$

has a unique positive solution. If we set $x = \cos A$ and $y = \cos B$, $A, B \in (0, \pi/2)$, then the uniqueness of the solution implies that z can be equal only to $\cos C$, where $A + B + C = \pi$, and the problem is solved.

15. The second equation is equivalent to

$$\frac{a^2}{yz} + \frac{b^2}{zx} + \frac{c^2}{xy} + \frac{abc}{xyz} = 4.$$

Let $x_1 = a/\sqrt{yz}$, $y_1 = b/\sqrt{zx}$, $z_1 = c/\sqrt{xy}$. Then $x_1^2 + y_1^2 + z_1^2 + x_1 y_1 z_1 = 4$, where $0 < x_1 < 2$, $0 < y_1 < 2$, $0 < z_1 < 2$. Thinking of this as an equation in $x_1/2$, $y_1/2$, and $z_1/2$, we obtain from the previous problem that $x_1 = 2\cos A$, $y_1 = 2\cos B$, and $z_1 = 2\cos C$, where A, B, and C are the angles of an acute triangle.

Adding the three equalities $2\sqrt{yz}\cos A = a$, $2\sqrt{zx}\cos B = b$, and $2\sqrt{xy}\cos C = c$, and using the fact that $x + y + z = a + b + c$ yields

$$x + y + z - 2\sqrt{yz}\cos A - 2\sqrt{zx}\cos B - 2\sqrt{xy}\cos C = 0.$$

We transform the left side into a sum of two squares. We have

$$x + y + z - 2\sqrt{yz}\cos A - 2\sqrt{zx}\cos B - 2\sqrt{xy}\cos C$$
$$= x + y + z - 2\sqrt{yz}\cos A - 2\sqrt{zx}\cos B$$
$$+ 2\sqrt{xy}(\cos A \cos B - \sin A \sin B)$$
$$= x(\sin^2 B + \cos^2 B) + y(\sin^2 A + \cos^2 A) + z$$
$$- 2\sqrt{yz}\cos A - 2\sqrt{zx}\cos B$$
$$+ 2\sqrt{xy}\cos A \cos B - 2\sqrt{xy}\sin A \sin B$$
$$= (\sqrt{x}\sin B - \sqrt{y}\sin A)^2 + (\sqrt{x}\cos B + \sqrt{y}\cos A - \sqrt{z})^2.$$

Since the sum of these two squares must be equal to zero, each of them must be zero. Therefore,

$$\sqrt{z} = \sqrt{x} \cdot \frac{b}{2\sqrt{zx}} + \sqrt{y} \cdot \frac{a}{2\sqrt{yz}} = \frac{b+a}{2\sqrt{z}},$$

and hence $z = (a+b)/2$. By symmetry, $y = (c+a)/2$ and $x = (b+c)/2$. It is easy to check that these three solutions satisfy the given system of equations.

(Submitted by the USA for the IMO, 1995; proposed by T. Andreescu)

Chapter 2

Algebra and Analysis

T. Andreescu and R. Gelca, *Mathematical Olympiad Challenges*, DOI: 10.1007/978-0-8176-4611-0_2, 171
© Birkhäuser Boston, a part of Springer Science+Business Media, LLC 2009

2.1   No Square is Negative

1. Let the numbers be $a_1, a_2, \ldots, a_n$. From the given conditions, we find

$$a_1^2 + a_2^2 + \cdots + a_n^2 = (a_1 + a_2 + \cdots + a_n)^2 - 2(a_1 a_2 + a_1 a_3 + \cdots + a_{n-1} a_n) = 0.$$

It follows that $a_1 = a_2 = \cdots = a_n = 0$, hence $a_1^3 + a_2^3 + \cdots + a_n^3 = 0$.
(Leningrad Mathematical Olympiad)

2. If the inequalities

$$a - b^2 > \frac{1}{4}, \quad b - c^2 > \frac{1}{4}, \quad c - d^2 > \frac{1}{4}, \quad d - a^2 > \frac{1}{4}$$

hold simultaneously, then by adding them we obtain

$$a + b + c + d - (a^2 + b^2 + c^2 + d^2) > 1.$$

Moving everything to the right side and completing the squares gives

$$\left(\frac{1}{2} - a\right)^2 + \left(\frac{1}{2} - b\right)^2 + \left(\frac{1}{2} - c\right)^2 + \left(\frac{1}{2} - d\right)^2 < 0,$$

a contradiction.
   (*Revista Matematică din Timişoara (Timişoara's Mathematics Gazette)*, proposed by T. Andreescu)

3. Assuming the contrary and summing up, we obtain

$$\left(\frac{1}{x} + \frac{1}{4-x}\right) + \left(\frac{1}{y} + \frac{1}{4-y}\right) + \left(\frac{1}{z} + \frac{1}{4-z}\right) < 3.$$

On the other hand,

$$\frac{1}{a} + \frac{1}{4-a} \geq 1$$

for all positive real numbers a less than 4, since this is equivalent to

$$(a-2)^2 \geq 0.$$

Hence the conclusion.
   (Hungarian Mathematical Olympiad, 2001)

4. *First solution:* As in the case of the previous problem, we add the three equations and rewrite the expression as a sum of squares. By summing and moving everything to the left side, we obtain

$$2x + 2y + 2z - \sqrt{4x - 1} - \sqrt{4y - 1} - \sqrt{4z - 1} = 0.$$

We want to write this expression as a sum of three squares, one depending on x only, one depending on y, and one depending on z. Let us divide by 2 and look at

$x + \sqrt{x - \frac{1}{4}}$. The presence of the $\frac{1}{4}$ under the square root suggests to us to add and subtract $\frac{1}{4}$. We have

$$x - \frac{1}{4} - \sqrt{x - \frac{1}{4}} + \frac{1}{4} = \left(\sqrt{x - \frac{1}{4}} - \frac{1}{2}\right)^2.$$

Returning to the original problem, we have

$$\left(\sqrt{x - \frac{1}{4}} - \frac{1}{2}\right)^2 + \left(\sqrt{y - \frac{1}{4}} - \frac{1}{2}\right)^2 + \left(\sqrt{z - \frac{1}{4}} - \frac{1}{2}\right)^2 = 0,$$

so each of these squares must be equal to 0. It follows that $x = y = z = \frac{1}{2}$ is the only solution of the given system.

Second solution: Square, and note that

$$0 = ((x+y)^2 - 4z + 1) + ((y+z)^2 - 4x + 1) + ((z+x)^2 - 4y + 1)$$
$$= (x+y-1)^2 + (y+z-1)^2 + (z+x-1)^2$$

Hence the equations give $x + y = y + z = z + x = 1$ or $x = y = z = 1/2$.

(Romanian mathematics contest, proposed by T. Andreescu, second solution by R. Stong)

5. Since $a > 0$ and $a \neq 1$, the equality can be rewritten as

$$\frac{1}{\log_a x} + \frac{1}{\log_a y} = \frac{4}{\log_a xy},$$

which is equivalent to

$$\frac{\log_a x + \log_a y}{\log_a x \log_a y} = \frac{4}{\log_a x + \log_a y}.$$

Eliminating the denominators, we obtain

$$(\log_a x + \log_a y)^2 = 4 \log_a x \log_a y,$$

which implies $(\log_a x - \log_a y)^2 = 0$, and this can hold only if $x = y$.

(Romanian mathematics contest, proposed by T. Andreescu)

6. If we try to complete a square involving the first two terms of the left side, we obtain $(x^2 - y^2)^2 + 2x^2y^2 + z^4 - 4xyz$. Of course the presence of $2x^2y^2$ and $-4xyz$ suggests the possibility of adding a $2z^2$ and then completing one more square. At this moment, it is not hard to see that the equation can be rewritten as

$$(x^2 - y^2)^2 + (z^2 - 1)^2 + 2(xy - z)^2 = 0.$$

This equality can hold only if all three squares are equal to zero. From $z^2 - 1 = 0$ we have $z = \pm 1$, and after a quick analysis we conclude that the solutions are $(1,1,1)$, $(-1,-1,1)$, $(-1,1,-1)$, and $(1,-1,-1)$.

(*Revista Matematică din Timişoara (Timişoara's Mathematics Gazette)*, proposed by T. Andreescu)

7. *First solution:* From the second inequality, we obtain $z \geq |x+y| - 1$. Plugging this into the first inequality yields

$$2xy - (1 - |x+y|)^2 \geq 1.$$

We have

$$
\begin{aligned}
2xy - (1 + |x+y|)^2 &= 2xy - |x+y|^2 + 2|x+y| - 1 \\
&= 2xy - x^2 - y^2 - 2xy + 2(\pm x \pm y) - 1 \\
&= -x^2 - y^2 + 2(\pm x \pm y) - 1
\end{aligned}
$$

for some choice of signs plus and minus. From the inequality deduced above, it follows that

$$0 \geq x^2 + y^2 - 2(\pm x \pm y) + 1 + 1 = (1 \pm x)^2 + (1 \pm y)^2.$$

The two squares must both be equal to zero. Hence x and y can only have the values 1 or -1. Moreover, we saw above that xy is positive, so x and y must have the same sign. For $x = y = 1$ or $x = y = -1$, we obtain $2 - z^2 \geq 1$ and $z - 2 \geq -1$; hence $z^2 \leq 1$ and $z \geq 1$. The only z satisfying both inequalities is $z = 1$; hence there are two solutions to our problem, $x = y = z = 1$ and $x = y = -1, z = 1$.

Second solution: Writing the second inequality as $z + 1 \geq |x+y|$, and squaring we obtain $(z+1)^2 \geq (x+y)^2$. Now adding twice the first inequality to this and rearranging gives $0 \geq (x-y)^2 + (z-1)^2$. Thus any solution has $x = y$ and $z = 1$. Plugging these in, the inequalities become $2x^2 \geq 2$ and $2 \geq 2|x|$, hence $x = 1$.

(T. Andreescu, second solution by R. Stong)

8. The solution is a quickie if we note that $x^4 + ax^3 + 2x^2 + bx + 1 = (x^2 + \frac{a}{2}x)^2 + (1 + \frac{b}{2}x)^2 + \frac{1}{4}(8 - a^2 - b^2)x^2$. In this case, the polynomial is strictly positive unless $a^2 + b^2 - 8 \leq 0$. Hence the conclusion.

(Communicated by I. Boreico)

9. Completing squares, we obtain

$$x^4 + ax^3 + bx^2 + cx + 1 = \left(x^2 + \frac{a}{2}x\right)^2 + \left(b - \frac{a^2 + c^2}{4}\right)x^2 + \left(\frac{c}{2}x + 1\right)^2.$$

The inequality follows now from the fact that $b \geq (a^2 + c^2)/4$.

(*Revista Matematică din Timişoara (Timişoara's Mathematics Gazette)*, proposed by T. Andreescu)

10. The desired inequality is equivalent to

$$\frac{1}{2}\left[(x-y)^2 + (y-z)^2 + (z-x)^2\right] \geq \frac{3}{4}(x-y)^2,$$

that is,

$$2\left[(y-z)^2 + (z-x)^2\right] \geq (x-y)^2.$$

Letting $a = y - z$ and $b = z - x$, this becomes $2(a^2 + b^2) \geq (a+b)^2$, which reduces to $(a-b)^2 \geq 0$.

11. We try to transform the equation into a sum of squares equal to zero. To this end, we multiply the equality by 2, move everything to the right side, and complete squares. We have

$$
\begin{aligned}
(x_1 + x_2 + \cdots &+ x_n) - 2\sqrt{x_1 - 1} - 4\sqrt{x_2 - 2^2} - \cdots - 2n\sqrt{x_n - n^2} \\
&= (x_1 - 1 - 2\sqrt{x_1 - 1} + 1) + (x_2 - 2^2 - 4\sqrt{x_2 - 2^2} + 2^2) + \cdots \\
&\quad + (x_n - n^2 - 2n\sqrt{x_n - n^2} + n^2) = (\sqrt{x_1 - 1} - 1)^2 + (\sqrt{x_2 - 2^2} - 2)^2 \\
&\quad + \cdots + (\sqrt{x_n - n^2} - n)^2.
\end{aligned}
$$

By hypothesis, the sum of those squares must be equal to 0, hence all squares are equal to 0. Thus $\sqrt{x_1 - 1} = 1$, $\sqrt{x_2 - 2^2} = 2$, ..., $\sqrt{x_n - n^2} = n$. The unique solution to the equation is $x_1 = 2, x_2 = 8, \ldots, x_n = 2n^2$.

(*Gazeta Matematică (Mathematics Gazette, Bucharest), proposed by T. Andreescu*)

12. (a) Squaring both sides, we obtain

$$a^2 b^2 + b^2 c^2 + c^2 a^2 + 2abc(a+b+c) \geq 3abc(a+b+c).$$

This is equivalent to

$$\frac{1}{2}\left[(ab - bc)^2 + (bc - ca)^2 + (ca - ab)^2\right] \geq 0.$$

(b) The inequality is equivalent to

$$\sqrt{12abc(a+b+c)} \leq (a+b+c)^2 - (a^2 + b^2 + c^2).$$

This is the same as

$$\sqrt{12(a+b+c)abc} \leq 2(ab + bc + ca),$$

which reduces to the one before.

(Part (b) appeared at the Austrian Mathematical Olympiad, 1984)

13. As in the example from the introductory essay, we will plug particular values for m, n, and k into the given equation. Letting $m = n = k = 0$, we obtain $2f(0) - f^2(0) \geq 1$; hence $0 \geq (f(0) - 1)^2$, which implies $f(0) = 1$. For $m = n = k = 1$, the same argument shows that $f(1) = 1$. For $m = n = 0$, we obtain $2 - f(k) \geq 1$; hence $f(k) \leq 1$ for all k. Also, for $k = 1$ and $m = 0$, we obtain $1 + f(n) - 1 \geq 1$, which implies that $f(n) \geq 1$ for all n. It follows that $f(n) = 1$ for all n.

(D.M. Bătineţu)

14.   Since in a right parallelepiped the diagonal is given by the formula $d = \sqrt{a^2 + b^2 + c^2}$, the inequality is equivalent to

$$(a^2b^2 + b^2c^2 + c^2a^2)^2 \geq 3a^2b^2c^2(a^2 + b^2 + c^2).$$

After regrouping terms, this becomes

$$\frac{c^4}{2}(a^2 - b^2)^2 + \frac{a^4}{2}(b^2 - c^2)^2 + \frac{b^4}{2}(c^2 - a^2)^2 \geq 0.$$

Note that the equality holds if and only if $a = b = c$, i.e., the parallelepiped is a cube.

(L. Pîrşan and C. G. Lazanu, *Probleme de algebră şi trigonometrie (Problems in algebra and trigonometry)*, Facla, Timişoara, 1983)

15. By using the addition formula for the cosine, we obtain

$$\sum_{i=1}^{n}\sum_{j=1}^{n} ij\cos(a_i - a_j) = \sum_{i=1}^{n}\sum_{j=1}^{n}(ij\cos a_i \cos a_j + ij\sin a_i \sin a_j)$$

$$= \sum_{i=1}^{n} i\cos a_i \sum_{j=1}^{n} j\cos a_j + \sum_{i=1}^{n} i\sin a_i \sum_{j=1}^{n} j\sin a_j$$

$$= \left(\sum_{i=1}^{n} i\cos a_i\right)^2 + \left(\sum_{i=1}^{n} i\sin a_i\right)^2 \geq 0.$$

2.2   Look at the Endpoints

1. The expression from the left side of the inequality is a linear function in each of the four variables. Its minimum is attained at one of the endpoints of the interval of definition. Thus we have only to check $a, b, c, d \in \{0, 1\}$. If at least one of them is 1, the expression is equal to $a + b + c + d$, which is greater than or equal to 1. If all of them are zero, the expression is equal to 1, which proves the inequality.

2. The inequality is equivalent to

$$a(k - b) + b(k - c) + c(k - a) \leq k^2.$$

If we view the left side as a function in a, it is linear. The conditions from the statement imply that the interval of definition is $[0, k]$. It follows that in order to maximize the left-hand side, we need to choose $a \in \{0, k\}$. Repeating the same argument for b and c, it follows that the maximum of the left-hand side is attained for some $(a, b, c) \in \{0, k\}^3$. Checking the eight possible situations, we obtain that this maximum is k^2, and we are done.

(All Union Mathematical Olympiad)

3. Let us fix $x_2, x_3, \ldots, x_n$ and then consider the function $f : [0, 1] \to \mathbf{R}$, $f(x) = x + x_2 + \cdots + x_n - xx_2 \cdots x_n$. This function is linear in x, hence attains its maximum

at one endpoint of the interval $[0,1]$. Thus in order to maximize the left side of the inequality, one must choose x_1 to be 0 or 1, and by symmetry, the same is true for the other variables. Of course, if all x_i are equal to 1, then we have equality. If at least one of them is 0, then their product is also zero, and the sum of the other $n-1$ terms is at most $n-1$, which proves the inequality.

(Romanian mathematics contest)

4. The expression is linear in each of the variables, so, as in the solutions to the previous problems, the maximum is attained for $a_k = \frac{1}{2}$ or 1, $k = 1,2,\ldots,n$. If $a_k = \frac{1}{2}$ for all k, then $S_n = n/4$. Let us show that the value of S_n cannot exceed this number. If exactly m of the a_k's are equal to 1, then m terms of the sum are zero. Also, at most m terms are equal to $\frac{1}{2}$, namely those of the form $a_k(1-a_{k+1})$ with $a_k = 1$ and $a_{k+1} = \frac{1}{2}$. Each of the remaining terms has both factors equal to $\frac{1}{2}$ and hence is equal to $\frac{1}{4}$. Thus the value of the sum is at most $m \cdot 0 + m/2 + (n-2m)/4 = n/4$, which shows that the maximum is $n/4$.

(Romanian IMO Team Selection Test, 1975)

5. Denote the left side of the inequality by $S(x_1,x_2,\ldots,x_n)$. This expression is linear in each of the variables x_i. As before, it follows that it is enough to prove the inequality when the x_i's are equal to 0 or 1.

If exactly k of the x_i's are equal to 0, and the others are equal to 1, then $S(x_1,x_2,\ldots,x_n) \le n-k$, and since the sum $x_1x_2 + x_2x_3 + \cdots + x_nx_1$ is at least $n-2k$, $S(x_1,x_2,\ldots,x_n)$ is less than or equal to $n-k-(n-2k) = k$. Thus the maximum of S is less than or equal to $\min(k,n-k)$, which is at most $\lfloor n/2 \rfloor$. It follows that for n even, equality holds when $(x_1,x_2,x_3,\ldots) = (1,0,1,0,\ldots,1,0)$ or $(0,1,0,1,\ldots0,1)$. For n odd, equality holds when all pairs (x_i,x_{i+1}), $i = 1,2,\ldots,n$ consist of a zero and a one, except for one pair that consists of two ones (with the convention $x_{n+1} = x_1$) or if $x_1,\ldots,x_n$ is a rotate of $0,1,0,1,\ldots,0,1,x$ where x is arbitrary (which corresponds to the case where a linear function is constant hence it attains its extremum on the whole interval).

(Bulgarian Mathematical Olympiad, 1995)

6. The sum we want to minimize is linear in each variable; hence the minimum is attained for some $a_i \in \{-98,98\}$. Since there is an odd number of indices, if we look at the indices mod 19, there exists an i such that a_i and a_{i+1} have the same sign. Hence the sum is at least $-18 \cdot 98^2 + 98^2 = -17 \cdot 98^2$. Equality is attained for example when $a_1 = a_3 = \cdots = a_{19} = -98$, $a_2 = a_4 = \cdots = a_{18} = 98$, but it should be observed that there are other choices that yield the same maximum.

7. For any nonnegative numbers α and β, the function

$$x \mapsto \frac{\alpha}{x+\beta}$$

is convex for $x \ge 0$. Viewed as a function in any of the three variables, the given expression is a sum of two convex functions and two linear functions, so it is convex. Thus when two of the variables are fixed, the maximum is attained when the third is at one of the endpoints of the interval, so the values of the expression are always less than the largest value obtained by choosing $a,b,c \in \{0,1\}$. An easy check of the eight possible cases shows that the value of the expression cannot exceed 1.

(USAMO, 1980)

8. If we fix four of the numbers and regard the fifth as a variable x, then the left side becomes a function of the form $\alpha x + \beta/x + \gamma$, with α, β, γ positive and x ranging over the interval $[p,q]$. This function is convex on the interval $[p,q]$, being the sum of a linear and a convex function, so it attains its maximum at one (or possibly both) of the endpoints of the interval of definition. As before, this shows that if we are trying to maximize the value of the expression, it is enough to let a,b,c,d,e take the values p and q.

If n of the numbers are equal to p, and $5 - n$ are equal to q, then the left side is equal to

$$n^2 + (5-n)^2 + n(5-n)\left(\frac{p}{q} + \frac{q}{p}\right) = 25 + n(5-n)\left(\sqrt{\frac{p}{q}} - \sqrt{\frac{q}{p}}\right)^2.$$

The maximal value of $n(5-n)$ is attained when $n = 2$ or 3, in which case $n(5-n) = 6$, and the inequality is proved.

(USAMO, 1977)

9. *First solution:* Using the AM–GM inequality, we can write

$$\sqrt[3]{\left(\sum_{k=1}^{n} x_k\right)\left(\sum_{k=1}^{n} \frac{1}{x_k}\right)^2} \le \frac{1}{3}\left(\sum_{k=1}^{n} x_k + \sum_{k=1}^{n} \frac{1}{x_k} + \sum_{k=1}^{n} \frac{1}{x_k}\right)$$

$$= \sum_{k=1}^{n} \frac{x_k + \frac{1}{x_k} + \frac{1}{x_k}}{3}.$$

The function $x + 2/x$ is convex on the interval $[1,2]$, so it attains its maximum at one of the endpoints of the interval. Also, the value of the function at each of the endpoints is equal to 3. This shows that

$$\sum_{k=1}^{n} \frac{x_k + \frac{1}{x_k} + \frac{1}{x_k}}{3} \le n,$$

and the inequality is proved.

Second solution: Here is a direct solution, more in the spirit of this section, which was pointed out to us by R. Stong. As a function of any one $x_i = x$, the left-hand side of the desired inequality is of the form $f(x) = C_1/x^2 + C_2/x + C_3 + C_4 x$, which is convex. Hence the maximum is attained when all the x_i are on the boundary. Suppose m of them are 2 and $n - m$ of them are 1. Then the left-hand side becomes

$$(n+m)\left(n - \frac{m}{2}\right)^2 = n^3 - \frac{(3n-m)m^2}{4} \le n^3,$$

with equality if and only if $m = 0$.

Let us point out that the same idea can be used to prove the more general form of this inequality, due to Gh. Sőllősy, which holds for $x_i \in [a,b]$, $i = 1, 2, \ldots, n$:

$$\left(\sum_{k=1}^{n} x_i\right)\left(\sum_{k=1}^{n} \frac{1}{x_i}\right)^{ab} \leq \left(\frac{a+b}{1+ab}n\right)^{1+ab}$$

(L. Panaitopol)

10. Assume that the inequality is not always true, and choose the smallest n for which it is violated by some real numbers. Consider the function one variable function

$$f(x_1) = \sum_{i=1}^{n}\sum_{j=1}^{n}|x_i + x_j| - n\sum_{i=1}^{n}|x_i|.$$

This function is not linear or convex, but it is piece-wise linear, meaning that it is linear on each of finitely many intervals that partition the real axis. In fact, the endpoints of the intervals are among the numbers $0, x_2, x_3, \ldots, x_n$. Note also that for $|x_1|$ sufficiently large compared to $|x_i|$ for $i \neq 1$,

$$f(x_1) = |x_1 + x_1| + \sum_{i=2}^{n}(|x_1| \pm x_i) - n|x_1| + \text{constant},$$

hence $\lim_{|x_1| \to \infty} f(x_1) = \infty$. Now assume that this function takes negative values. Then it must be negative at some endpoint of one of the intervals on which it is linear. Thus it must be negative when $x_1 = 0$, or $x_1 = -x_i$ for some i. In the former case

$$f(0) = 2\sum_{i=2}^{n}|x_i| + \sum_{i=2}^{n}\sum_{j=2}^{n}|x_i + x_j| - n\sum_{i=2}^{n}|x_i|$$

$$= \sum_{i=2}^{n}\sum_{j=2}^{n}|x_i + x_j| - (n-2)\sum_{i=2}^{n}|x_i|$$

$$\geq \sum_{i=2}^{n}\sum_{j=2}^{n}|x_i + x_j| - (n-1)\sum_{i=2}^{n}|x_i|,$$

which however is nonnegative by the minimality of n. If $x_1 = -x_i$ for some i, say $x_1 = -x_2$, then

$$f(x_1) = 2\sum_{i>2}|x_i + x_1| + 2\sum_{i>2}|x_i - x_2| + 4|x_1| + \sum_{i>2}\sum_{j>2}|x_i + x_j|$$

$$-2n|x_1| - n\sum_{i>2}|x_i|.$$

On the one hand $\sum_{i>2}\sum_{j>2}|x_i + x_j| \geq (n-2)\sum_{i>2}|x_i|$. On the other hand,

$$2\sum_{i>2}|x_i + x_1| + 2\sum_{i>2}|x_i - x_1| \geq 2\sum_{i>2}(|x_i| + |x_1|).$$

Hence

$$2 \sum_{i>2} |x_i + x_1| + 2 \sum_{i>2} |x_i - x_2| + 4|x_1| + \sum_{i,j>2} |x_i + x_j| - 2n|x_1| - n \sum_{i>2} |x_i|$$
$$\geq 2n|x_1| + 2 \sum_{i>2} |x_i| - 2n|x_1| - 2 \sum_{i>2} |x_i|$$
$$+ \sum_{i,j>2} |x_i + x_j| - (n-2) \sum_{i>2} |x_j|,$$

and this is nonnegative. It follows that $f(x_1)$ is nonnegative at all the endpoints of the intervals, hence it is nonnegative everywhere, and we are done.

(Mathematical Olympiad Summer Program, 2006)

11. The function $f(x,y,z) = x^2 + y^2 + z^2 - xyz - 2$ is quadratic in each of x, y, z, and having positive dominant coefficient, it attains the maximal value at the endpoints of the interval, thus for $x, y, z \in \{0, 1\}$. It is easy to check that in this case the inequality holds.

(Romanian Team Selection Test, 2006)

12. Let (x_1, y_1), (x_2, y_2), (x_3, y_3) be the vertices of the triangle inside the square of vertices $(0,0)$, $(1,0)$, $(0,1)$, $(1,1)$. Then $x_1, x_2, x_3, y_1, y_2, y_3 \in [0,1]$. The area of the triangle is half the absolute value of the determinant

$$\begin{vmatrix} 1 & 1 & 1 \\ x_1 & x_2 & x_3 \\ y_1 & y_2 & y_3 \end{vmatrix},$$

that is, half of

$$|x_2 y_3 - x_3 y_2 + x_3 y_1 - x_1 y_3 + x_1 y_2 - x_2 y_1|.$$

This is a convex function in each variable, so arguing as before we find that it has a maximal value that is attained when $x_1, x_2, x_3, y_1, y_2, y_3 \in \{0, 1\}$, that is, when the vertices of the triangle are vertices of the square. And in this case the area is $1/2$ (or 0, but of course this is not a maximum).

13. How can we possibly solve such a problem using the endpoint method? The intuitive reason is simple: the area (and the sum/difference of areas) is a linear function in the height or the length of the base, and a linear function attains its extremes at the endpoints.

Let us define $f(P) = \sum S_i - 2S$. Let also $l_i = A_i A_{i+1}$ and V_i be the third vertex of T_i. Note that V_i is uniquely determined unless there is a side l_j parallel to l_i, in this case V_i being any of its endpoints.

For $n = 3$ the assertion is clear, and for $n = 4$ we have $S_1 + S_2 + S_3 + S_4 \geq [A_1 A_2 A_3] + [A_2 A_3 A_4] + [A_3 A_4 A_1] + [A_4 A_1 A_2] = 2S$, where $[\quad]$ denotes the area. Next we shall use induction of step 2.

We are now going to apply the following operation, for as long as we can:

Choose a side $A_i A_{i+1}$ that is not parallel to any of the other sides of the polygon. Next, we try to move $X = A_i$ on the line $A_{i-1} A_i$, ensuring that while we move it, the polygon still remains convex and also line A_iA_{i+1} never becomes parallel to any of the other sides. We claim that f is linear in XA_{i-1}. Note that for any side l_k, V_k remains unchanged. To see this, note that if V_k changes, then there must be some intermediate step with V_k ambiguous, hence this intermediate has a side parallel to A_kA_{k+1}. However the only side whose direction is changing is A_iA_{i+1}, and we ensure that A_iA_{i+1} is never parallel to other sides of the polygon. Therefore T_k is either constant or has one vertex $X = A_i$ and the other two vertices fixed. In any case, S_k is clearly linear in A_iX, and obviously so is S. Therefore f, as a linear function, takes its minimal values at the extremities. What could the extremities be? We could have one of the following cases:

(a) $A_i = A_{i-1}$ in which the polygon degenerates into $A_1 \ldots A_{i-1}A_{i+1} \ldots A_n$, and we use induction.

(b) A_i goes to infinity, which only occurs if A_iA_{i-1} is parallel to $A_{i+1}A_{i+2}$, and in this case the inequality is easy to prove.

(c) A_i becomes collinear with $A_{i+1}A_{i+2}$, in which the polygon degenerates into $A_1 \ldots A_iA_{i+2} \ldots A_n$, and we use induction again.

(d) A_iA_{i+1} becomes parallel to one of the sides of the polygon. In this case, the number of pairs of parallel sides in P increases.

We are done in the cases (a), (b), (c). If we encounter case (d), repeat the operation and so on. Eventually we reach a polygon in which all sides are divided into pairs of parallel ones.

In this case, we can deduce that n is even and l_i parallel to $l_{i+\frac{n}{2}}$ (we work modulo n). Assume now that $n \geq 6$ because for $n = 4$ actually we have equality. Let $m = \frac{n}{2}$. We can see that $S_i + S_{i+m} = [A_iA_{i+1}A_{i+m}A_{i+m+1}]$, so $f(P) = \sum[A_iA_{i+1}A_{m+i}A_{m+i+1}] - 2S$.

Because opposite sides are parallel, we have that $\angle A_1A_2A_3 + \angle A_2A_3A_4 > 180$, that is A_1A_2 and A_4A_3 intersect at a point X. Let $A_{m+1}A_{m+2}$ and $A_{m+4}A_{m+3}$ intersect at point Y. We claim that

$$f(P) \geq f(A_1XA_4 \ldots A_{m+1}YA_{m+4} \ldots A_n).$$

This would provide the final step in the problem, since the polygon on the right has $n - 2$ sides, so we can apply the induction hypothesis.

Now

$$f(P) - f(A_1XA_4 \ldots A_{m+1}YA_{m+4} \ldots A_n)$$
$$= 2[A_2XA_3] + 2[A_{m+2}YA_{m+3}] + [A_2A_3A_{m+2}A_{m+3}] + [A_1A_2A_{m+1}A_{m+2}]$$
$$+ [A_3A_4A_{m+3}A_{m+4}] - [A_1XA_{m+1}Y] - [A_4XA_{m+4}Y],$$

and

$$[A_1XA_{m+1}Y] - [A_1A_2A_{m+1}A_{m+2}] = [A_2XA_{m+2}Y] = [A_2YX] + [XYA_{m+2}],$$

and analogously for $[A_4XA_{m+4}Y]$. The problem reduces then to

$$2[A_2XA_3] + 2[A_{m+2}YA_{m+3}] + [A_2A_3A_{m+2}A_{m+3}] - [A_2YX]$$
$$- [XYA_{m+2}] - [A_3YX] - [XYA_{m+3}] \geq 0.$$

However

$$[A_2YX] + [XYA_{m+2}] + [A_3YX] + [XYA_{m+3}]$$

$$= [A_2YA_{m+2}X] + +([XA_3A_{m+2}] + \frac{1}{2}(\overline{XA_3}, \overline{A_{m+2}Y}))$$

$$+ ([A_2YA_{m+3}] + \frac{1}{2}(\overline{YA_{m+3}}, \overline{A_2X}))$$

$$= [A_2A_3A_{m+2}A_{m+3}] + [XA_2A_3] + [YA_{m+2}A_{m+3}] + \frac{1}{2}(\overline{XA_3}, \overline{A_{m+2}Y})$$

$$+ \frac{1}{2}(\overline{YA_{m+3}}, \overline{A_2X}).$$

Here $(\overline{u}, \overline{v})$ denotes the dot product of $\overline{u}$ and $\overline{v}$. We are left to prove that

$$[XA_2A_3] + [YA_{m+2}A_{m+3}] \geq \frac{1}{2}\overline{XA_3}, \overline{A_{m+2}Y}) + \frac{1}{2}(\overline{YA_{m+2}}, \overline{A_2X}).$$

However as triangles $XA_2A_3, YA_{m+2}A_{m+3}$ are similar, we deduce that both $\frac{1}{2}(\overline{XA_3}, \overline{A_{m+2}Y})$ and $\frac{1}{2}(\overline{YA_{m+2}}, \overline{A_2X})$ are equal to $\sqrt{[XA_2A_3][YA_{m+2}A_{m+3}]}$, and we complete the solution by applying the AM–GM inequality.

(Communicated by I. Boreico)

14. Let $f(t) = t^2(1-y) - z^2t + y^2 + z^2 - y^2z - 1$. We must prove that $f(t) \leq 0$. The function $f(t)$ is convex (not strictly if $y = 1$) so the maximum is attained at the endpoints. Repeating the argument for y and z, we may assume $x, y, z \in 0, 1$ and check cases. The inequality follows.

15. The function $f(x) = x^{12}$ is convex, thus if $a < b < c < d$ with $a + d = b + c$, then $f(a) + f(d) \geq f(b) + f(c)$ (another way of proving it is to see that the function $g(x) = (t + x)^{12} + (t - x)^{12}$ is increasing). Thus, if we have two numbers that are not at the ends of the interval $[-\frac{1}{\sqrt{3}}; \sqrt{3}]$, then we can push them away increasing the value of the expression. We can push them until one becomes an endpoint of the interval. Thus we can assume that at most one number is not $-\frac{1}{\sqrt{3}}$ or $\sqrt{3}$. Assume we have k numbers equal to $-\frac{1}{\sqrt{3}}$, $1996 - k$ equal to $\sqrt{3}$, and one last number, say x. As the sum of all the numbers is $-318\sqrt{3}$, we have $-k\frac{1}{\sqrt{3}} + (1996 - k)\sqrt{3} + x = -318\sqrt{3}$. Multiplying by $\sqrt{3}$ we get $-k + 3(1996 - k) + \sqrt{3}x = -954$ so $4k = 6942 + \sqrt{3}x$. As $\sqrt{3}x$ is between -1 and 3, $4k$ is between 6941 and 6945, and since k is an integer, we get $k = 1736$. Thus the maximal set of numbers consists of 1736 numbers equal to $-\frac{1}{\sqrt{3}}$, 260 numbers equal to $\sqrt{3}$, and one number equal to $-318\sqrt{3} + \frac{1736}{\sqrt{3}} - 260\sqrt{3} = \frac{2}{\sqrt{3}}$. It follows that the greatest possible value of our expression equals $\frac{1736}{3^6} + 260 \cdot 3^6 + \frac{2^{12}}{3^6}$.

(Communicated by I. Boreico)

2.3   Telescopic Sums and Products in Algebra

1. We can write

$$\sum_{k=1}^{n} k!(k^2 + k + 1) = \sum_{k=1}^{n} [(k+1)^2 - k]k!$$

$$= \sum_{k=1}^{n} [(k+1)!(k+1) - k!k] = (n+1)!(n+1) - 1.$$

2. We have

$$\sum_{k=1}^{n} \frac{1}{a_k a_{k+1}} = \frac{1}{d} \sum_{k=1}^{n} \frac{a_{k+1} - a_k}{a_k a_{k+1}} = \frac{1}{d} \sum_{k=1}^{n} \left(\frac{1}{a_k} - \frac{1}{a_{k+1}} \right).$$

Hence the sum is equal to $(1/d)(a_{n+1} - a_1)/(a_{n+1}a_1)$. Since $a_{n+1} - a_1 = nd$, this is equal to $n/((a_1 + nd)a_1)$.

3. Trying a "partial fraction" decomposition

$$\frac{6^k}{(3^k - 2^k)(3^{k+1} - 2^{k+1})} = \frac{A}{3^k - 2^k} - \frac{B}{3^{k+1} - 2^{k+1}},$$

we find

$$(3^{k+1} - 2^{k+1})A - (3^k - 2^k)B = 6^k.$$

We can try either

$$3^k(3A - B) = 6^k$$
$$2^k(2A - B) = 0$$

or

$$3^k(3A - B) = 0$$
$$2^k(2A + B) = 6^k.$$

The first system gives $A = 2^k, B = 2^{k+1}$ with the decomposition

$$\frac{6^k}{(3^k - 2^k)(3^{k+1} - 2^{k+1})} = \frac{2^k}{3^k - 2^k} - \frac{2^{k+1}}{3^{k+1} - 2^{k+1}}$$

and the second gives $A = 3^k, B = 3^{k+1}$ with the decomposition

$$\frac{6^k}{(3^k - 2^k)(3^{k+1} - 2^{k+1})} = \frac{3^k}{3^k - 2^k} - \frac{3^{k+1}}{3^{k+1} - 2^{k+1}}.$$

In both cases the sum telescopes, and we find that it equals

$$\sum_{k=1}^{\infty} \frac{6^k}{(3^k - 2^k)(3^{k+1} - 2^{k+1})} = \frac{2}{3-2} - \lim_{k \to \infty} \frac{2^{k+1}}{3^{k+1} - 2^{k+1}} = 2.$$

(44th W.L. Putnam Mathematical Competition, 1984)

4. We will use the recurrence relation of the sequence to telescope the sum. Since $x_{k+1} = x_k^2 + x_k$, we obtain

$$\frac{1}{x_{k+1}} = \frac{1}{x_k(x_k + 1)} = \frac{1}{x_k} - \frac{1}{x_k + 1},$$

so that

$$\frac{1}{x_k + 1} = \frac{1}{x_k} - \frac{1}{x_{k+1}}.$$

Therefore,

$$\frac{1}{x_1 + 1} + \frac{1}{x_2 + 1} + \cdots + \frac{1}{x_{100} + 1} = \frac{1}{x_1} - \frac{1}{x_{101}}.$$

Since $x_1 = \frac{1}{2}$ and $0 < 1/x_{101} < 1$, the integer part of the sum is 1.

(Tournament of the Towns, Autumn 1985; proposed by A. Andjans)

5. By using the recurrence formula for the Fibonacci sequence, we obtain the following chains of equalities:

(a) $$\sum_{n=2}^{\infty} \frac{F_n}{F_{n-1}F_{n+1}} = \sum_{n=2}^{\infty} \frac{F_{n+1} - F_{n-1}}{F_{n-1}F_{n+1}} = \sum_{n=2}^{\infty} \left(\frac{1}{F_{n-1}} - \frac{1}{F_{n+1}} \right)$$

$$= \lim_{N \to \infty} \left(\frac{1}{F_1} + \frac{1}{F_2} - \frac{1}{F_N} - \frac{1}{F_{N+1}} \right) = \frac{1}{F_1} + \frac{1}{F_2} = 2;$$

(b) $$\sum_{n=2}^{\infty} \frac{1}{F_{n-1}F_{n+1}} = \sum_{n=2}^{\infty} \frac{F_n}{F_{n-1}F_nF_{n+1}} = \sum_{n=2}^{\infty} \frac{F_{n+1} - F_{n-1}}{F_{n-1}F_nF_{n+1}}$$

$$= \sum_{n=2}^{\infty} \left(\frac{1}{F_{n-1}F_n} - \frac{1}{F_nF_{n+1}} \right)$$

$$= \lim_{n \to \infty} \left(\frac{1}{F_1F_2} - \frac{1}{F_NF_{N+1}} \right)$$

$$= \frac{1}{F_1F_2} = 1.$$

6. For a positive integer n, we can write

$$1 + \frac{1}{n^2} + \frac{1}{(n+1)^2} = \frac{n^2(n+1)^2 + (n+1)^2 + n^2}{n^2(n+1)^2} = \frac{(n^2 + n + 1)^2}{n^2(n+1)^2}.$$

Therefore,

$$\sqrt{1+\frac{1}{n^2}+\frac{1}{(n+1)^2}} = \frac{n^2+n+1}{n^2+n} = 1+\frac{1}{n(n+1)}.$$

Hence the given sum is equal to

$$\sum_{n=1}^{1999}\left(1+\frac{1}{n(n+1)}\right) = \sum_{n=1}^{1999}\left(1+\frac{1}{n}-\frac{1}{n+1}\right) = 2000-\frac{1}{2000}.$$

7. There are some terms missing to make this sum telescope. However, since the left-hand side is greater than

$$\frac{1}{\sqrt{3}+\sqrt{5}}+\frac{1}{\sqrt{7}+\sqrt{9}}+\cdots+\frac{1}{\sqrt{9999}+\sqrt{10001}},$$

the inequality will follow from

$$\frac{1}{\sqrt{1}+\sqrt{3}}+\frac{1}{\sqrt{3}+\sqrt{5}}+\frac{1}{\sqrt{5}+\sqrt{7}}+\cdots+\frac{1}{\sqrt{9999}+\sqrt{10001}} > 48.$$

Now we are able to telescope. Rationalize the denominators and obtain the equivalent inequality

$$\frac{\sqrt{3}-\sqrt{1}}{2}+\frac{\sqrt{5}-\sqrt{3}}{2}+\frac{\sqrt{7}-\sqrt{5}}{2}+\cdots+\frac{\sqrt{10001}-\sqrt{9999}}{2} > 48.$$

The left side is equal to $(\sqrt{10001}-1)/2$, and an easy check shows that this is larger than 48.

(Ukrainian mathematics contest)

8. Since for a positive integer k, $(\sqrt{k+1}-\sqrt{k})(\sqrt{k+1}+\sqrt{k})=1$, we have

$$2(\sqrt{k+1}-\sqrt{k}) = \frac{2}{\sqrt{k+1}+\sqrt{k}} < \frac{1}{\sqrt{k}}$$

and

$$\frac{1}{\sqrt{k}} < \frac{2}{\sqrt{k}+\sqrt{k-1}} = 2(\sqrt{k}-\sqrt{k-1}).$$

Combining the two yields

$$2(\sqrt{k+1}-\sqrt{k}) < \frac{1}{\sqrt{k}} < 2(\sqrt{k}-\sqrt{k-1}).$$

By adding all these inequalities for k between m and n, we obtain

$$2(\sqrt{n+1}-\sqrt{m}) < \frac{1}{\sqrt{m}}+\frac{1}{\sqrt{m+1}}$$

$$+\cdots+\frac{1}{\sqrt{n-1}}+\frac{1}{\sqrt{n}} < 2(\sqrt{n}-\sqrt{m-1}),$$

and the double inequality is proved.

9.   The idea is first to decrease the denominator of a_n, replacing $k^{4/3}$ by $(k-1)^{2/3}(k+1)^{2/3}$, and then to rationalize it. We have

$$a_n < \frac{k}{(k-1)^{4/3}+(k-1)^{2/3}(k+1)^{2/3}+(k+1)^{4/3}}$$

$$= \frac{k((k+1)^{2/3}-(k-1)^{2/3})}{(k+1)^2-(k-1)^2} = \frac{1}{4}((k+1)^{2/3}-(k-1)^{2/3}).$$

It follows that

$$\sum_{n=1}^{999} a_n < \frac{1}{4}\sum_{n=1}^{999}((k+1)^{2/3}-(k-1)^{2/3})$$

$$= \frac{1}{4}(1000^{2/3}+999^{2/3}-1^{2/3}-0^{2/3})$$

$$< \frac{1}{4}(100+100-1) < 50.$$

(T. Andreescu)

10. It is natural to transform the terms of the sum as

$$\frac{1}{\sqrt{n}(n+1)} = \frac{\sqrt{n}}{n(n+1)} = \frac{\sqrt{n}}{n} - \frac{\sqrt{n}}{n+1}.$$

This allows us to rewrite the sum as

$$1 + \sum_{n=2}^{\infty} \frac{\sqrt{n}-\sqrt{n-1}}{n}.$$

The sum does not telescope, but it is bounded from above by

$$1 + \sum_{n=2}^{\infty} \frac{\sqrt{n}-\sqrt{n-1}}{\sqrt{n}\sqrt{n-1}} = 1 + \sum_{k=2}^{\infty}\left(\frac{1}{\sqrt{n-1}}-\frac{1}{\sqrt{n}}\right),$$

which telescopes to 2. This proves the inequality.
   (Romanian college admission exam)

11. By induction,

$$F_{2m}F_{m-1} - F_{2m-1}F_m = (-1)^m F_m, \quad m \geq 1.$$

Setting $m = 2^{n-1}$ yields

$$F_{2^n}F_{2^{n-1}-1} - F_{2^n-1}F_{2^{n-1}} = F_{2^{n-1}}, \quad n \geq 2,$$

or

$$\frac{1}{F_{2^n}} = \frac{F_{2^{n-1}-1}}{F_{2^{n-1}}} - \frac{F_{2^n-1}}{F_{2^n}}, \quad n \geq 2.$$

Thus

$$\sum_{n=0}^{\infty} \frac{1}{F_{2^n}} = \frac{1}{F_1} + \frac{1}{F_2} + \lim_{N\to\infty}\left(\frac{F_1}{F_2} - \frac{F_{2^N-1}}{F_{2^N}}\right) = 3 - \frac{1}{\frac{\sqrt{5}+1}{2}} = \frac{7-\sqrt{5}}{2}.$$

12. We have

$$\prod_{n=2}^{\infty} \frac{n^3-1}{n^3+1} = \lim_{N\to\infty}\prod_{n=2}^{N} \frac{n^3-1}{n^3+1} = \lim_{N\to\infty}\prod_{n=2}^{N} \frac{(n-1)(n^2+n+1)}{(n+1)(n^2-n+1)}$$

$$= \lim_{n\to\infty}\prod_{n=2}^{N} \frac{n-1}{n+1} \prod_{n=2}^{N} \frac{(n+1)^2-(n+1)+1}{n^2-n+1}$$

$$= \lim_{N\to\infty} \frac{1\cdot 2\cdot((N+1)^2-(N+1)+1)}{3N(N+1)} = \frac{2}{3}.$$

(W.L. Putnam Mathematical Competition)

13. We can write

$$\prod_{n=0}^{\infty}\left(1+\frac{1}{2^{2^n}}\right) = 2\lim_{N\to\infty}\left(1-\frac{1}{2^{2^0}}\right)\prod_{n=0}^{N}\left(1+\frac{1}{2^{2^n}}\right).$$

Since

$$\left(1-\frac{1}{2^{2^n}}\right)\left(1+\frac{1}{2^{2^n}}\right) = \left(1-\frac{1}{2^{2^{n+1}}}\right),$$

the latter product telescopes, and is equal to $1-1/2^{2^{N+1}}$. It follows that the answer to the problem is 2.

14. The Fibonacci and Lucas sequences satisfy the identities $F_{n+1}+F_{n-1}=L_{n+1}$ and $F_{2n}=F_{n+1}^2-F_{n-1}^2$ for all $n\geq 1$. Then

$$F_{2n} = (F_{n+1}+F_{n-1})(F_{n+1}-F_{n-1}) = L_{n+1}F_n.$$

Thus $L_{n+1}=F_{2n}/F_n$, $n\geq 1$. Therefore,

$$\prod_{k=1}^{m} L_{2^k+1} = \prod_{k=1}^{m} \frac{F_{2^{k+1}}}{F_{2^k}} = \frac{F_{2^{m+1}}}{F_2} = F_{2^{m+1}},$$

and we are done.
(T. Andreescu)

2.4   On an Algebraic Identity

1. *First solution:* We have $2^{2^n-2} + 1 = 1 + \frac{m^4}{4}$, where $m = 2^{2^{n-2}}$. We can factor

$$1 + \frac{m^4}{4} = \left(1 + m + \frac{1}{2}m^2\right)\left(1 - m + \frac{1}{2}m^2\right),$$

and the conclusion follows.

Second solution: Observe that the exponent is congruent to 2 modulo 4 and by Fermat's little theorem $2^4 \equiv 1 \pmod{5}$. Therefore

$$2^{2^n-2} + 1 \equiv 2^2 + 1 \equiv 5 \pmod{5}.$$

It follows that all our numbers are divisible by 5, and they are of course greater than 5 because for $n > 2$,

$$2^{2^n-2} + 1 > 2^{2^2-2} + 1 = 5.$$

2. Observe that factoring $X^4 + 1$ yields

$$X^4 + 1 = 4((X/\sqrt{2})^4 + \frac{1}{4})$$
$$= 4((X/\sqrt{2})^2 + X/\sqrt{2} + \frac{1}{2})((X/\sqrt{2})^2 - X/\sqrt{2} + \frac{1}{2})$$
$$= (X^2 - \sqrt{2}X + 1)(X^2 + \sqrt{2}X + 1).$$

This shows that the roots α and β of the characteristic equation of the sequence $X^2 - \sqrt{2}X + 1 = 0$ are roots of $X^4 + 1$, so they are eighth roots of unity. The general term of the sequence is of the form $x_n = a\alpha^n + b\beta^n$ for some a and b, and hence the sequence is periodic of period 8.

(*Kvant (Quantum)*)

3. Since $k^4 + \frac{1}{4} = (k^2 - k + \frac{1}{2})(k^2 + k + \frac{1}{2})$, we have

$$\sum_{k=1}^{n} \frac{k^2 - \frac{1}{2}}{k^4 + \frac{1}{4}} = \sum_{k=1}^{n} \left(\frac{k - \frac{1}{2}}{k^2 - k + \frac{1}{2}} - \frac{k + \frac{1}{2}}{k^2 + k + \frac{1}{2}}\right)$$
$$= \sum_{k=1}^{n} \left(\frac{k - \frac{1}{2}}{k^2 - k + \frac{1}{2}} - \frac{(k+1) - \frac{1}{2}}{(k+1)^2 - (k+1) + \frac{1}{2}}\right).$$

This is a telescopic sum, equal to $1 - (2n+1)/(2n^2 + 2n + 1)$.

(T. Andreescu)

4. We will use the factorization

$$m^4 + \frac{1}{4} = \left(m^2 + m + \frac{1}{2}\right)\left(m^2 - m + \frac{1}{2}\right).$$

The product becomes

$$\prod_{k=1}^{n} \frac{(2k-1)^4 + \frac{1}{4}}{(2k)^4 + \frac{1}{4}}$$
$$= \prod_{k=1}^{n} \frac{((2k-1)^2 + (2k-1) + \frac{1}{2})((2k-1)^2 - (2k-1) + \frac{1}{2})}{((2k)^2 + 2k + \frac{1}{2})((2k)^2 - 2k + \frac{1}{2})}.$$

Since $m^2 - m + \frac{1}{2} = (m-1)^2 + (m-1) + \frac{1}{2}$, the factors in the numerator cancel those in the denominator, except for $1^2 - 1 + \frac{1}{2}$ in the numerator and $(2n)^2 + 2n + \frac{1}{2}$ in the denominator. Hence the answer is $1/(8n^2 + 4n + 1)$.
   (Communicated by S. Savchev)

5. If we choose $a = 4k^4$, with $k > 1$, then

$$n^4 + 4k^4 = (n^2 + 2nk + 2k^2)(n^2 - 2nk + 2k^2).$$

Since $n^2 + 2nk + 2k^2 > k > 1$ and $n^2 - 2nk + 2k^2 = (n-k)^2 + k^2 > k^2 > 1$, none of the numbers $n^4 + 4k^4$ is prime.
   (11th IMO, 1969)

6. If n is even, the number is clearly divisible by 2. If n is odd, say $n = 2k + 1$, then by applying the Sophie Germain identity for $X = n$ and $Y = 2^{k+1}$, we obtain

$$n^4 + 4^n = n^4 + \frac{1}{4} 4^{2k+2} = (n^2 + 2^{k+1}n + 2^{2k+1})(n^2 - 2^{k+1}n + 2^{2k+1}).$$

If $n > 1$, both factors are greater than 1, which proves that in this situation the number is composite. Of course, when $n = 1$, $4^n + n^4 = 5$, which is prime.

7. We have

$$P(X^4) = X^{16} + 6X^8 - 4X^4 + 1 = (X^4 - 1)^4 + 4(X^3)^4.$$

This can be factored as

$$\left[(X^4 - 1)^2 + 2(X^4 - 1)X^3 + 2(X^3)^2\right] \left[(X^4 - 1)^2 - 2(X^4 - 1)X^3 + 2(X^3)^2\right]$$

and we are done.

8. Using the Sophie Germain identity, we factor $n^{12} + 64$ as $(n^6 - 4n^3 + 8)$ $(n^6 + 4n^3 + 8)$. On the other hand, $n^{12} + 64$ is the sum of two cubes; hence it factors as $(n^4 + 4)(n^8 - 4n^4 + 16)$. By the same identity, $n^4 + 4 = (n^2 - 2n + 2)(n^2 + 2n + 2)$. The polynomials $n^2 - 2n + 2$ and $n^2 + 2n + 2$ are irreducible over the ring of polynomials with integer coefficients; hence they divide in some order the polynomials $n^6 - 4n^3 + 8$ and $n^6 + 4n^3 + 8$. Checking cases yields

$$n^6 + 4n^3 + 8 = (n^2 - 2n + 2)(n^4 + 2n^3 + 2n^2 + 4n + 4)$$

and

$$n^6 - 4n^3 + 8 = (n^2 + 2n + 2)(n^4 - 2n^3 + 2n^2 - 4n + 4).$$

Thus

$$n^{12} + 64 = (n^2 - 2n + 2)(n^2 + 2n + 2)(n^4 - 2n^3 + 2n^2 - 4n + 4)$$
$$\times (n^4 + 2n^3 + 2n^2 + 4n + 4).$$

The four factors are strictly increasing in that order, so they are distinct.
   (T. Andreescu)

   9. Summing as a geometric progression, we obtain

$$\sum_{k=0}^{m}(-4)^k n^{4(m-k)} = n^{4m}\sum_{k=0}^{m}\left(-\frac{4}{n}\right)^k = \frac{(n^4)^{m+1} + 4^{m+1}}{n^4 + 4}$$
$$= \frac{\left(n^{m+1}\right)^4 + 4\left(2^{m/2}\right)^4}{n^4 + 4}.$$

Using the Sophie Germain identity, the numerator can be written as the product of $n^{2(m+1)} + 2^{m/2+1}n^{m+1} + 2^{m+1}$ and $n^{2(m+1)} - 2^{m/2+1}n^{m+1} + 2^{m+1}$. Because $m \geq 2$, the denominator is less than any of these two factors, so after cancellations, the remaining number is still a product of two numbers greater than one (one coming from the first factor, one from the second), and the problem is solved.
   (T. Andreescu)

   10. We will show that the least number is 16. First we have to check that the numbers from 1 to 15 don't work. To this end, we apply the Eisenstein criterion of irreducibility:

   Given a polynomial $P(x) = a_n x^n + a_{n-1}x^{n-1} + \cdots + a_0$ with integer coefficients, suppose that there exists a prime number p such that a_n is not divisible by p, a_k is divisible by p for $k = 0, 1, \ldots, n-1$, and a_0 is not divisible by p^2. Then $P(x)$ is irreducible over $\mathbf{Z}[x]$ (meaning that it cannot be written in a nontrivial way as a product of two polynomials with integer coefficients).

   For $n = 10, 11, 12, 13, 14$, and 15, we apply this criterion for the primes 5, 11, 3, 13, 7, and 5, respectively.
   In the cases $n = 8$ or 9, if the polynomial could be factored in the desired fashion, then the factorization would have a linear term. But it is easy to check that these polynomials have no integer roots. Thus we proved that $n \geq 16$.
   For $n = 16$, this reduces to Problem 8 above.
   (*Mathematical Reflections*, proposed by T. Andreescu)

2.5   Systems of Equations

   1. It is not difficult to guess that $x = y = z = 0$ is a solution. Let us see whether there are other solutions. If $x > 0$, then $\log(x + \sqrt{x^2 + 1}) > 0$, and from the first equation we deduce $y > x > 0$. From the second and the third equations we obtain $x > z > y > x > 0$, which is impossible.

If $x < 0$, then

$$x + \sqrt{x^2 + 1} = \frac{1}{-x + \sqrt{x^2 + 1}} < 1.$$

Hence $y < x < 0$, and consequently $x < z < y < x < 0$, which is again impossible. Therefore, the only solution is $x = y = z = 0$.
(Israeli Mathematical Olympiad, 1995)

2. We have $\log(2xy) = \log 2 + \log x + \log y$. By moving the logarithms containing variables to the right and adding 1 to each side of the three equations, we obtain

$$\log 20 = (\log x - 1)(\log y - 1),$$
$$1 = (\log y - 1)(\log z - 1),$$
$$\log 20 = (\log z - 1)(\log x - 1).$$

Multiplying all equations and taking the square root yields

$$\pm \log 20 = (\log x - 1)(\log y - 1)(\log z - 1).$$

This, combined with the equality $\log 20 = (\log x - 1)(\log y - 1)$, shows that $\log z - 1 = \pm 1$. The other equations give $\log x - 1 = \pm \log 20$ and $\log y - 1 = \pm 1$, and we obtain the two solutions to the system $(200, 100, 100)$ and $\left(\frac{1}{2}, 1, 1\right)$.
(*Revista Matematică din Timisoara (Timişoara's Mathematics Gazette)*, proposed by T. Andreescu)

3. Let $\sqrt[3]{xyz} = a$. From the AM–GM inequality

$$12 = xy + yz + zx \geq 3a^2$$

and

$$a^3 = 2 + x + y + z \geq 2 + 3a.$$

Therefore $a^2 \leq 4$ and $a^3 - 3a - 2 = (a - 2)(a + 1)^2 \geq 0$. Hence all equalities hold, $a = 2$, and $x = y = z$. Thus $(x, y, z) = (2, 2, 2)$ is the only solution.
(British Mathematical Olympiad, 1998)

4. The solution is very similar to the one we gave for problem 1. We start by observing that the function $f : [0, \infty) \to [0, \infty)$, $f(t) = 4t^2 / (4t^2 + 1)$ is strictly increasing. Hence if $x < y$, then $f(x) < f(y)$, so $y < z$. Repeating the argument, we obtain $z < x$; hence $x < y < z < x$, which is impossible. Similarly, $x > y$ leads to a contradiction. Therefore, $x = y = z$. Solving the equation $4t^2 / (4t^2 + 1) = t$ yields $t = 0$ or $t = \frac{1}{2}$. Hence the only triples that satisfy the system are $(0, 0, 0)$ and $\left(\frac{1}{2}, \frac{1}{2}, \frac{1}{2}\right)$.
(Canadian Mathematical Olympiad, 1996)

5. For $n = 2$ and $n = 3$, the identity

$$(ax^n + by^n)(x + y) - (ax^{n-1} + by^{n-1})xy = ax^{n+1} + y^{n+1}$$

leads to the equations

$$7(x+y) - 3xy = 16 \text{ and } 16(x+y) - 7xy = 42.$$

Solving these two equations simultaneously yields

$$x+y = -14 \text{ and } xy = -38.$$

Applying the recurrence identity for $n = 4$ gives

$$ax^5 + by^5 = (42)(-14) - (16)(-38) = -588 + 608 = 20.$$

(AIME, 1990)

6. The only solutions are $x = y = z = 1$ and $x = y = z = -1$. The given equalities imply that

$$(x - y^{-1}) + (y - z^{-1}) + (z - x^{-1}) = xyz - (xyz)^{-1},$$

which factors as

$$(x - y^{-1})(y - z^{-1})(z - x^{-1}) = 0.$$

Thus one of $x - y^{-1}, y - z^{-1}, z - x^{-1}$ is zero, but the given equalities imply that all three are zero. Thus $xy = yz = zx = 1$, $(xyz)^2 = 1$, and so $x = y = z = 1$ or $x = y = z = -1$.
(Math Horizons, April 2000, proposed by T. Andreescu)

7. The two equalities imply

$$24 = (x+y+z)^3 - (x^3+y^3+z^3) = 3\sum x^2 y + 6xyz.$$

Dividing by 3 and factoring, we get

$$(x+y)(x+z)(y+z) = 8,$$

That is

$$(3-x)(3-y)(x+y) = 8.$$

All factors are integers that divide 8, and an easy check yields the solutions

$$(x,y,z) = (1,1,1),(4,4,-5),(4,-5,4),(-5,4,4).$$

8. Let (x,y,z) be a solution. Clearly, if one of these numbers is positive, the other two must be positive as well. Multiplying by -1 if necessary, we may assume that $x,y,z > 0$.
Adding the three equations, we obtain

$$x+y+z = 2\left(\frac{1}{x} + \frac{1}{y} + \frac{1}{z}\right).$$

Also, applying the AM–GM inequality to each equation of the system yields $2x \geq 2\sqrt{2}$, $2y \geq 2\sqrt{2}$, $2z \geq 2\sqrt{2}$. This shows that in the above equation, the left side is greater than or equal to $3\sqrt{2}$, whereas the right side is less than or equal $3\sqrt{2}$. To obtain equality, we must have $x = y = z = \sqrt{2}$, which gives one solution. The other solution is obtained by changing sign and is $x = y = z = -\sqrt{2}$.

Remark. This is a system of the form $y = f(x)$, $z = f(y)$, $x = f(z)$, where $f(t) = \frac{1}{2}(t + 2/t)$. The sequence given by

$$t_0 \in \mathbf{R}, \quad t_{n+1} = f(t_n), \, n \geq 0,$$

is traditionally used to compute $\sqrt{2}$ with great precision because it converges really rapidly to it. No matter what $t_0 \in \mathbf{R}$ is, each subsequent term is greater than or equal to $\sqrt{2}$ in absolute value. If, for definiteness, $t_0 > 0$, then $t_n \geq \sqrt{2}$ for $n \geq 1$ and also $t_1 \geq t_2 \geq \cdots$. A term in this sequence can repeat only if it is exactly $\sqrt{2}$. There is no difficulty in solving the analogous system with any number of variables.

9. Subtracting the second equation from the first, we obtain

$$(x - z)((x+y)^2 + (x+y)(y+z) + (y+z)^2) = z - x.$$

Since $(x+y)^2 + (x+y)(y+z) + (y+z)^2 > 0$, we obtain $x = z$. By symmetry $y = z$, and we are left with solving the equation $8x^3 = x$. This equation has the solutions $x = 0$ and $x = \pm\frac{1}{2}$. It follows that the solutions to the given system of equations are $x = y = z = 0$, $x = y = z = 1/(2\sqrt{2})$, and $x = y = z = -1/(2\sqrt{2})$.

(Tournament of the Towns, 1985)

10. Let (x, y, z) be a solution. If $xyz \neq 0$, then, since the absolute value is positive, we obtain $x^2 > |yz|$, $y^2 > |zx|$, and $z^2 > |xy|$, which by multiplication gives $x^2y^2z^2 > x^2y^2z^2$, a contradiction. Thus one of the numbers is zero, and using the equation that contains it on the left side, we obtain that another of the three numbers must be zero as well. The third one can be only 0 or ± 1. Thus the solutions are $(0,0,0)$, $(1,0,0)$, $(0,1,0)$, $(0,0,1)$, $(-1,0,0)$, $(0,-1,0)$, and $(0,0,-1)$.

(T. Andreescu)

11. Adding the three equations, we obtain $2x + 2y + 2z = 6.6$; hence $x + y + z = 3.3$. Subtracting from this the initial equations gives the equivalent system

$$\{y\} + \lfloor z \rfloor = 2.2,$$
$$\{x\} + \lfloor y \rfloor = 1.1,$$
$$\{z\} + \lfloor x \rfloor = 0.$$

The first equation gives $\lfloor z \rfloor = 2$, $\{y\} = 0.2$, the second $\lfloor y \rfloor = 1$, $\{x\} = 0.1$, and the third $\lfloor x \rfloor = 0$ and $\{z\} = 0$. Hence the solution is $x = .1$, $y = 1.2$, and $z = 2$.

(Romanian mathematics contest, 1979; proposed by T. Andreescu)

12. Let $s = x_1 + x_2 + x_3 + x_4$. The system becomes

$$(s - x_4)x_4 = a,$$
$$(s - x_3)x_3 = a,$$
$$(s - x_2)x_2 = a,$$
$$(s - x_1)x_1 = a.$$

This is equivalent to $x_k^2 - sx_k + a = 0$, $k = 1, 2, 3, 4$. It follows that x_1, x_2, x_3, x_4 are solutions to the equation $u^2 - su + a = 0$.

The rest of the work is routine and inevitable. Instead of analyzing the 16 possible cases separately, we proceed as follows. If $x_1 = x_2 = x_3 = x_4$, then each x_i equals $s/4$. Plugging this into any of the equations yields $s = \pm 4\alpha$, where α is one of the solutions to the equation $3x^2 = a$ (remember that a is complex, so the notation $\sqrt{a}$ does not make sense). This case leads to the two solutions $(\alpha, \alpha, \alpha, \alpha)$ and $(-\alpha, -\alpha, -\alpha, -\alpha)$.

If two x_i's are distinct, say $x_1 \neq x_2$, then they are the two roots of the equation $u^2 - su + a = 0$, so their sum is s. Then $x_3 + x_4 = 0$, and it suffices to consider two cases.

If $x_3 \neq x_4$, the same argument shows that $x_3 + x_4 = s$; hence $s = 0$, and the quadruple (x_1, x_2, x_3, x_4) is of the form $(\beta, -\beta, \beta, -\beta)$, where β is one of the solutions of the equation $x^2 + a = 0$. From symmetry, we obtain the six solutions $(\beta, -\beta, \beta, -\beta)$, $(\beta, \beta, -\beta, -\beta)$, $(\beta, -\beta, -\beta, \beta)$, $(-\beta, \beta, \beta, -\beta)$, $(-\beta, \beta, -\beta, \beta)$, and $(-\beta, -\beta, \beta, \beta)$.

If $x_1 \neq x_2$ and $x_3 = x_4$, then $x_3 = x_4 = 0$. This implies that three x_i's are zero and the fourth is s. This is, however, possible if and only if $a = 0$, in which case we obtain the additional solutions $(s, 0, 0, 0)$, $(0, s, 0, 0)$, $(0, 0, s, 0)$, and $(0, 0, 0, s)$, with s any complex number.

(Romanian IMO selection test, 1976; proposed by I. Cuculescu)

13. *First solution:* Note that $(0, 0, 0, 0, 0)$ is a solution. Let us assume that x_1, x_2, x_3, x_4, x_5 is a nontrivial solution. It follows that $\sum(ak - k^3)x_k = 0$ and $\sum(ak^3 - k^5)x_k = 0$. We have

$$\sum_{k^2 \leq a}(a - k^2)kx_k = \sum_{k^2 > a}(a - k^2)kx_k,$$
$$\sum_{k^2 \leq a}(a - k^2)k^3 x_k = \sum_{k^2 > a}(a - k^2)k^3 x_k.$$

But

$$\sum_{k^2 \leq a}(a - k^2)k^3 x_k \leq a \sum_{k^2 \leq a}(a - k^2)kx_k = a \sum_{k^2 > a}(a - k^2)kx_k$$
$$\leq \sum_{k^2 > a}(a - k^2)k^3 x_k.$$

Since the first and the last terms are equal, all inequality signs are in fact equalities. We have

$$\sum_{k^2 > a} a(a - k^2)kx_k = \sum_{k^2 > a} k^2(a - k^2)kx_k.$$

But for $k^2 > a$, we have $a(k^2 - a)kx_k > k^2(k^2 - a)kx_k$, which combined with the inequality above shows that for $k^2 > a$, $x_k = 0$. A similar argument shows that $x_k = 0$ if $k^2 < a$. Thus for the system to admit a nontrivial solution, a must be equal to one of the perfect squares $1, 4, 9, 16, 25$. Note that if $a = m^2$ for some $m = 1, 2, 3, 4$, or 5, then $x_k = 0$ for $k \neq m$, and $x_m = m$ is a solution.

Second solution: As before, let x_1, x_2, x_3, x_4, x_5 be a nontrivial solution. From the equations of the system, it follows that

$$\left(\sum_{k=1}^{5} k^3 x_k \right)^2 = \left(\sum_{k=1}^{5} k x_k \right) \left(\sum_{k=1}^{5} k^5 x_k \right).$$

On the other hand, the Cauchy–Schwarz inequality applied to the sequences $\{\sqrt{k x_k}\}_{k=1,\dots,5}$ and $\{\sqrt{k^5 x_k}\}_{k=1,\dots,5}$ gives

$$\left(\sum_{k=1}^{5} k^3 x_k \right)^2 \leq \left(\sum_{k=1}^{5} k x_k \right) \left(\sum_{k=1}^{5} k^5 x_k \right).$$

The relation we deduced above shows that we have equality in the Cauchy–Schwarz inequality, and hence the two sequences are proportional. For $x_k \neq 0$ we have $\sqrt{k^5 x_k}/\sqrt{k x_k} = k^2$, and since all these values are distinct, it follows that $x_k \neq 0$ for exactly one k. As before, we conclude that the only possible values for a are $1, 4, 9, 16, 25$.

Third solution: Note that

$$0 \leq \sum_{k=1}^{5} k(k^2 - a)^2 x_k = \sum_{k=1}^{5} k^5 x_k - 2a \sum_{k=1}^{5} k^3 x_k + a^2 \sum_{k=1}^{5} k x_k$$
$$= a^3 - 2a^3 + a^3 = 0.$$

Hence $(k^2 - a)^2 x_k = 0$ for all k. Thus a nontrivial solution must have $a = k^2$ for some k and only one nonzero x_k.
(21st IMO, 1979)

14. The system can be rewritten as

$$(y - 3)^3 = y^3 - x^3,$$
$$(z - 3)^3 = z^3 - y^3,$$
$$(x - 3)^3 = x^3 - z^3.$$

Adding these gives

$$(x - 3)^3 + (y - 3)^3 + (z - 3)^3 = 0.$$

Without loss of generality, we may assume that $x \geq 3$. From the third equation of the initial system, we obtain $z^3 - 27 = 9x(x-3)$; hence $z \geq 3$. Similarly, $y \geq 3$. The above equality implies that $x = y = z = 3$ is the only possible solution.

15. Add the third equation to the first and subtract the second to obtain

$$2ax = (x-y)^2 + (z-x)^2 - (y-z)^2 = 2(x^2 - xy - xz + yz).$$

Factoring this gives

$$ax = (x-y)(x-z).$$

In a similar manner we obtain

$$by = (y-z)(y-x) \text{ and } cz = (z-x)(z-y).$$

Now let (x,y,z) be a solution. Without loss of generality, we may assume $x \geq y \geq z$. Then $by = (y-z)(y-x) \leq 0$ and $cz = (z-x)(z-y) \geq 0$, and the conditions $b > 0$, $c > 0$ imply $y \leq 0 \leq z \leq y$. Thus $y = z = 0$ and $ax = x^2$. Thus the solutions in this case are $(0,0,0)$ and $(a,0,0)$. By symmetry, all solutions are $(0,0,0)$, $(a,0,0)$, $(0,b,0)$, and $(0,0,c)$.

(Balkan Mathematical Olympiad, 1984)

16. Squaring each equation and subtracting the product of the other two yields

$$a^2 - bc = x(x^3 + y^3 + z^3 - 3xyz),$$
$$b^2 - ca = y(x^3 + y^3 + z^3 - 3xyz),$$
$$c^2 - ab = z(x^3 + y^3 + z^3 - 3xyz).$$

Let $k = x^3 + y^3 + z^3 - 3xyz$. Then

$$(a^2 - bc)^2 - (b^2 - ca)(c^2 - ab) = k^2(x^2 - yz) = k^2 a.$$

The same computation that produced the system above shows that the expression on the left is $a(a^3 + b^3 + c^3 - 3abc)$, and the latter is positive by the AM–GM inequality. Hence

$$k = \pm\sqrt{a^3 + b^3 + c^3 - 3abc},$$

and the two solutions to the system (one for each choice of k) are

$$x = \frac{a^2 - bc}{k}, \quad y = \frac{b^2 - ca}{k}, \quad z = \frac{c^2 - ab}{k}.$$

(Proposed by K. Kedlaya for the USAMO, 1998)

2.6 Periodicity

1. One expects the period to be related to ω, so a good idea is to iterate the given relation. First note that $f(x) = 2$ implies $f(x+\omega) = 3$. But 3 is not in the range of f, so 2 is not in the range as well. Similarly, one shows that f never assumes the value 1. Successively, we obtain

$$f(x+2\omega) = \frac{f(x+\omega) - 5}{f(x+\omega) - 3} = \frac{2f(x) - 5}{f(x) - 2},$$

$$f(x+3\omega) = \frac{2f(x+\omega) - 5}{f(x+\omega) - 2} = \frac{3f(x) - 5}{f(x) - 1},$$

$$f(x+4\omega) = \frac{3f(x+\omega) - 5}{f(x+\omega) - 1} = f(x).$$

Hence the function has period 4ω.

(*Gazeta Matematică (Mathematics Gazette, Bucharest)*, proposed by T. Andreescu)

2. Substituting $x = y = 0$, we obtain $f(0)^2 = f(0)$, so $f(0)$ is equal to 0 or 1. If $f(0) = 0$, then letting $x = x_0$ and $y = 0$, we obtain $-1 = f(x_0) = 0 \cdot f(x_0) = 0$, which cannot happen. Thus $f(0) = 1$.

For $x = y = x_0$, we obtain $f(2x_0) = 1$. This suggests that $2x_0$ might be a period for f. Let us show that this is indeed the case.

Replace x by $x + 2x_0$ and y by $x - 2x_0$ to obtain

$$f(2x) + f(4x_0) = 2f(x+2x_0)f(x-2x_0).$$

Since $f(2x) = 2f^2(x) - 1$ and $f(4x_0) = 2f^2(2x_0) - 1 = 1$, the above relation becomes

$$f(x+2x_0)f(x-2x_0) = f(x)^2.$$

Similarly, for x arbitrary and $y = 2x_0$, we obtain

$$f(x+2x_0) + f(x-2x_0) = 2f(x).$$

Since the sum and product of two numbers completely determine the numbers, we have $f(x-2x_0) = f(x+2x_0) = f(x)$, so f has period $2x_0$.

(M. Martin)

3. Since f is not injective, there exist two distinct numbers α and β with $f(\alpha) = f(\beta)$. It is natural to expect that $\beta - \alpha$ is a period for f. By replacing x first by α and then by β, we obtain

$$f(\alpha + y) = g(f(\alpha), y) = g(f(\beta), y) = f(\beta + y).$$

For $y = z - \alpha$, this implies $f(z) = f(z + \beta - \alpha)$ for all $z \in \mathbf{R}$.

(*Gazeta Matematică (Mathematics Gazette, Bucharest)*, proposed by D.M. Bătineţu)

4. (a) *First solution:* As in problem 1, we expect the period to be related to a. Iterating the relation from the statement gives

$$f(x+2a) = \frac{1}{2} + \sqrt{f(x+a) - f(x+a)^2}$$

$$= \frac{1}{2} + \sqrt{\frac{1}{2} + \sqrt{f(x) - f(x)^2} - \frac{1}{4} - \sqrt{f(x) - f(x)^2 - (f(x) - f(x)^2)}}$$

$$= \frac{1}{2} + \sqrt{\left(\frac{1}{2} - f(x)\right)^2} = \frac{1}{2} + \left|f(x) - \frac{1}{2}\right|.$$

The defining relation shows that $f(x) \geq \frac{1}{2}$ for all x. Hence the above computation implies $f(x+2a) = f(x)$ for all x, which proves that f is periodic.

Second solution: An alternative solution that avoids the use of square roots was suggested to us by R. Stong. Rewrite and square to obtain $f(x) - f^2(x) + f(x+a) - f^2(x+a) = 1/4$. Replacing x by $x+a$ in this formula gives $f(x+a) - f^2(x+a) + f(x+2a) - f^2(x+2a) = 1/4$. Subtracting gives

$$0 = f(x) - f^2(x) - f(x+2a) + f^2(x+2a) = [f(x+2a) - f(x)][1 - f(x) - f(x+2a)].$$

Since the original defining equation gives $f(x) \geq 1/2$, the second factor is nonzero unless $f(x+2a) = f(x) = 1/2$. Either because both are $1/2$ or by canceling, we get $f(x+2a) = f(x)$.

(b) An example of such a function is

$$f(x) = \begin{cases} \frac{1}{2}, & 2n \leq x < 2n+1, \\ 1, & 2n+1 \leq x < 2n+2, \end{cases}$$

where $n \in \mathbf{Z}$. Another example is the constant function $f(x) = \frac{1}{2} + \frac{1}{2\sqrt{2}}$.

(10th IMO, 1968)

5. One computes

$$a_2 = \frac{1 - a_1}{a_0}, \quad a_3 = \frac{a_0 + a_1 - 1}{a_0 a_1}, \quad a_4 = \frac{1 - a_0}{a_1}, \quad a_5 = a_0, \quad a_6 = a_1.$$

The sequence is periodic, hence bounded.

(T. Andreescu)

6. If $\alpha = 2$, then the sequence $x_n = n$ satisfies the given recurrence relation and is clearly not periodic. Thus let us assume that $\alpha \neq 2$. In this case, the equation $x^2 - \alpha x + 1 = 0$ has two distinct solutions, r and r^{-1}, and the general term of the sequence can be written in exponential form as $x_n = Ar^n + Br^{-n}$, where A and B are determined by the first two terms of the sequence.

If the sequence is periodic, then r must have absolute value equal to one, for otherwise the absolute value of the general term would tend to infinity. Thus $x_n = Ar^n + Br^{-n}$. The sequence $\overline{x_n}$ must also be periodic, which by addition implies that $(A + \bar{B})r^n + (\bar{A} + B)r^{-n}$ is periodic. Hence the sequence $\text{Re}((A + \bar{B})r^n)$ is periodic (here Re z denotes the real part of z). Writing $r = \cos \pi t + i \sin \pi t$, we conclude that $\mathrm{Re}(A + \bar{B}) \cos n\pi t$ is periodic. This implies that t is rational. Indeed, if t were not rational, then the numbers of the form $2m\pi + n\pi t$, $m, n \in \mathbf{Z}$, $n > 0$, would be dense in $\mathbf{R}$, so $\cos n\pi t$, $n > 0$, would be dense in $[0, 1]$, and it could not be periodic. It follows that α must be of the form $2 \cos \pi t$, with t rational.

(Mathematical Olympiad Summer Program, 1996)

7. Let T be a period of f. Assume by way of contradiction that $T = p/q$, where p and q are relatively prime positive integers. Then $qT = p$ is also a period of f. Let $n = kp + r$, where k and r are integers and $0 < r < p - 1$.

Then $f(n) = f(kp+r) = f(r)$, so $f(n) \in \{f(1), f(2), \ldots, f(p-1)\}$ for all positive integers n, in contradiction to the fact that $\{f(n) \mid n \in \mathbf{N}\}$ has infinitely many elements. The proof is complete.

(Revista Matematică din Timişoara (Timişoara's Mathematics Gazette), 1981; proposed by D. Andrica)

8. Assume by way of contradiction that $g : \mathbf{R} \to \mathbf{R}$, $g(x) = \sin f(x)$ is periodic. In that case, $g'(x) = f'(x) \cos f(x)$ is also periodic, and since it is continuous (as both f and f' are continuous), $g'(x)$ is bounded.

Consider the sequence $y_n = (4n + 1)\frac{\pi}{2}$. Because f is continuous and $\lim_{n \to \infty} f(x) = \infty$, there is some positive integer n_0 such that if $n \geq n_0$, there is x_n such that $f(x_n) = y_n$. Note that $\lim_{x_n \to \infty} x_n = \infty$. We obtain

$$\lim_{n \to \infty} g'(x_n) = \sin(4n+1)\frac{\pi}{2} \cdot \lim_{n \to \infty} f'(x_n) = 1 \cdot \infty = \infty.$$

This contradicts the fact that g is bounded. Hence our assumption was false, and g is not a periodic function.

9. We will denote the last digit of a positive integer n by $l(n)$. The sequence $\{l(n)\}_n$ is obviously periodic of period 10. Also, for a fixed $a \in \mathbf{N}$, the sequence $\{l(a^n)\}_n$ is periodic, and the period is equal to 1 if a ends in 0, 1, 5, or 6; 2 if a ends in 4 or 9; and 4 if a ends in 2, 3, 7, or 8.

Since the least common multiple of 10 and 4 is 20, if we let

$$m = (n+1)^{n+1} + (n+2)^{n+2} + \cdots + (n+20)^{n+20},$$

then $l(m)$ does not depend on n. Thus let us compute the last digit of $1^1 + 2^2 + 3^3 + \cdots + 20^{20}$. Because of the periodicity of sequences of the form $\{l(a^n)\}_n$, the last digit of this number is the same as that of

$$1 + 2^2 + 3^3 + 4^2 + 5 + 6 + 7^3 + 8^4 + 9^1$$
$$+ 1 + 2^4 + 3^1 + 4^2 + 5 + 6 + 7^1 + 8^2 + 9^1.$$

An easy computation shows that the last digit is 4. Consequently, the last digit of a sum of the form

$$(n+1)^{n+1} + (n+2)^{n+2} + \cdots + (n+100)^{n+100}$$

is equal to $l(4 \cdot 5) = l(20) = 0$; hence b_n is periodic, with period 100.

(Romanian IMO Team Selection Test, 1980)

10. (a) Suppose $u_n > a$. If u_n is even, then $u_{n+1} = u_n/2 < u_n$. If u_n is odd, $u_{n+2} = (u_n + a)/2 < u_n$. Hence for each term greater than a, there is a smaller subsequent term. These form a decreasing subsequence, which must eventually terminate, and this happens only if $u_n \leq a$.

(b) We will show that infinitely many terms of the sequence are less than $2a$. Suppose this is not true, and let u_m be the largest with this property. If u_m is even, then $u_{m+1} = u_m/2 < 2a$. If u_m is odd, then $u_{m+1} = u_m + a$ is even; hence $u_{m+2} = (u_m + a)/2 < 3a/2 < 2a$, which is again impossible. This shows that there are infinitely many terms less than $2a$. An application of the pigeonhole principle with infinitely many pigeons shows that some term u_n repeats, leading to a periodic sequence.

(French Mathematical Olympiad, 1996)

11. Since the sequence is bounded, some terms must repeat infinitely many times. Let K be the largest number that occurs infinitely many times in the sequence, and let N be a natural number such that $a_i \leq K$ for $i \geq N$. Choose some $m \geq N$ such that $a_m = K$. We will prove that m is a period of the sequence, i.e., $a_{i+m} = a_i$ for all $i \geq N$.

Assume first that $a_{i+m} = K$ for some i. Since $a_i + a_m$ is divisible by $a_{i+m} = K$, we have that $a_i = K = a_{i+m}$.

Otherwise, if $a_{i+m} < K$, choose $j \geq N$ such that $a_{i+j+m} = K$. We obtain $a_{i+m} + a_j < 2K$. Since $a_{i+m} + a_j$ is divisible by $a_{i+j+m} = K$, it follows that $a_{i+m} + a_j = K$, and therefore $a_j < K$. Since $a_{i+j+m} = K$, the above argument implies $a_{i+j+m} = a_{i+j} = K$, so K divides $a_i + a_j$. It follows that $a_i + a_j = K$, since $a_i \leq K$ and $a_j < K$; hence $a_{i+m} = K - a_j = a_i$.

(Leningrad Mathematical Olympiad, 1988)

12. We reduce the terms of the sequence modulo 2006 and examine the sequence of residues, which we still call $\{x_k\}_k$, $k \geq 1$. This sequence is periodic because there are only finitely many possible sequences of 2005 consecutive residues modulo 2006, and once such a sequence is repeated, every subsequent value is repeated. Moreover, writing the recursion backwards as $x_{k-2005} = x_{k+1} - x_k$, we see that the sequence extends to a doubly infinite sequence $\{x_k\}_k$, $k \in \mathbf{Z}$, which is periodic.

It thus suffices to find 2005 consecutive residues that are equal to zero in the doubly infinite sequence. Running the recursion backwards, we easily find

$$x_1 = x_0 = \cdots = x_{-2004} = 1$$
$$x_{-2005} = \cdots = x_{-4009} = 0,$$

and the conclusion follows.

(67th W.L. Putnam Mathematical Competition, 2006)

13. Consider the function $F : \mathbf{R} \to \mathbf{R}$, $F(x) = \int_0^x f(t)dt$. Then F satisfies $F(x + \sqrt{3}) - F(x) = ax + b$ and $F(x + \sqrt{2}) - F(x) = cx + d$, for all $x \in \mathbf{R}$. We can find two polynomial functions f_1 and f_2 of second degree such that $f_1(x + \sqrt{3}) - f_1(x) = ax + b$ and $f_2(x + \sqrt{2}) - f_2(x) = cx + d$, for all $x \in \mathbf{R}$. From these equalities, we can derive that the functions $g_i = F - f_i$, $i = 1, 2$, are periodic, with periods $\sqrt{3}$ and $\sqrt{2}$, respectively. It follows that for all x, $f_1(x) - f_2(x) + g_1(x) - g_2(x) = 0$. But since g_1 and g_2 are continuous and periodic, they are bounded; hence $f_1 - f_2$ must be constant.

It follows that $g_1(x) = g_2(x) + c$, for some $c \in \mathbf{R}$. Hence, along with the period $\sqrt{3}$, g_1 also has period $\sqrt{2}$. It follows that all numbers of the form $r\sqrt{3} + s\sqrt{2}$, with $r, s \in \mathbf{Z}$, are periods of g_1, and since the set of all numbers of this form is dense in $\mathbf{R}$, g_1 is constant on a dense set. By continuity, g_1 is constant on $\mathbf{R}$. The same argument shows that g_2 is constant.

We found out that $F(x) = mx^2 + nx + p$ for some $m, n, p \in \mathbf{R}$. Standard results about integrals imply that the function F is differentiable at each point x where f is continuous, and at these points $F'(x) = 2mx + n = f(x)$. Since f is monotonic, the set of such points is dense in $\mathbf{R}$. Thus f coincides with a linear function on a dense subset of $\mathbf{R}$. A squeezing argument using the monotonicity of f shows that f coincides everywhere with the linear function, and the problem is solved.

(Romanian Mathematical Olympiad, 1999; proposed by M. Piticari)

14. The answer to the problem is negative. Arguing by contradiction, let us assume that some polynomial $P(x)$, not identically equal to zero, and function $f(x)$ satisfy the functional equation from the statement. We want to eliminate the function from the functional equation and obtain an equation in the polynomial only. Let

$$\phi(x) = \frac{3x - 3}{3 + x}.$$

Then, if we denote by $\phi^{(n)}$ the function ϕ composed with itself n times, we have

$$\phi^{(2)}(x) = \frac{x - 3}{x + 1}, \quad \phi^{(3)}(x) = -\frac{3}{x},$$

$$\phi^{(4)}(x) = \frac{x + 3}{1 - x}, \quad \phi^{(5)}(x) = \frac{3x + 3}{3 - x}, \quad \phi^{(6)}(x) = x.$$

The functional equation from the statement can be written as

$$f(x) - \frac{x^2}{3} f(\phi(x)) = P\left(\phi^{(5)}(x)\right)$$

and iterating we obtain

$$f(\phi(x)) - \frac{(\phi(x))^2}{3} f\left(\phi^{(2)}(x)\right) = P(x)$$

$$f(\phi^{(2)}(x)) - \frac{(\phi^{(2)}(x))^2}{3} f\left(\phi^{(3)}(x)\right) = P(\phi(x))$$

$$f\left(\phi^{(3)}(x)\right) - \frac{\left(\phi^{(3)}(x)\right)}{3} f\left(\phi^{(4)}(x)\right) = P\left(\phi^{(2)}(x)\right)$$

$$f\left(\phi^{(4)}(x)\right) - \frac{\left(\phi^{(4)}(x)\right)}{3} f\left(\phi^{(5)}(x)\right) = P\left(\phi^{(3)}(x)\right)$$

$$f\left(\phi^{(5)}(x)\right) - \frac{\left(\phi^{(5)}(x)\right)}{3} f(x) = P\left(\phi^{(4)}(x)\right).$$

Note that

$$x\phi(x)\phi^{(2)}(x)\phi^{(3)}(x)\phi^{(4)}(x)\phi^{(5)}(x) = -27.$$

Hence if we multiply the second relation from the above by $x^2/3$, the third by $(x\phi(x))^2/9$, the fourth by $(x\phi(x)\phi^{(2)}(x))^2/27$, the fifth by $(x\phi(x)\phi^{(2)}(x)\phi^{(3)}(x))^2/81$, and the sixth by $(x\phi(x)\phi^{(2)}(x)\phi^{(3)}(x)\phi^{(4)}(x))^2/243$, and then add everything to first, the terms on the left cancel out, and we obtain

$$0 = P(\phi^{(5)}(x)) + \frac{x^2}{3}P(x) + \frac{(x\phi(x))^2}{9}P(\phi(x)) + \frac{\left(x\phi(x)\phi^{(2)}(x)\right)^2}{27}P(\phi^{(2)}(x))$$

$$+ \frac{\left(x\phi(x)\phi^{(2)}(x)\phi^{(3)}(x)\right)^2}{81}P(\phi^{(3)}(x))$$

$$+ \frac{\left(x\phi(x)\phi^{(2)}(x)\phi^{(3)}(x)\phi^{(4)}(x)\right)^2}{243}P(\phi^{(4)}(x)).$$

If the polynomial is non-constant, then the first term in this expression is a rational function whose denominator is divisible by $3 - x$ whereas the numerator is not. The factor $3 - x$ does not appear in any other denominator, hence after we bring everything to the common denominator, we obtain a rational function whose numerator is of the form $R(x) + (3 - x)Q(x)$ with $R(x)$ and $Q(x)$ polynomials and $R(x)$ not divisible by $3 - x$. We conclude that this numerator cannot be identically zero, so the equality in the last equation cannot hold.

If the polynomial $P(x)$ is constant, then after dividing by this constant we obtain

$$0 = 1 + \frac{x^2}{3} + \frac{(x\phi(x))^2}{9} + \frac{\left(x\phi(x)\phi^{(2)}(x)\right)^2}{27}$$

$$+ \frac{\left(x\phi(x)\phi^{(2)}(x)\phi^{(3)}(x)\right)^2}{81} + \frac{\left(x\phi(x)\phi^{(2)}(x)\phi^{(3)}(x)\phi^{(4)}(x)\right)^2}{243},$$

which is not true, for example because the right-hand side is a sum of squares. The contradiction proves that there is no polynomial with the required property.
(Proposed by R. Gelca for the USAMO, 2008)

2.7 The Abel Summation Formula

1. (a) Applying the Abel summation formula, we obtain

$$1 + 2q + 3q^2 + \cdots + nq^{n-1}$$

$$= (1 - 2) + (2 - 3)(1 + q) + (3 - 4)(1 + q + q^2) + \cdots$$

$$+ ((n-1) - n)(1 + 2 + \cdots + q^{n-2}) + n(1 + q + q^2 + \cdots + q^{n-1})$$

$$= -\left(\frac{q-1}{q-1} + \frac{q^2-1}{q-1} + \frac{q^3-1}{q-1} + \cdots + \frac{q^{n-1}-1}{q-1}\right) + n\frac{q^n-1}{q-1}$$

$$= -\frac{1}{q-1}(1+q+q^2+\cdots+q^{n-1}-n)+n\frac{q^n-1}{q-1}$$

$$= -\frac{1}{q-1}\left(\frac{q^n-1}{q-1}-n\right)+n\frac{q^n-1}{q-1}=\frac{nq^n}{q-1}-\frac{q^n-1}{(q-1)^2}.$$

(b) Applying the Abel summation formula again, we have

$$1+4q+9q^2+\cdots+n^2q^{n-1}$$
$$= (1-4)+(4-9)(1+q)+(9-16)(1+q+q^2)+\cdots$$
$$+((n-1)^2-n^2)(1+q+\cdots+q^{n-2})+n^2(1+q+\cdots+q^{n-1})$$

Summing the geometric series yields

$$-\left(3\frac{q-1}{q-1}+5\frac{q^2-1}{q-1}+7\frac{q^3-1}{q-1}+\cdots+(2n-1)\frac{q^{n-1}-1}{q-1}\right)+n^2\frac{q^n-1}{q-1}$$

$$=\left(\frac{q-1}{q-1}+\frac{q^2-1}{q-1}+\frac{q^3-1}{q-1}+\cdots+\frac{q^{n-1}-1}{q-1}\right)$$
$$-2\left(2\frac{q-1}{q-1}+3\frac{q^2-1}{q-1}+4\frac{q^3-1}{q-1}+\cdots+n\frac{q^{n-1}-1}{q-1}\right)+n^2\frac{q^n-1}{q-1}$$

$$=\frac{1}{q-1}(1+q+\cdots+q^{n-1}-n)$$
$$-\frac{2}{q-1}\left(1+2q+\cdots+nq^{n-1}-\frac{n(n+1)}{2}\right)+n^2\frac{q^n-1}{q-1}.$$

Using part (a) of the problem, we obtain that this is equal to

$$\frac{1}{q-1}\left(\frac{q^n-1}{q-1}-n\right)-\frac{2}{q-1}\left(\frac{nq^n}{q-1}-\frac{q^n-1}{(q-1)^2}-\frac{n(n+1)}{2}\right)$$
$$+n^2\frac{q^n-1}{q-1}$$
$$=\frac{n^2q^n}{q-1}-\frac{(2n-1)q^n+1}{(q-1)^2}+\frac{2q^n-2}{(q-1)^3}.$$

(A.M. Yaglom and I.M. Yaglom, *Neelementarnye zadaci v elementarnom izlozenii (Non-elementary problems in an elementary exposition)*, Gosudarstv. Izdat. Tehn.-Teor. Lit., Moscow, 1954)

2. We can write

$$a_i^k-b_i^k=(a_i-b_i)(a_i^{k-1}+a_i^{k-2}b_i+\cdots+a_ib_i^{k-2}+b_i^{k-1}).$$

To simplify computations, set $c_i=a_i-b_i$ and $d_i=a_i^{k-1}+a_i^{k-2}b_i+\cdots+a_ib_i^{k-2}+b_i^{k-1}$. The hypothesis implies $c_1+c_2+\cdots+c_j\geq 0$ for all j and $d_i\geq d_{i+1}>0$, the latter since a_i and b_i are decreasing positive sequences. Hence

$$a_1^k - b_1^k + a_2^k - b_2^k + \cdots + a_n^k - b_n^k = c_1 d_1 + c_2 d_2 + \cdots + c_n d_n$$
$$= (d_1 - d_2)c_1 + (d_2 - d_3)(c_1 + c_2) + \cdots + d_n(c_1 + c_2 + \cdots + c_n) \geq 0,$$

and the inequality is proved.

(D. Buşneag and I.V. Maftei, *Teme pentru cercurile şi concursurile de matematică ale elevilor (Lectures for student mathematics circles and competitions)*, Scrisul românesc, Craiova, 1983)

3. We will prove a more general statement.

If $a_1, a_2, \ldots, a_n$ are positive, $0 \leq b_1 \leq b_2 \leq \cdots \leq b_n$, and for all $k \leq n$, $a_1 + a_2 + \cdots + a_k \leq b_1 + b_2 + \cdots + b_k$, then

$$\sqrt{a_1} + \sqrt{a_2} + \cdots + \sqrt{a_n} \leq \sqrt{b_1} + \sqrt{b_2} + \cdots + \sqrt{b_n}.$$

The special case of the original problem is obtained for $n = 4$, by setting $b_k = k^2$, $k = 1, 2, 3, 4$.

Let us prove the above result. We have

$$\frac{a_1}{\sqrt{b_1}} + \frac{a_2}{\sqrt{b_2}} + \cdots + \frac{a_n}{\sqrt{b_n}}$$

$$= a_1 \left(\frac{1}{\sqrt{b_1}} - \frac{1}{\sqrt{b_2}} \right) + (a_1 + a_2) \left(\frac{1}{\sqrt{b_2}} - \frac{1}{\sqrt{b_3}} \right)$$

$$+ (a_1 + a_2 + a_3) \left(\frac{1}{\sqrt{b_3}} - \frac{1}{\sqrt{b_4}} \right) + \cdots + (a_1 + a_2 + \cdots + a_n) \frac{1}{\sqrt{b_n}}.$$

The differences in the parentheses are all positive. Using the hypothesis, we obtain that this expression is less than or equal to

$$b_1 \left(\frac{1}{\sqrt{b_1}} - \frac{1}{\sqrt{b_2}} \right) + (b_1 + b_2) \left(\frac{1}{\sqrt{b_2}} - \frac{1}{\sqrt{b_3}} \right)$$

$$+ \cdots + (b_1 + b_2 + \cdots + b_n) \frac{1}{\sqrt{b_n}}$$

$$= \sqrt{b_1} + \sqrt{b_2} + \cdots + \sqrt{b_n}.$$

Therefore

$$\frac{a_1}{\sqrt{b_1}} + \frac{a_2}{\sqrt{b_2}} + \cdots + \frac{a_n}{\sqrt{b_n}} \leq \sqrt{b_1} + \sqrt{b_2} + \cdots + \sqrt{b_n}.$$

Using this result and the Cauchy–Schwarz inequality, we obtain

$$(\sqrt{a_1} + \sqrt{a_2} + \cdots + \sqrt{a_n})^2$$

$$= \left(\sqrt[4]{b_1} \cdot \sqrt{\frac{a_1}{\sqrt{b_1}}} + \sqrt[4]{b_2} \cdot \sqrt{\frac{a_2}{\sqrt{b_2}}} + \cdots + \sqrt[4]{b_n} \cdot \sqrt{\frac{a_n}{\sqrt{b_n}}} \right)^2$$

$$\leq (\sqrt{b_1} + \sqrt{b_2} + \cdots + \sqrt{b_n}) \left(\frac{a_1}{\sqrt{b_1}} + \frac{a_2}{\sqrt{b_2}} + \cdots + \frac{a_n}{\sqrt{b_n}} \right)$$

$$\leq (\sqrt{b_1} + \sqrt{b_2} + \cdots + \sqrt{b_n})^2.$$

This gives

$$\sqrt{a_1} + \sqrt{a_2} + \cdots + \sqrt{a_n} \le \sqrt{b_1} + \sqrt{b_2} + \cdots + \sqrt{b_n}.$$

(Romanian IMO Team Selection Test, 1977; proposed by V. Cârtoaje)

4. We have

$$a_1 + a_2 + \cdots + a_n = \left(1 - \frac{1}{2}\right)(1 \cdot 2a_1) + \left(\frac{1}{3} - \frac{1}{4}\right)(3 \cdot 4a_2)$$

$$+ \cdots + \left(\frac{1}{2n-1} - \frac{1}{2n}\right)((2n-1) \cdot 2na_n)$$

$$= \left(1 - \frac{1}{2} - \frac{1}{3} + \frac{1}{4}\right)(1 \cdot 2a_1) + \left(\frac{1}{3} - \frac{1}{4} - \frac{1}{5} + \frac{1}{6}\right)(1 \cdot 2a_1 + 3 \cdot 4a_2)$$

$$+ \cdots + \left(\frac{1}{2n-1} - \frac{1}{2n}\right)(1 \cdot 2a_1 + 3 \cdot 4a_2 + \cdots + (2n-1) \cdot 2na_n).$$

Using the AM–GM inequality and the hypothesis, we obtain that $1 \cdot 2a_1 \ge 1$, $1 \cdot 2a_1 + 3 \cdot 4a_2 \ge 2$, ..., $1 \cdot 2a_1 + 3 \cdot 4a_2 + \cdots + (2n-1) \cdot 2na_n \ge n$. Hence

$$a_1 + a_2 + \cdots + a_n \ge \left(1 - \frac{1}{2} - \frac{1}{3} + \frac{1}{4}\right) + 2\left(\frac{1}{3} - \frac{1}{4} - \frac{1}{5} + \frac{1}{6}\right)$$

$$+ \cdots + n\left(\frac{1}{2n-1} - \frac{1}{2n}\right)$$

$$= 1 - \frac{1}{2} + \frac{1}{3} - \frac{1}{4} + \frac{1}{5} - \frac{1}{6} + \cdots + \frac{1}{2n-1} - \frac{1}{2n}$$

$$= 1 + \frac{1}{2} + \frac{1}{3} + \frac{1}{4} + \frac{1}{5} + \cdots + \frac{1}{2n-1} + \frac{1}{2n}$$

$$- 2\left(\frac{1}{2} + \frac{1}{4} + \frac{1}{6} + \cdots + \frac{1}{2n}\right)$$

$$= \frac{1}{n+1} + \frac{1}{n+2} + \cdots + \frac{1}{2n},$$

and we are done.

5. We want to reduce the inequalities involving products to inequalities involving sums. For this we use the AM–GM inequality. We have

$$\frac{x_1}{y_1} + \frac{x_2}{y_2} + \cdots + \frac{x_k}{y_k} \ge k \sqrt[k]{\frac{x_1}{y_1} \frac{x_2}{y_2} \cdots \frac{x_k}{y_k}} \ge k,$$

where the last inequality follows from the hypothesis.

Returning to the original inequality, we have

$$x_1 + x_2 + \cdots + x_n = \frac{x_1}{y_1}y_1 + \frac{x_2}{y_2}y_2 + \cdots + \frac{x_n}{y_n}y_n = \frac{x_1}{y_1}(y_1 - y_2)$$

$$+ \left(\frac{x_1}{y_1} + \frac{x_2}{y_2}\right)(y_2 - y_3) + \cdots + \left(\frac{x_1}{y_1} + \frac{x_2}{y_2} + \cdots + \frac{x_n}{y_n}\right)y_n.$$

By using the inequalities deduced at the beginning of the solution for the first factor in each term, we obtain that this expression is greater than or equal to

$$1(y_1 - y_2) + 2(y_2 - y_3) + \cdots + ny_n = y_1 + y_2 + \cdots + y_n,$$

and we are done.

6. We start by proving another inequality, namely that if $a_1, a_2, \ldots, a_n$ are positive and $b_1 \geq b_2 \geq \cdots \geq b_n \geq 0$ and if for all $k \leq n$, $a_1 + a_2 + \cdots + a_k \geq b_1 + b_2 + \cdots + b_k$, then

$$a_1^2 + a_2^2 + \cdots + a_n^2 \geq b_1^2 + b_2^2 + \cdots + b_n^2.$$

This inequality is the same as the one in Problem 2, in the particular case where the exponent is 2, but with a weaker hypothesis. Using the Abel summation formula, we can write

$$a_1 b_1 + a_2 b_2 + \cdots + a_n b_n = a_1(b_1 - b_2) + (a_1 + a_2)(b_2 - b_3)$$
$$+ (a_1 + a_2 + a_3)(b_3 - b_4) + \cdots + (a_1 + a_2 + \cdots + a_n)b_n.$$

The inequalities in the statement show that this is greater than or equal to

$$b_1(b_1 - b_2) + (b_1 + b_2)(b_2 - b_3) + \cdots + (b_1 + b_2 + \cdots + b_n)b_n$$
$$= b_1^2 + b_2^2 + \cdots + b_n^2.$$

Combining this with the Cauchy–Schwarz inequality, we obtain

$$(a_1^2 + a_2^2 + \cdots + a_n^2)(b_1^2 + b_2^2 + \cdots + b_n^2) \geq (a_1 b_1 + a_2 b_2 + \cdots + a_n b_n)^2$$
$$\geq (b_1^2 + b_2^2 + \cdots + b_n^2)^2,$$

and the proof is complete.

Returning to our problem, note first that

$$\sqrt{n} - \sqrt{n-1} > \frac{1}{2\sqrt{n}}.$$

Indeed, multiplying by the rational conjugate of the left side, this becomes $n - (n-1) > (\sqrt{n} + \sqrt{n-1})/(2\sqrt{n})$. After eliminating the denominator and canceling out terms, this becomes $\sqrt{n} > \sqrt{n-1}$.

The conclusion of the problem now follows from the inequality proved in the beginning by choosing $b_n = \sqrt{n} - \sqrt{n-1}$.

(USAMO, 1995)

7. *First solution:* The inequality can be rewritten as

$$\sum_{k=1}^{n} \frac{1}{k} \left(\frac{\phi(k)}{k} - 1 \right) \geq 0.$$

If we set $\lambda_k = \phi(k)/k$, then the injectivity of ϕ implies $\lambda_1 \lambda_2 \cdots \lambda_k \geq 1$ for all k. From the AM–GM inequality, we obtain

$$\sum_{i=1}^{k} \lambda_i \geq k \sqrt[k]{\lambda_1 \lambda_2 \cdots \lambda_k} \geq k.$$

Applying the Abel summation formula yields

$$\sum_{k=1}^{n} \frac{1}{k}(\lambda_k - 1) = \sum_{k=1}^{n-1} \left(\frac{1}{k} - \frac{1}{k+1} \right) (\lambda_1 + \lambda_2 + \cdots + \lambda_k - k)$$
$$+ \frac{1}{n}(\lambda_1 + \cdots + \lambda_n - n).$$

Each term in the sum above is positive, so we are done.

Second solution: First note that $\sum_{k=1}^{m} \phi(k) \geq \frac{m(m+1)}{2}$ since the $\phi(k)$ are distinct. Hence $\sum_{k=1}^{m} (\phi(k) - k) \geq 0$ for all m. Then Abel summation gives

$$\sum_{k=1}^{n} \frac{\phi(k) - k}{k^2} = \frac{1}{n^2} \sum_{k=1}^{n} (\phi(k) - k) + \sum_{m=1}^{n-1} \left(\frac{1}{m^2} - \frac{1}{(m+1)^2} \right) \sum_{k=1}^{m} (\phi(k) - k),$$

which is nonnegative. Rearranging gives the desired result.

(20th IMO, 1978; proposed by France, second solution by R. Stong)

8. This problem is an application of Abel's criterion for convergence of series. The proof for the general case is similar to the one below. Write

$$S_n = a_1 + \frac{a_2}{2} + \frac{a_3}{3} + \frac{a_4}{4} + \cdots + \frac{a_n}{n} = a_1 \left(1 - \frac{1}{2} \right) + (a_1 + a_2) \left(\frac{1}{2} - \frac{1}{3} \right)$$
$$+ (a_1 + a_2 + a_3) \left(\frac{1}{3} - \frac{1}{4} \right) + \cdots + (a_1 + a_2 + \cdots + a_n) \frac{1}{n}.$$

Note that

$$S_{n+1} - S_n = (a_1 + a_2 + \cdots + a_n) \left(\frac{1}{n} - \frac{1}{n+1} \right) + \frac{a_{n+1}}{n+1},$$

which has order size of $\frac{|L|}{n(n+1)}$, where $L = a_1 + a_2 + \cdots + a_n$. Since the series

$$\sum_{n=1}^{\infty} \frac{|L|}{n(n+1)}$$

converges, our series converges as well.

Observe that Abel summation applies for infinite sums provided $\lim_{n \to \infty} a_n \sum_{k=1}^{n} b_k = 0$. In your case, with $a_n = 1/n$ and (alas) $b_k = a_k$, this condition holds.

(11th W.L. Putnam Mathematical Competition, 1951)

9. (a) Let $S_k = (x_1 - y_1) + (x_2 - y_2) + \cdots + (x_k - y_k)$ and $z_k = \frac{1}{x_k y_k}$. Then, we have $S_k \geq 0$ and $z_k - z_{k+1} > 0$, for any $k = 1, 2, \ldots, n-1$. It follows that

$$\frac{1}{x_1} + \frac{1}{x_2} + \cdots + \frac{1}{x_n} - \frac{1}{y_1} - \frac{1}{y_2} - \cdots - \frac{1}{y_n}$$

$$= \left(\frac{1}{x_1} - \frac{1}{y_1}\right) + \left(\frac{1}{x_2} - \frac{1}{y_2}\right) + \cdots + \left(\frac{1}{x_n} - \frac{1}{y_n}\right)$$

$$= \frac{y_1 - x_1}{x_1 y_1} + \frac{y_2 - x_2}{x_2 y_2} + \cdots + \frac{y_n - x_n}{x_n y_n}$$

$$= -S_1 z_1 - (S_2 - S_1)z_2 - \cdots - (S_n - S_{n-1})z_n$$

$$= -S_1(z_1 - z_2) - S_2(z_2 - z_3) - \cdots - S_{n-1}(z_{n-1} - z_n) - S_n z_n \leq 0,$$

with equality if and only if $S_k = 0$, $k = 1, 2, \ldots, n$, that is, when $x_k = y_k$, $k = 1, 2, \ldots, n$.

(b) We can assume without loss of generality that $a_1 < a_2 < \cdots < a_n$. From the hypothesis, it follows that for any partition of the set $\{a_1, a_2, \ldots, a_k\}$ into two subsets, the sum of the elements of the first subset is different from the sum of the elements of the second subset. Since we can perform such a partition in 2^k ways, it follows that $a_1 + a_2 + \cdots + a_k \geq 2^k$. We now apply (a) to the numbers $a_1, a_2, \ldots, a_n$ and $1, 2, 2^2, \ldots, 2^{n-1}$ (whose sum is $2^n - 1$). It follows that

$$\frac{1}{a_1} + \frac{1}{a_2} + \cdots + \frac{1}{a_n} \leq \frac{1}{1} + \frac{1}{2} + \cdots + \frac{1}{2^{n-1}} = 2 - \frac{1}{2^{n-1}} < 2.$$

(Romanian Mathematical Olympiad, 1999)

10. The sequence $\{u_n\}_n$ is well-known, and its basic property is a straightforward consequence of the definition:

$$\frac{1}{u_1} + \frac{1}{u_2} + \cdots + \frac{1}{u_n} + \frac{1}{u_1 u_2 \cdots u_n} = 1 \text{ for } n = 1, 2, 3, \ldots.$$

Thus, we need to prove that if $x_1, x_2, \ldots, x_n$ are positive integers satisfying

$$\frac{1}{x_1} + \frac{1}{x_2} + \cdots + \frac{1}{x_n} < 1,$$

then

$$\frac{1}{x_1} + \frac{1}{x_2} + \cdots + \frac{1}{x_n} \leq \frac{1}{u_1} + \frac{1}{u_2} + \cdots + \frac{1}{u_n}.$$

The proof is by induction on n. Everything is clear for $n = 1$. Assume that the claim holds for each $k = 1, 2, \ldots n-1$, and consider positive integers $x_1, x_2, \ldots, x_n$ such that

$$\frac{1}{x_1} + \frac{1}{x_2} + \cdots + \frac{1}{x_n} < 1.$$

Let

$$\frac{1}{x_1} + \frac{1}{x_2} + \cdots + \frac{1}{x_i} = X_i, \quad \frac{1}{u_1} + \frac{1}{u_2} + \cdots + \frac{1}{u_i} = U_i, \quad i = 1, 2, \ldots, n.$$

We have $x_1 x_2 \cdots x_n X_n \leq x_1 x_2 \cdots x_n - 1$, or $X_n \leq 1 - 1/(x_1 x_2 \cdots x_n)$.

Now, assume on the contrary that $X_n > U_n$. This, combined with $X_n \leq 1 - 1/(x_1 x_2 \cdots x_n)$ and $U_n = 1 - 1/(u_1 u_2 \cdots u_n)$, implies

$$x_1 x_2 \cdots x_n > u_1 u_2 \cdots u_n.$$

On the other hand, $X_i < 1$ for $i = 1, 2, \ldots, n-1$, so by the inductive hypothesis,

$$X_1 \leq U_1, \quad X_2 \leq U_2, \quad \ldots, \quad X_{n-1} \leq U_{n-1}.$$

Consider the sum $\sum_{i=1}^{n} x_i / u_i$ and apply the Abel summation formula to obtain

$$\sum_{i=1}^{n} \frac{x_i}{u_i} = \sum_{i=i}^{n-1} U_i(x_i - x_{i+1}) + U_n x_n.$$

We may assume, without loss of generality, that $x_1 \leq x_2 \leq \cdots \leq x_n$, so $x_i - x_{i+1} \leq 0$ for $i = 1, 2, \ldots, n-1$. Then, since $x_1 x_2 \cdots x_n > u_1 u_2 \cdots u_n$, $X_k \leq U_k$, $1 \leq k \leq n-1$, and because of the assumption $X_n > U_n$. We obtain

$$\sum_{i=1}^{n} \frac{x_i}{u_i} < \sum_{i=1}^{n-1} X_i(x_i - x_{i+1}) + X_n x_n = \sum_{i=1}^{n} \frac{x_i}{x_i} = n.$$

Now, by the AM–GM inequality,

$$n > \sum_{i=1}^{n} \frac{x_i}{u_i} \geq n \sqrt[n]{\frac{x_1 x_2 \cdots x_n}{u_1 u_2 \cdots u_n}},$$

so $u_1 u_2 \cdots u_n > x_1 x_2 \cdots x_n$, in contradiction to the inequality deduced above for $\sum_{i=1}^{n} x_i / u_i$.

2.8 $x + 1/x$

1. The equality

$$x^2 + \frac{1}{x^2} = \left(x - \frac{1}{x}\right)^2 - 2$$

shows that $x^2 + 1/x^2$ can be written as a polynomial in $x - 1/x$.

The conclusion of the problem follows by induction from the identity

$$x^{2n+1} - \frac{1}{x^{2n+1}} = \left(x^2 + \frac{1}{x^2}\right)\left(x^{2n-1} - \frac{1}{x^{2n-1}}\right) - \left(x^{2n-3} - \frac{1}{x^{2n-3}}\right).$$

(20th W.L. Putnam Mathematical Competition, 1959)

2. Using

$$x^{n+1} + y^{n+1} = (x+y)(x^n + y^n) - xy(x^{n-1} + y^{n-1})$$

and the equalities from the statement, we see that $1 = x + y - xy$. Factoring, we obtain $(x-1)(y-1) = 0$; hence one of the numbers x and y is equal to 1. Assume $x = 1$. Then $y^n + 1 = y^{n+1} + 1$, and since $y > 0$, we must have $y = 1$ as well; hence $x = y$.
   (M. Mogoşanu)

3. We recognize right away the golden ratio $(1 + \sqrt{5})/2$. It is well-known that the golden ratio appears in a "golden rectangle," but, more important to us right now is that it appears in a "golden triangle." This is the isosceles triangle with angles $\pi/5$, $2\pi/5$, $2\pi/5$.

Let us therefore take a close look at the triangle ABC with $\angle B = \angle C = 2\pi/5$ (see Figure 2.8.1). If CD is the bisector of $\angle ACB$, then the triangles DAC and CBD are isosceles (just compute their angles). Moreover, the triangles ABC and CBD are similar, and since $BC = CD = AD$, it follows that $AB/BC = BC/(AB - BC)$. Therefore, $AB/BC = (1 + \sqrt{5})/2$, the golden ratio.

Now let us apply the law of sines in triangle ABC. We obtain

$$\frac{AB}{BC} = \frac{\sin C}{\sin A} = \frac{\sin \frac{2\pi}{5}}{\sin \frac{\pi}{5}} = 2\cos \frac{\pi}{5}.$$

Therefore, $\cos 2\pi/5 = (1 + \sqrt{5})/4$.

Returning to the problem, we deduce that x satisfies $x + 1/x = 2\cos \pi/5$. This gives

$$x^{2000} + \frac{1}{x^{2000}} = 2\cos \frac{2000\pi}{5} = 2\cos 400\pi = 2.$$

(T. Andreescu)

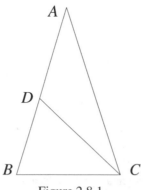

Figure 2.8.1

4. Denote $|z + 1/z|$ by r. From the hypothesis,

$$\left| \left(z + \frac{1}{z} \right)^3 \right| = \left| z^3 + \frac{1}{z^3} + 3\left(z + \frac{1}{z} \right) \right| \le \left| z^3 + \frac{1}{z^3} \right| + \left| 3\left(z + \frac{1}{z} \right) \right| \le 2 + 3r.$$

Hence $r^3 \le 2+3r$, which by factorization gives $(r-2)(r+1)^2 \le 0$. This implies $r \le 2$, as desired.

(*Revista Matematică din Timişoara (Timişoara's Mathematics Gazette)*, proposed by T. Andreescu)

5. Assume that $\cos 1°$ is a rational number. Consider the complex number $z = \cos 1° + i \sin 1°$. Since $z + 1/z = 2\cos 1°$ is rational, as in the introduction we conclude that $z^{45} + 1/z^{45}$ is rational as well. But $z^{45} + 1/z^{45} = 2\cos 45° = \sqrt{2}$, a contradiction. This shows that $\cos 1°$ is irrational.

6. Suppose that $\max\{|a|, |b|, |c|\} = |a|$. Solving for b yields

$$b = \frac{ac \pm \sqrt{(a^2-4)(c^2-4)}}{2}.$$

Since b is a real number, it follows that $(a^2-4)(c^2-4) \ge 0$; but $a^2-4 > 0$, so $c^2 \ge 4$. Similarly $b^2 \ge 4$. Then the equations $x^2 - ax + 1 = 0$, $x^2 - bx + 1 = 0$, $x^2 - cx + 1 = 0$ have real solutions. Let u be a solution to $x^2 - ax + 1 = 0$ and v a solution to $x^2 - bx + 1 = 0$. Then $u + \frac{1}{u} = a$, $v + \frac{1}{v} = b$, and

$$c = \frac{ab \pm \sqrt{(a^2-4)(b^2-4)}}{2}$$

$$= \frac{1}{2}\left[\left(u+\frac{1}{u}\right)\left(v+\frac{1}{v}\right) \pm \left(u-\frac{1}{u}\right)\left(v-\frac{1}{v}\right)\right].$$

The "+" choice gives $c = uv + \frac{1}{uv}$, hence there exist $w = \frac{1}{uv}$ such that $c = w + \frac{1}{w}$ and $uvw = 1$. The "−" choice gives $c = \frac{u}{v} + \frac{v}{u}$, and writing $a = \frac{1}{u} + u$, $b = v + \frac{1}{v}$, one has $\frac{1}{u}v\frac{u}{v} = 1$ as desired.

7. We will show by induction on n that

$$\frac{x^n - x^{-n}}{n} < \frac{x^{n+1} - x^{-n-1}}{n+1}.$$

The inequality is clearly true for $n = 1$, since if $x > 1$, then $x - x^{-1} > 0$, so $x - x^{-1} < (x^2 - x^{-2})/2$ reduces to $2 < x + x^{-1}$. The latter is true by the AM–GM inequality (since $x \ne 1$).

Assume that

$$\frac{x^{k-1} - x^{-(k-1)}}{k-1} < \frac{x^k - x^{-k}}{k},$$

and let us show that

$$\frac{x^k - x^{-k}}{k} < \frac{x^{k+1} - x^{-(k+1)}}{k+1}.$$

We have $x^{k+1} - x^{-(k+1)} = (x + x^{-1})(x^k - x^{-k}) - (x^{k-1} - x^{-(k-1)})$. Since $x + x^{-1} > 2$ using the inductive hypothesis we obtain

$$x^{k+1} - x^{-(k+1)} > 2(x^k - x^{-k}) - \frac{k-1}{k}(x^k - x^{-k}) = \frac{k+1}{k}(x^k - x^{-k}),$$

and the conclusion follows.

(USAMO, 1986; proposed by C. Rousseau)

8. We have $(1 + \sqrt{2}) + (1 - \sqrt{2}) = 2$, hence $(1 + \sqrt{2})^{2n} + (1 - \sqrt{2})^{2n}$ is an integer for all n. Note also that $(1 - \sqrt{2})^{2n} < 1$ for all n. Hence

$$\{(1 + \sqrt{2})^{2n}\} = 1 - (1 - \sqrt{2})^{2n}.$$

Passing to the limit in this equality, we obtain

$$\lim_{n \to \infty} \{(1 + \sqrt{2})^{2n}\} = 1 - \lim_{n \to \infty} (1 - \sqrt{2})^{2n} = 1.$$

(Russian Mathematical Olympiad, 1977–1978)

9. *First solution:* The number $(\sqrt{3} + 1)^{2n+1} - (\sqrt{3} - 1)^{2n+1}$ is an integer, since after expanding with the binomial formula, all terms containing $\sqrt{3}$ cancel out. Since $(\sqrt{3} - 1)^{2n+1}$ is less than 1, it follows that $\lfloor(\sqrt{3} + 1)^{2n+1}\rfloor = (\sqrt{3} + 1)^{2n+1} - (\sqrt{3} - 1)^{2n+1}$.

We will prove by induction on n that $(\sqrt{3} + 1)^{2n+1} - (\sqrt{3} - 1)^{2n+1}$ is divisible by 2^{n+1} but not by 2^{n+2}. For $n = 0$, we obtain $\lfloor(\sqrt{3} + 1)\rfloor = 2$, so the property holds in this case. Assume that the property holds for all numbers less than n and let us show that it holds for n.

Set $x = ((\sqrt{3} + 1)/\sqrt{2})$. Then $x^{-1} = ((\sqrt{3} - 1)/\sqrt{2})$. An easy computation shows that $x^2 + x^{-2} = 4$. From the recurrence formula exhibited in the solution to problem 1, we obtain

$$x^{2n+1} - \frac{1}{x^{2n+1}} = 4\left(x^{2n-1} - \frac{1}{x^{2n-1}}\right) - \left(x^{2n-3} - \frac{1}{x^{2n-3}}\right).$$

Hence, after multiplying both sides by $(\sqrt{2})^{2n+1}$, we have

$$(\sqrt{3} + 1)^{2n+1} - (\sqrt{3} - 1)^{2n+1}$$
$$= (\sqrt{2})^2 \cdot 4\left((\sqrt{3} + 1)^{2n-1} - (\sqrt{3} - 1)^{2n-1}\right)$$
$$- (\sqrt{2})^4\left((\sqrt{3} + 1)^{2n-3} - (\sqrt{3} - 1)^{2n-3}\right)$$
$$= 4\left(2\left((\sqrt{3} + 1)^{2n-1} - (\sqrt{3} - 1)^{2n-1}\right)\right.$$
$$\left. - \left((\sqrt{3} + 1)^{2n-3} - (\sqrt{3} - 1)^{2n-3}\right)\right).$$

By the induction hypothesis, the first term in parentheses is divisible by 2^n, and the second is divisible by 2^{n-1} but not by 2^n. This shows that the whole expression in parentheses is divisible by 2^{n-1} and is not divisible by 2^n. Multiplying by the 4 up front, we obtain that $(\sqrt{3} + 1)^{2n+1} - (\sqrt{3} - 1)^{2n+1}$ is divisible by 2^{n+1} but not by 2^{n+2}, and the problem is solved.

Second solution: Here is another approach that was suggested to us by R. Stong. Taking $x = (u/v)^{1/2}$ in the identity of Problem 1 from this section and clearing denominators gives

$$u^{2n+1} - v^{2n+1} = (u^2 + v^2)(u^{2n-1} - v^{2n-1}) - u^2 v^2 (u^{2n-3} - v^{2n-3}).$$

This identity can also be easily checked directly. Taking $u = \sqrt{3} + 1$ and $v = \sqrt{3} - 1$, we see that $b_n = u^{2n+1}$ $v^{2n+1} = (\sqrt{3}+1)^{2n+1} - (\sqrt{3}-1)^{2n+1}$ satisfies the recursion $b_n = 8b_{n-1} - 4b_{n-2}$. Since $b_{-1} = -1$ and $b_0 = 2$, it follows that b_n is an integer for all $n \geq 0$. Since $(\sqrt{3}+1)^{2n+1} - 1 < b_n < (\sqrt{3}+1)^{2n+1}$, (for $n > 0$) it follows that $a_n = b_n$. Now it follows immediately from the recursion and induction that $u_n - b_n$ is exactly divisible by 2^{n+1}.

10. Since the number $\cos a + \sin a$ is rational, its square must be rational as well. Thus $1 + 2\cos a \sin a = (\cos a + \sin a)^2$ is rational. This shows that both the sum and the product of $\cos a$ and $\sin a$ are rational and the fact that $\cos^n a + \sin^n a$ is rational can be proved inductively using the formula

$$\cos^{n+1} a + \sin^{n+1} a = (\cos a + \sin a)(\cos^n a + \sin^n a)$$
$$- \cos a \sin a(\cos^{n-1} a + \sin^{n-1} a).$$

11. *First solution:* Of course, $a_1 = \frac{1}{2}$. The identity

$$\left(X^2 + \frac{1}{2}X + 1\right)\left(X^2 - \frac{1}{2}X + 1\right) = X^4 + \frac{7}{4}X^2 + 1$$

yields $a_2 = \frac{7}{4}$.

We have

$$X^n + \frac{1}{X^n} = \left(X + \frac{1}{X}\right)\left(X^{n-1} + \frac{1}{X^{n-1}}\right) - \left(X^{n-2} + \frac{1}{X^{n-2}}\right)$$
$$= \left(X + \frac{1}{X} + \frac{1}{2}\right)\left(X^{n-1} + \frac{1}{X^{n-1}}\right) - \frac{1}{2}\left(X^{n-1} + \frac{1}{X^{n-1}} + a_{n-1}\right)$$
$$- \left(X^{n-2} + \frac{1}{X^{n-2}} + a_{n-2}\right) + \frac{a_{n-1}}{2} + a_{n-2}.$$

If we define recursively the numbers a_n by $a_n = -a_{n-1}/2 - a_{n-2}$, then

$$X^n + \frac{1}{X^n} + a_n = (X + \frac{1}{X} + \frac{1}{2})P(X) \cdot \frac{1}{X^{n-1}}$$

for some polynomial P with rational coefficients.

Consequently,

$$X^{2n} + a_n X^n + 1 = \left(X^2 + \frac{1}{2}X + 1\right)P(X),$$

and the problem is solved.

Second solution: Note that if x is either of the two solutions to $x^2 + (1/2)x + 1 = 0$, then $x + 1/x = -1/2$ and hence $x^n + 1/x^n = P_n(-1/2)$ or $x^{2n} - P_n(-1/2)x^n + 1 = 0$. Hence taking $a_n = -P_n(-1/2)$ suffices.

(Romanian competition, proposed by T. Andreescu, second solution by R. Stong)

12. The conclusion is trivial for $ac = 0$, so we may assume that $a \neq 0$ and $c \neq 0$. Then $ar^2 + br + c = 0$ implies that $r \neq 0$ and that $ar + c/r = -b$ is a rational number.

Using now the identity

$$a^{n+1}r^{n+1} + \frac{c^{n+1}}{r^{n+1}} = \left(ar + \frac{c}{r} \right) \left(a^n r^n + \frac{c^n}{r^n} \right) - ac \left(a^{n-1}r^{n-1} + \frac{c^{n-1}}{r^{n-1}} \right)$$

it follows by induction that for all positive integers n, $a^n r^n + c^n/r^n$ is a rational number, say $-b_n$. Then $a^n(r^n)^2 + b_n r^n + c^n = 0$, and the problem is solved.

(T. Andreescu)

13. Rephrase the solution as follows. The relation from the statement implies that $A(aI_n - A) = I_n$, which shows that A is invertible. Thus the given relation can be rewritten as

$$A + A^{-1} = aI_n.$$

Since $A_m + A^{-m} = P_m(A + A^{-1})$, where P_m is the polynomial that expresses $x^m + 1/x^m$ in terms of $x + 1/x$, it follows that $a_m = T_m(a)$ satisfies the requirement of the problem.

(D. Andrica)

2.9 Matrices

1. Since $AB - A - B = 0_n$, by adding I_n to both sides and factoring, we obtain $(I_n - A)(I_n - B) = I_n$. It follows that $I_n - A$ is invertible, and its inverse is $I_n - B$. Hence $(I_n - B)(I_n - A) = I_n$, which implies $BA - A - B = 0_n$. Consequently, $BA = A + B = AB$.

2. Complete the matrices with zeros to obtain the 5×5 matrices A' and B'. The matrix $A'B'$ is equal to AB, while $B'A'$ has BA in the left upper corner, and is zero elsewhere. The identity about determinants discussed at the beginning of the section implies $\det(I_5 - A'B') = \det(I_5 - B'A')$. We have $\det(I_5 - A'B') = \det(I_5 - AB) = (-1)^5 \det(AB - I_5)$. Also, since the only nonzero element in the last row of $I_5 - B'A'$ is a 1 in the lower right corner, and the corresponding minor is $\det(I_4 - BA)$, we obtain $\det(I_5 - B'A') = \det(I_4 - BA) = (-1)^4 \det(AB - I_4)$, and the conclusion follows.

3. Assuming that $X + Y + Z = XY + YZ + ZX$, we see that $XYZ = XZ - ZX$ is equivalent to

$$XYZ + X + Y + Z = XZ - ZX + XY + YZ + ZX.$$

Then

$$(X - I_n)(Y - I_n)(Z - I_n) = XYZ - XY - YZ - XZ + X + Y + Z - I_n,$$
$$= -I_n,$$

and the matrices $X - I_n$, $Y - I_n$, and $Z - I_n$ are invertible. Taking a circular permutation of the factors (i.e., by multiplying to the right by the factor and to the left by its inverse), we obtain, for example,

$$(Z - I_n)(X - I_n)(Y - I_n) = -I_n.$$

Thus

$$ZXY - XY - ZY - ZX + X + Y + Z = \mathcal{O}_n,$$

which is equivalent to $ZXY = ZY - YZ$. This proves that the first equality from the group of three implies the last. Permuting the letters, we obtain that the three equalities are equivalent.

(Romanian mathematics contest, 1985; proposed by T. Andreescu and I.V Maftei)

4. The equality

$$\begin{bmatrix} I_n & A \\ B & I_n \end{bmatrix} \cdot \begin{bmatrix} I_n & -A \\ 0_n & I_n \end{bmatrix} = \begin{bmatrix} I_n & 0_n \\ B & I_n - AB \end{bmatrix}$$

shows that the $2n \times 2n$ matrix from the statement can be written as the product of a matrix with determinant equal to one and a matrix with determinant equal to $\det(I_n - AB)$. Therefore,

$$\det \begin{bmatrix} I_n & A \\ B & I_n \end{bmatrix} = \det \begin{bmatrix} I_n & 0_n \\ B & I_n - AB \end{bmatrix} \det \begin{bmatrix} I_n & -A \\ 0_n & I_n \end{bmatrix}^{-1} = \det(I_n - AB).$$

5. *First solution:* We have

$$(A + I)(A^{n-1} - A^{n-2} + \cdots + (-1)^{n-1}I_n - \alpha I_n) = A^n + (-1)^{n-1}I_n - \alpha A - \alpha I_n$$
$$= ((-1)^{n-1} - \alpha)I_n,$$

which shows that $A + I_n$ is invertible.

Second solution: View A as a matrix with complex entries, and denote by $\sigma(A)$ the set of complex eigenvalues of A. This set is called the spectrum of A, a name motivated by quantum physics. Note that the matrix $\lambda I - A$ is invertible if and only if λ lies in the complement of the spectrum.

The spectral mapping theorem states that for any polynomial p, one has $p(\sigma(A)) = \sigma(p(a))$. In particular for $p(z) = z^n - \alpha z$, $p(\sigma(A)) = \sigma(p(A)) = \sigma(\mathcal{O}_n) = 0$. Thus the eigenvalues of A are zeros of p. Since -1 is not a zero of p, the matrix $-I_n - A$ is invertible, so $A + I_n$ is invertible, and we are done.

(Romanian Mathematical Olympiad, 1990; proposed by C. Cocea)

6. We have $(A^2 + B^2)(A - B) = A^3 - A^2B + AB^2 - B^3 = 0_n$. Since $A \neq B$, this shows that $A^2 + B^2$ has a zero divisor. Hence it is not invertible, so its determinant is 0.

(51st W.L. Putnam Mathematical Competition, 1991)

7. Write $A^2 + I_n = (A + iI_n)(A - iI_n)$, where i is the imaginary unit. Hence

$$\det(A^2 + I_n) = \det(A + iI_n)\det(A - iI_n) = |\det(A + iI_n)|^2 \geq 0.$$

8. By the previous problem, we have $\det(I_n + A^{2p}) \geq 0$ and $\det(I_n + B^{2q}) \geq 0$. From $AB = 0_n$, we obtain $A^{2p}B^{2q} = 0_n$; thus

$$
\begin{aligned}
\det(I_n + A^{2p} + B^{2q}) &= \det(I_n + A^{2p} + B^{2q} + A^{2p}B^{2q}) \\
&= \det((I_n + A^{2p})(I_n + B^{2q})) \\
&= \det(I_n + A^{2p})\det(I_n + B^{2q}) \geq 0_n.
\end{aligned}
$$

(M. and S. Rădulescu)

9. Let $\omega \neq 1$ be a third root of unity. Since A, B, C commute and $ABC = 0_n$, we can write

$$
\begin{aligned}
A^3 + B^3 + C^3 &= A^3 + B^3 + C^3 - 3ABC \\
&= (A + B + C)(A^2 + B^2 + C^2 - AB - BC - CA) \\
&= (A + B + C)(A + \omega B + \omega^2 C)(A + \omega^2 B + \omega C) \\
&= (A + B + C)(A + \omega B + \omega^2 C)\overline{(A + \omega B + \omega^2 C)}.
\end{aligned}
$$

Hence

$$
\begin{aligned}
&\det(A^3 + B^3 + C^3)\det(A + B + C) \\
&= \det((A + B + C)^2)\det(A + \omega B + \omega^2 C)\det\overline{(A + \omega B + \omega^2 C)} \\
&= (\det(A + B + C))^2\det(A + \omega B + \omega^2 C)\overline{\det(A + \omega B + \omega^2 C)} \\
&= (\det(A + B + C))^2|\det(A + \omega B + \omega^2 C)|^2 \geq 0.
\end{aligned}
$$

10. Assume $X^2 + pX + qI_n = O_n$ for some $n \times n$ matrix X. This equality can be written in the form

$$\left(X + \frac{p}{2}I_n\right)^2 = \frac{p^2 - 4q}{4}I_n.$$

Taking determinants on both sides and using the fact that $\det(AB) = \det A \cdot \det B$, we obtain

$$\left(\det\left(X + \frac{p}{2}I_n\right)\right)^2 = \left(\frac{p^2 - 4q}{4}\right)^n.$$

The left side is nonnegative, and the right side is strictly negative by hypothesis. This contradiction ends the proof.

(Romanian Mathematical Olympiad, proposed by T. Andreescu and I.D. Ion)

11. We start by observing that if we let $p(x) = \det(xI_n - B)$ be the characteristic polynomial of B, then by the Cayley–Hamilton theorem, $p(B) = 0_n$.

Since C commutes with both A and B, we have $AB^2 - B^2A = (AB - BA)B + B(AB - BA) = BC + CB = 2BC$. We show by induction that for any $k > 0$,

$$AB^k - B^kA = kB^{k-1}C.$$

For $k = 1, 2$ it is true. Assuming that it is true for $k - 1$, we have

$$AB^k - B^kA = (AB - BA)B^{k-1} + B(AB^{k-1} - B^{k-1}A)$$
$$= CB^{k-1} - B(k-1)B^{k-2}C = kB^{k-1}C,$$

which proves the claim. As a consequence, for any polynomial q, $Aq(B) - q(B)A = q'(B)C$, where q' is the derivative of q. In particular, $0_n = Ap(B) - p(B)A = p'(B)C$. Hence $Ap'(B)C - p'(B)AC = p''(B)C^2 = 0_n$. Inductively we obtain $p^{(k)}(B)C^k = 0_n$; in particular, $n!C^n = 0_n$. Thus $C^n = 0_n$, and we are done.

(Jacobson's lemma)

2.10 The Mean Value Theorem

1. The function

$$f(x) = \sum_{i=0}^{n} \frac{a_i}{i+1} x^{i+1}$$

has its derivative equal to $\sum_{i=0}^{n} a_i x^i$. Since $f(x)$ is differentiable and $f(0) = f(1) = 0$, Rolle's theorem applied on the interval $[0, 1]$ proves the existence in this interval of a zero of the function $\sum_{i=0}^{n} a_i x^i$.

2. Let us show that, moreover, there are no distinct pairs of real numbers with this property. Without loss of generality, we can assume $x > u \geq v > y$ (the case $u > x \geq y > v$ follows by symmetry).

Set $x_1 = x^3, y_1 = y^3, u_1 = u^3, v_1 = v^3$ and rewrite the system as

$$x_1^{2/3} - u_1^{2/3} = v_1^{2/3} - y_1^{2/3}$$
$$x_1 - u_1 = v_1 - y_1.$$

Applying the mean value theorem to the function $f(t) = t^{2/3}$ on the intervals $[u_1, x_1]$ and $[y_1, v_1]$, we deduce that there exist $t_1 \in (u_1, x_1)$ and $t_2 \in (y_1, v_1)$ such that

$$x_1^{2/3} - u_1^{2/3} = \frac{2}{3} t_1^{-1/3}(x_1 - u_1),$$
$$v_1^{2/3} - y_1^{2/3} = \frac{2}{3} t_2^{-1/3}(v_1 - y_1).$$

Consequently, $t_1 = t_2$, which is impossible, since (u_1, x_1) and (y_1, v_1) are disjoint intervals. We conclude that such numbers do not exist.

3. Fermat's little theorem implies that $a^p - a \equiv 0 \pmod{p}$, $b^p - b \equiv 0 \pmod{p}$, $c^p - c \equiv 0 \pmod{p}$, and $d^p - d \equiv 0 \pmod{p}$. Hence $(a - c) + (b - d) \equiv 0 \pmod{p}$. If we prove that $a + b \neq c + d$, then the conclusion of the problem follows from

$$|a - b| + |c - d| \geq |(a - c) + (b - d)|.$$

If $a + b = c + d$, then we may assume that $a > c > d > b$. Applying the mean value theorem to the function $f(t) = t^p$ on the intervals $[c, a]$ and $[b, d]$, we obtain $t_1 \in (c, a)$ and $t_2 \in (b, d)$ such that $pt_1^{p-1}(a - c) = pt_2^{p-1}(b - d)$. But, because $a - c = b - d$, this implies $t_1 = t_2$, a contradiction since they lie in different intervals. This completes the solution.

(*Revista Matematică din Timişoara (Timişoara's Mathematics Gazette)*, proposed by T. Andreescu)

4. If we apply Cauchy's theorem to the functions $f(x)/x$ and $1/x$, we conclude that there exists $c \in (a, b)$ with

$$\left(\frac{f(b)}{b} - \frac{f(a)}{a} \right) \left(-\frac{1}{c^2} \right) = \left(\frac{1}{b} - \frac{1}{a} \right) \left(\frac{cf'(c) - f(c)}{c^2} \right).$$

Hence

$$\frac{af(b) - bf(a)}{a - b} = f(c) - cf'(c).$$

5. Since f is positive, the function $\ln f(x)$ is well-defined and satisfies the hypothesis of the mean value theorem. Hence there exists $c \in (a, b)$ with

$$\frac{\ln f(b) - \ln f(a)}{b - a} = \frac{f'(c)}{f(c)}.$$

This implies

$$\ln \frac{f(b)}{f(a)} = (b - a) \frac{f'(c)}{f(c)},$$

and the conclusion follows by exponentiation.

6. The function $h = f/g$ satisfies the conditions in the hypothesis of Rolle's theorem; hence there exists $c \in (a, b)$ with $h'(c) = 0$. Since

$$h'(x) = \frac{f'(x)g(x) - f(x)g'(x)}{(g(x))^2},$$

we have $f'(c)g(c) - f(c)g'(c) = 0$; hence $f(c)/g(c) = f'(c)/g'(c)$.

7. Consider the function $g : [a, b] \to \mathbf{R}$ defined by

$$g(x) = (x - a)(x - b)f(x).$$

We see that g is continuous on $[a,b]$, differentiable on (a,b), and that $g(a) = g(b) = 0$. Applying Rolle's theorem, we obtain a $\theta \in (a,b)$ with $g'(\theta) = 0$. But $g'(x) = (x-b)f(x) + (x-a)f(x) + (x-a)(x-b)f'(x)$. We have thus obtained a θ with $(\theta - b)f(\theta) + (\theta - a)f(\theta) + (\theta - a)(\theta - b)f'(\theta) = 0$. Dividing by $(\theta - a)(\theta - b)f(\theta)$ yields

$$\frac{f'(\theta)}{f(0)} = \frac{1}{a-\theta} + \frac{1}{b-\theta}.$$

(*Gazeta Matematică (Mathematics Gazette, Bucharest)*, 1975; proposed by D. Andrica)

8. Let $n < N$ be two natural numbers. By applying the mean value theorem to the function $g(x) = xf(x)$ on the interval $[n,N]$, we deduce that there exists $c_n \in (n,N)$ such that $g'(c_n) = (Nf(N) - nf(n))/(N-n)$. Since

$$\lim_{N \to \infty} \frac{Nf(N) - nf(n)}{N-n} = a$$

for sufficiently large n, $|g'(c_n) - a| < 1/n$. Clearly, $\lim_{n \to \infty} c_n = \infty$. Also, since $g'(x) = f(x) + xf'(x)$, it follows that $g'(x)$ has a limit at infinity. It follows that $\lim_{x \to \infty} g'(x) = \lim_{n \to \infty} g'(c_n) = a$; hence $\lim_{x \to \infty} xf'(x) = \lim_{x \to \infty} g'(x) - \lim_{x \to \infty} f(x) = a - a = 0$.

9. The limit we must compute is clearly nonnegative. We will show that it is equal to zero by bounding the sequence from above with a sequence converging to zero.

Let $f : [n, n+1] \to \mathbf{R}$, $f(x) = (1 + 1/x)^x$, where n is a natural number. Applying the mean value theorem to the function f we deduce that there exists a number $c_n \in (n, n+1)$ such that $f(n+1) - f(n) = f'(c_n)$. We want to compute $\lim_{n \to \infty} \sqrt{n} f'(c_n)$. Note that

$$f'(c_n) = \left(1 + \frac{1}{c_n}\right)^{c_n} \left(\ln\left(\frac{1}{c_n} + 1\right) - \frac{1}{c_n + 1}\right).$$

Since $n < c_n < n+1$, this implies that

$$f'(c_n) < \left(1 + \frac{1}{c_n}\right)^{c_n} \left(\ln\left(\frac{1}{n} + 1\right) - \frac{1}{n+2}\right).$$

Using the fact that the sequence c_n tends to infinity, we obtain

$$0 \leq \lim_{n \to \infty} \sqrt{n} f'(c_n) \leq \lim_{n \to \infty} \sqrt{n} \left(1 + \frac{1}{c_n}\right)^{c_n} \left(\ln\left(\frac{1}{n} + 1\right) - \frac{1}{n+2}\right) = 0.$$

Here we have used that

$$\lim_{n \to \infty} \left(1 + \frac{1}{c_n}\right)^{c_n} = e$$

and

$$\lim_{n \to \infty} \sqrt{n} \ln\left(\frac{1}{n} + 1\right) = \lim_{n \to \infty} \frac{\sqrt{n}}{n+2} = 0.$$

The squeezing principle implies that the original limit is also equal to 0.

(*Gazeta Matematică (Mathematics Gazette, Bucharest)*, proposed by I.V. Maftei)

10. From $\int_0^a f(x)dx = 0$, by making the change of variable $x = at$, we have $\int_0^1 f(at)dt = 0$. Let $M = \sup_{x \in [0,1]} |f'(x)|$. For a fixed x, consider the function $g(t) = f(tx)$. The mean value theorem applied to this function on the interval $[a, 1]$ shows that $|f(x) - f(ax)| \leq (1-a)xM$. Integrating, we obtain

$$\left| \int_0^1 f(x)dx \right| = \left| \int_0^1 (f(x) - f(ax))dx \right|$$

$$\leq (1-a)M \int_0^1 x \, dx = \frac{1}{2}(1-a)M.$$

Equality is attained for $f(x) = \pm M(x - a/2)$.

(Romanian Mathematical Olympiad, 1984; proposed by R. Gologan)

11. *First solution:* Let $M = \max_{0 \leq x \leq 1} |f'(x)|$. With the change of variables $x = \alpha t$, we have

$$\left| \int_0^\alpha f(x)dx \right| = \left| \alpha \int_0^1 f(\alpha t)dt \right| = \left| \alpha \int_0^1 (f(\alpha t) - f(t))dt \right|$$

$$\leq \alpha \int_0^1 |f(t) - f(\alpha t)|dt,$$

since by hypothesis $\int_0^1 f(t)dt = 0$. Using the mean value theorem, we can find $c \in (\alpha t, t)$ such that $f(t) - f(\alpha t) = f'(c)(t - \alpha t)$, hence $|f(t) - f(\alpha t)| \leq Mt(1 - \alpha)$. We therefore have

$$\left| \int_0^\alpha f(x)dx \right| \leq \alpha(1-\alpha)M \int_0^1 t \, dt$$

The inequality from the statement follows from $\alpha(1-\alpha) \leq \frac{1}{4}$ if $\alpha \in [0,1]$ and $\int_0^1 t \, dt = \frac{1}{2}$.

Second solution: A short solution without the use of the mean value theorem is also possible. As before, set $M = \max_{0 \leq x \leq 1} |f'(x)|$. Consider the function

$$F(x) = \int_0^x f(t)dt + \frac{M}{2}x^2.$$

Note that $F'(x) = f(x) + Mx$ and $F''(x) = f'(x) + M \geq 0$. Hence $F(x)$ is a convex function. Thus for $\alpha \in [0,1]$ we have

$$F(\alpha) \leq (1-\alpha)F(0) + \alpha F(1),$$

that is

$$\int_0^\alpha f(t)dt + \frac{M}{2}\alpha^2 \leq 0 + \alpha \left(0 + \frac{M}{2} \right).$$

We thus obtain

$$\int_0^\alpha f(t)dt \le \frac{M}{2}(\alpha - \alpha^2) \le \frac{M}{2}.$$

Replacing f by $-f$ yields the inequality for the absolute value.
   (W.L. Putnam Mathematical Competition, 2007, proposed by T. Andreescu)

12. Showing that f is linear amounts to showing that its derivative is constant. We will prove that the derivative at any point is equal to the derivative at zero. Fix x, which we assume to be positive, the case of x negative being completely analogous. Define the set

$$M = \{t | t \ge 0 \text{ and } f'(t) = f'(x)\}.$$

Clearly, M is bounded from below, so let t_0 be the infimum of M. The relation in the statement implies that

$$\lim_{x \to 0} f'(x) = \lim_{x \to 0} \frac{f(x) - f(x/2)}{x/2} = 2\lim_{x \to 0} \frac{f(x) - f(0)}{x} - \lim_{x \to 0} \frac{f(x/2) - f(0)}{x/2}$$
$$= 2f'(0) - f'(0) = f'(0),$$

which shows that f' is continuous at 0. It obviously is continuous everywhere else, being the ratio of two continuous functions. This implies that M is closed, and hence t_0 is in M.
   We want to prove that $t_0 = 0$. Suppose, by way of contradiction, that $t_0 > 0$. Since

$$\frac{f(t_0) - f(t_0/2)}{t_0/2} = f'(t_0),$$

the mean value theorem applied on the interval $[t_0/2, t_0]$ proves the existence of a $c \in (t_0/2, t_0)$ such that $f'(c) = f'(t_0) = f'(x)$, which contradicts the minimality of t_0. This shows that for all x, $f'(x) = f'(0) = $ constant, and the problem is solved.
   (Romanian Mathematical Olympiad, 1999; proposed by M. Piticari and S. Rădulescu)

Chapter 3

Number Theory and Combinatorics

T. Andreescu and R. Gelca, *Mathematical Olympiad Challenges*, DOI: 10.1007/978-0-8176-4611-0_3, 223
© Birkhäuser Boston, a part of Springer Science+Business Media, LLC 2009

3.1 Arrange in Order

1. Let a,b,c,d,e,f be the six digits of the number, arranged in increasing order of their value. Choose for the first three digits f,c,a and for the last three digits e,d,b. Then $f+c+a-e-d-b=(f-e)+(c-d)+(a-b)\leq f-e\leq 9$, since $c-d$ and $a-b$ are negative or zero. Also, $e+d+b-f-c-a=(e-f)+(b-c)+(d-a)\leq d-a\leq 9$. Hence the number $fcaedb$ satisfies the desired property.

2. The cut $x=y$ separates the points for which $x<y$ from those for which $x>y$. The same holds for the other cuts. Hence each piece corresponds to an ordering of x,y,z. There are six such orderings; hence there are six pieces.
   (AHSME, 1987)

3. Arrange the numbers in increasing order $a_1\leq a_2\leq\cdots\leq a_{2n+1}$. By hypothesis, we have

$$a_{n+2}+a_{n+3}+\cdots+a_{2n+1}<a_1+a_2+\cdots+a_{n+1}.$$

This implies

$$a_1>(a_{n+2}-a_2)+(a_{n+3}-a_3)+\cdots+(a_{2n+1}-a_{n+1}).$$

Since all the differences in the parentheses are nonnegative, it follows that $a_1>0$, and we are done.
   (*Kőzépiskolai Matematikai Lapok (Mathematics Gazette for High Schools, Budapest)*)

4. Arrange the numbers in increasing order $1\leq a_1<a_2<\cdots<a_7\leq 1706$. If $a_{i+1}<4a_i-a_1$ for some $i=2,3,\cdots 6$, then $a_i<a_1+a_{i+1}<4a_i$ and we are done. If $a_{i+1}\geq 4a_i-a_1$ for all $i=2,3,\ldots,6$, then

$$a_3\geq 4a_2-a_1\geq 4(a_1+1)-a_1=3a_1+4,$$
$$a_4\geq 4a_3-a_1\geq 4(3a_1+4)-a_1=11a_1+16,$$
$$a_5\geq 4a_4-a_1\geq 4(11a_1+16)-a_1=43a_1+64,$$
$$a_6\geq 4a_5-a_1\geq 4(43a_1+64)-a_1=171a_1+256,$$
$$a_7\geq 4a_6-a_1\geq 4(171a_1+256)-a_1=683a_1+1024\geq 1707,$$

a contradiction.

5. The solution is rather simple if one gets the idea to arrange the numbers in order. Let $a=a_1<a_2<a_3<\cdots<a_n=A$ be the numbers, m their least common multiple, and d their greatest common divisor. Then on the one hand, $m/a_n<m/a_{n-1}<\cdots<m/a_1=m/a$, and on the other hand, $a_1/d<a_2/d<\cdots<a_n/d=A/d$. Since the numbers in these two sequences are positive integers, both m/a and A/d must be at least n, which implies $m\geq na$ and $d\leq A/n$.

6. The natural approach is via the pigeonhole principle to show that

$$\frac{\binom{2n}{2}}{n^2-1}>2.$$

However, this approach fails. The solution proceeds as follows. We may assume that $a_1 < a_2 < a_3 < \cdots < a_{2n}$. Consider the differences $a_2 - a_1, a_3 - a_2, \cdots, a_{2n} - a_{2n-1}$. If no three are equal, then

$$(a_2 - a_1) + (a_3 - a_2) + \cdots + (a_{2n} - a_{2n-1})$$
$$\geq 1 + 1 + 2 + 2 + \cdots + (n-1) + (n-1) + n = n^2.$$

On the other hand, this sum is equal to $a_{2n} - a_1$, hence is less than or equal to $n^2 - 1$. This is a contradiction, and the problem is solved.

(Proposed by G. Heuer for the USAMO)

7. From the $2n + 3$ points, choose two such that the line they determine has all other points on one side. Consider the sets that are intersections of the half-plane containing all points with the disk determined by the two points and a third from the remaining $2n + 1$ points (see Figure 3.1.1) ordered with respect to inclusion $U_1 \subset U_2 \subset \cdots \subset U_{2n+1}$.

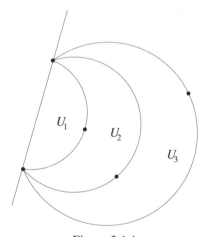

Figure 3.1.1

The set U_{n+1} contains exactly n of the points in its interior; hence the corresponding disk contains n points inside and n outside.

(Proposed for the IMO, 1993)

8. There exist finitely many lines joining pairs of given points. Choose a line l not parallel to any of these such that all points are on one side of l. Start moving l parallel to itself; it will meet the points one at a time. Label the points $A_1, A_2, \ldots, A_{4n}$, in the order they have been encountered. Then $A_1A_2A_3A_4, A_5A_6A_7A_8, \ldots, A_{4n-3}A_{4n-2}A_{4n-1}A_{4n}$ are nonintersecting quadrilaterals.

9. Let the numbers be $1 \leq a_1 < a_2 < \cdots < a_{69} \leq 100$. Clearly, $a_1 \leq 32$. Consider the sequences

$$a_3 + a_1 < a_4 + a_1 < \cdots < a_{69} + a_1$$

and

$$a_3 - a_2 < a_4 - a_2 < \cdots < a_{69} - a_2.$$

Each of their terms is a positive integer not exceeding $100 + 32 = 132$. Since the two sequences have jointly $67 + 67 = 134$ terms, there must exist indices $i, j \in \{3, 4, \ldots, 69\}$ such that $a_i + a_1 = a_j - a_2$. We have $a_1 < a_2 < a_i$, and since $a_1 + a_2 + a_i = a_j$, the first part is done.

A counterexample for the second part is given by the set $\{33, 34, 35, \ldots, 100\}$.

10. Assume that there exist 25 numbers for which the property does not hold. Arrange them in increasing order $0 < x_1 < x_2 < \cdots < x_{25}$. Since x_{25} is the largest, the sum of $x_{25} + x_k$ is not in the set for any k. Thus the difference is in the set, and we can only have $x_{25} - x_1 = x_{24}, x_{25} - x_2 = x_{23}, \ldots, x_{25} - x_{12} = x_{13}$. It follows that for $k > 1$, $x_{24} + x_k > x_{25}$; hence $x_{24} - x_k$ is in the set. But $x_{24} - x_2 < x_{23}$, since $x_2 + x_{23} = x_{25}$. Thus $x_{24} - x_2 \le x_{22}, x_{24} - x_3 \le x_{21}, \ldots, x_{24} - x_{12} \le x_{11}, \ldots, x_{24} - x_{22} \le x_1$. It is important to remark that here we have used the fact that if $x_{24} - x_{12} = x_{12}$, then the sum and the difference of x_{24} and x_{12} are not equal to any of the remaining numbers. It follows that neither the sum nor the difference of x_{24} and x_{23} is in the set, a contradiction that solves the problem.

11. Let $a_0 > a_1 > \cdots > a_9$ be the selected numbers arranged in order. Then at most 20 numbers exceed a_0 (the largest and the second largest in each row), so $a_0 \ge 80$. Similarly $a_1 \ge 72$ (this time the largest and the second largest in each row, and the elements in the row containing a_0 may exceed a_1). Hence

$$a_0 + a_1 + \cdots + a_9 \ge 80 + 72 + (a_9 + 7) + (a_9 + 6)$$
$$+ \cdots + a_9 = 8a_9 + 180.$$

Meanwhile, the row containing a_9 has sum at most

$$100 + 99 + a_9 + \cdots + (a_9 - 7) = 8a_9 + 171,$$

which is less than the sum of the a_i's.

(St. Petersburg City Mathematical Olympiad, 1998)

12. Let the numbers be $a_1 < a_2 < \cdots < a_n$. For $i = 1, 2, \ldots, n$ let $s_i = a_1 + a_2 + \cdots + a_i$, $s_0 = 0$. All the sums in question are less than or equal to s_n, and if σ is one of them, we have

$$s_{i-1} < \sigma \le s_i$$

for an appropriate i. Divide the sums into n groups by letting C_i denote the group of sums satisfying the above inequality. We claim that these groups have the desired property. To establish this, it suffices to show that this inequality implies

$$\frac{1}{2}s_i < \sigma \le s_i.$$

Suppose $s_{i-1} < \sigma \le s_i$. The inequality $a_1 + a_2 + \cdots + a_{i-1} = s_{i-1} < \sigma$ and the arrangement $a_1 < a_2 < \cdots < a_n$ show that the sum σ contains at least one term a_k with $k \ge i$. Then, since $a_k \ge a_i$, we have

$$s_i - \sigma < s_i - s_{i-1} = a_i \le a_k \le \sigma,$$

which together with $\sigma \le s_i$ proves the claim.

(25th USAMO, 1996; proposed by T. Andreescu)

3.2 Squares and Cubes

1. We have $(x+y+z\pm1)^2 = x^2 + y^2 + z^2 + 2xy + 2x(z\pm1) + 2y(z\pm1) \pm 2z + 1$. It follows that

$$(x+y+z-1)^2 < x^2 + y^2 + z^2 + 2xy + 2x(z-1) + 2y(z+1)$$
$$< (x+y+z+1)^2.$$

Hence $x^2 + y^2 + z^2 + 2xy + 2x(z-1) + 2y(z+1)$ can be equal only to $(x+y+z)^2$. This implies $x = y$, hence the solutions to the given equation are of the form (m,m,n), $m,n \in \mathbf{N}$.

(T. Andreescu)

2. The nth term of this sequence is

$$\frac{11\ldots1077\ldots7811\ldots11}{3} = \frac{1}{3}\left(1 + 10 + \cdots + 10^n + 8 \cdot 10^{n+1}\right.$$
$$\left. + 7(10^{n+2} + \cdots + 10^{2n+1}) + 10^{2n+3} + \cdots + 10^{3n+2}\right)$$
$$= \frac{10^{n+1} - 1}{27} + \frac{8 \cdot 10^{n+1}}{3} + 7 \cdot 10^{n+2} \cdot \frac{10^n - 1}{27} + 10^{2n+3} \cdot \frac{10^n - 1}{27}.$$

This is further equal to

$$= \frac{10^{n+1} - 3 \cdot 10^{2n+2} + 3 \cdot 10^{n+1} - 1}{3} = \left(\frac{10^{n+1} - 1}{3}\right)^3$$
$$= [3(1 + 10 + \cdots + 10^n)]^3.$$

(L. Panaitopol, D. Şerbănescu, *Probleme de Teoria Numerelor şi Combinatorica pentru Juniori (Problems in Number Theory and Combinatorics for Juniors)*, GIL, 2003)

3. If the four consecutive positive integers are $k, k+1, k+2, k+3$, then

$$k(k+1)(k+2)(k+3) = (k(k+3))((k+1)(k+2))$$
$$= (k^2 + 3k)(k^2 + 3k + 2)$$
$$= ((k^2 + 3k + 1) - 1)((k^2 + 3k + 1) + 1)$$
$$= (k^2 + 3k + 1)^2 - 1.$$

This number lies between the consecutive squares $(k^2 + 3k)^2$ and $(k^2 + 3k + 1)^2$, hence cannot be a perfect square.

(Russian Mathematical Olympiad, 1979–1980)

4. There are more numbers *not* of this form. Let $n = k^2 + m^3$, with $k, m, n \in \mathbf{N}$ and $n < 1{,}000{,}000$. Clearly $k \leq 1000$ and $m \leq 100$. Therefore there cannot be more numbers in the desired form than the 100,000 pairs (k, m).

(Russian Mathematical Olympiad, 1996)

5. The solution is easy if one notes the identity

$$1 + (2n)^2 + (2n^2)^2 = (2n^2 + 1)^2.$$

Note that the conclusion also follows from the property that any number that is not congruent to 2 modulo 4 can be written as a difference of two squares, applied to $x^2 - 1$, when x is even.

(W. Sierpiński, *250 problems in elementary number theory*, Państwowe Wydawnictwo Naukowe, Warszawa, 1970)

6. Let $d = c + 1$. The equality $a - b = a^2 c - b^2(c + 1)$ implies

$$(a - b)[c(a + b) - 1] = b^2.$$

But $c(a + b) - 1$ and $a - b$ are relatively prime. Indeed, if p is a prime dividing $a - b$, then the above equality shows that p also divides b, so p will divide $a + b = (a - b) + 2b$. Hence p cannot divide $c(a + b) - 1$. It follows that $|a - b|$ is a perfect square as well.

Note that the first nontrivial example is $18 - 22 = 18^2(-3) - 22^2(-2)$.

(Proposed by T. Andreescu for the USAMO, 1999)

7. The interval $[s_n, s_{n+1})$ contains a perfect square if and only if $[\sqrt{s_n}, \sqrt{s_{n+1}})$ contains an integer, so it suffices to prove $\sqrt{s_{n+1}} - \sqrt{s_n} \geq 1$ for each $n \geq 1$. Substituting $s_n + k_n$ for s_{n+1}, we get the equivalent inequality $s_n + k_{n+1} \geq (\sqrt{s_n} + 1)^2$, or $k_{n+1} \geq 2\sqrt{s_n} + 1$.

Since $k_{m+1} - k_m \geq 2$ for each m, we have

$$
\begin{aligned}
s_n &= k_n + k_{n-1} + k_{n-2} + \cdots + k_1 \\
&\leq k_n + (k_n - 2) + (k_n - 4) + \cdots + \begin{cases} 2 & \text{if } k_n \text{ is even} \\ 1 & \text{if } k_n \text{ is odd} \end{cases} \\
&= \begin{cases} \frac{k_n(k_n+2)}{4} & \text{if } k_n \text{ is even} \\ \frac{(k_n+1)^2}{4} & \text{if } k_n \text{ is odd} \end{cases} \leq \frac{(k_n+1)^2}{4}.
\end{aligned}
$$

Hence $(k_n + 1)^2 \geq 4s_n$, and $k_{n+1} \geq k_n + 2 \geq 2\sqrt{s_n} + 1$, as desired.

(USAMO, 1994; proposed by T. Andreescu)

8. Let $b_n = (a_n + 1)/6$ for $n \geq 0$. Then $b_0 = b_1 = 1$ and

$$b_{n+1} = 98b_n - b_{n-1} - 16$$

for $n \geq 1$ (so in particular, b_n is always an integer). In addition, for $n \geq 2$,

$$b_{n+1} = 98b_n - b_{n-1} - (98b_{n-1} - b_n - b_{n-2}) = 99b_n - 99b_{n-1} + b_{n-2}.$$

Also, let $c_0 = c_1 = 1$, and for $n \geq 2$, set $c_n = 10c_{n-1} - c_{n-2}$. We will show that $b_n = c_n^2$ for all n. This holds for $n = 0, 1, 2$ (since $b_2 = 98 - 17 = 81$ and $c_2 = 10 - 1 = 9$); and assuming the claim for $b_1, \ldots, b_n$, we have

$$b_{n+1} = 99c_n^2 - 99c_{n-1}^2 + c_{n-2}^2$$
$$= 99c_n^2 - 99c_{n-1}^2 + (10c_{n-1} - c_n)^2$$
$$- 100c_n^2 - 20c_n c_{n-1} + c_{n-1}^2$$
$$= (10c_n - c_{n-1})^2 - c_{n+1}^2,$$

Thus the claim follows by induction, and the proof is complete.

(Proposed by T. Andreescu for the IMO, 1998)

9. Let us first solve the easy case $c = 0$. If $b \neq 0$, choose $n = p$ a prime that does not divide b. If $b = 0$, choose n such that $n + a$ is not a perfect square.

Now assume that c is not equal to zero. If a is even, say $a = 2a_0$, then choose n to be a perfect square $n = m^2$. By completing the square, we get

$$m^6 + 2a_0 m^4 + bm^2 + c = (m^3 + a_0 m)^2 + (b - a_0^2)m^2 + c.$$

For m large enough, this is between $(m^3 + a_0 m - 1)^2$ and $(m^3 + a_0 m + 1)^2$. The equality

$$m^6 + 2a_0 m^4 + bm^2 + c = (m^3 + a_0 m)^2$$

implies $(a_0^2 - b)m^2 = c$. Since $c \neq 0$, this can happen for at most two values of m. Hence we can find infinitely many m for which $m^6 + 2a_0 m^4 + bm^2 + c$ is not a perfect square.

The case where a is odd can be reduced to this one, since the change of variable $n \mapsto n + 1$ gives $n^3 + (a+3)n^2 + (3 + 2a + b)n + 1 + a + b + c$, and the parities of a and $a + 3$ are different.

(59th W.L. Putnam Mathematical Competition, 1998)

10. For $n = 10, 11, \ldots, 15$, the statement is satisfied by $3^3 = 27$. If $n \geq 16$, then

$$n > (2.5)^3 = \frac{1}{(1.4 - 1)^3} > \frac{1}{(\sqrt[3]{3} - 1)^3}.$$

Hence $\sqrt[3]{n} \geq 1/(\sqrt[3]{3} - 1)$. This implies $\sqrt[3]{3n} - \sqrt[3]{n} > 1$, so between $\sqrt[3]{n}$ and $\sqrt[3]{3n}$ there is at least one integer, and the conclusion follows.

(*Gazeta Matematică (Mathematics Gazette, Bucharest)*, proposed by T. Andreescu)

11. Suppose that A is such a number, and that it has n digits, i.e., $10^{n-1} \leq A < 10^n$. Then $10^n A + A$ is a perfect square. For this to be possible, $10^n + 1$ must be of the form $M^2 N$ with $M > 1$. Indeed, if no perfect square divides $10^n + 1$, then all of its prime divisors appear without multiplicity. Since $(10^n + 1)A$ is a perfect square, all these prime divisors must reappear in the decomposition of A, which implies that $A \geq 10^n + 1$, which is impossible.

Thus let us find a number of the form $10^n + 1$ that is not square free. One can check that $10^{11} + 1$ is divisible by 121, so we have the decomposition

$$10^{11} + 1 = 11^2 \times \frac{10^{11} + 1}{121}.$$

We cannot choose $(10^{11} + 1)/121$ to be equal to A, since this number is not greater than 10^{n-1}, but by multiplying it by an appropriate power of 9 (a perfect square), we get a number between 10^{n-1} and 10^n, and we choose A to be this number.

As a final remark, note that there exist infinitely many A with the desired property. Indeed, since $10^{(2k+1)11} + 1$ is divisible by $10^{11} + 1$ and so by 121, for all $k \in \mathbf{N}$, we can choose

$$A = 3^{2m} \frac{10^{11(2k+1)} + 1}{121}$$

with m such that $10^{(2k+1)11-1} \le A < 10^{(2k+1)11}$.

(Kvant (Quantum), proposed by B. Kukushkin)

12. The approach parallels that of the example in the introduction to Section 3.2. Let (a,b) be a pair of positive integers satisfying the condition, and let k be the ratio from the statement. Because $k > 0$, we have $2ab^2 - b^3 + 1 > 0$, so $a > b/2 - 1/2b^2$, and hence $a \ge b/2$. Using this and the fact that $k \ge 1$, we deduce that $a^2 \ge b^2(2a - b) + 1$, so $a^2 > b^2(2a - b) \ge 0$. Hence either $a > b$ or $2a = b$.

Now consider the two solutions a_1, a_2 to the quadratic equation

$$a^2 - 2kb^2 a + k(b^3 - 1) = 0,$$

for fixed positive integers k and b, and assume that one of them is an integer. Then the other is also an integer because $a_1 + a_2 = 2kb^2$. We may assume that $a_1 \ge a_2$, and we have $a_1 \ge kb^2 > 0$. Furthermore, since $a_1 a_2 = k(b^3 - 1)$, we obtain

$$0 \le a_2 = \frac{k(b^3 - 1)}{a_1} \le \frac{k(b^3 - 1)}{kb^2} < b.$$

From the above it follows that either $a_2 = 0$ or $2a_2 = b$.

If $a_2 = 0$, then $b^3 - 1 = 0$, and hence $a_1 = 2k$, $b = 1$.

If $a_2 = b/2$, then $k = b^2/4$, and $a_1 = b^4/2 - b/2$.

Therefore the only pairs of integers for which the ratio from the statement is a positive integer are $(a,b) = (2l, 1)$ or $(l, 2l)$ or $(8l^4 - l, 2l)$, for some positive integer l. All of these pairs satisfy the given condition.

(44th IMO, 2003, proposed by Bulgaria)

13. The reader familiar with Fibonacci numbers might recall the identity

$$F_{n+1}^2 - F_{n+1}F_n - F_n^2 = (-1)^n.$$

Hence we expect the answer to consist of Fibonacci numbers.

Note that $(m,n) = (1,1)$ satisfies the relation $(n^2 - mn - m^2)^2 = 1$. Also, if a pair (m,n) satisfies this relation and $0 < m < n$, then $m < n < 2m$, and by completing the square we get

$$
\begin{aligned}
(n^2 - mn - m^2)^2 &= ((n-m)^2 + mn - 2m^2)^2 \\
&= ((n-m)^2 + m(n-m) - m^2)^2 \\
&= (m^2 - m(n-m) - (n-m)^2)^2,
\end{aligned}
$$

which shows that $(n-m, m)$ satisfies the same relation and $0 < n - m < m$.

The transformation $(m,n) \to (n-m, m)$ must terminate after finitely many steps, and it terminates only when $m = n = 1$. Hence all pairs of numbers satisfying the relation are obtained from $(1,1)$ by applying the inverse transformation $(m,n) \mapsto (n, m+n)$ several times. Therefore, all pairs consist of consecutive Fibonacci numbers. The largest Fibonacci number less than 1981 is $F_{16} = 1597$, so the answer to the problem is $F_{15}^2 + F_{16}^2 = 3514578$.

(22nd IMO, 1981; proposed by The Netherlands)

14. Completing the cube, we obtain

$$
\begin{aligned}
x^3 - 3xy^2 + y^3 &= 2x^3 - 3x^2y - x^3 + 3x^2y - 3xy^2 + y^3 \\
&= 2x^3 - 3x^2y + (y-x)^3 \\
&= (y-x)^3 - 3(y-x)(-x)^2 + (-x)^3.
\end{aligned}
$$

This shows that if (x,y) is a solution, then so is $(y-x, -x)$. The two solutions are distinct, since $y - x = x$ and $-x = y$ lead to $x = y = 0$. Similarly,

$$
\begin{aligned}
x^3 - 3xy^2 + y^3 &= x^3 - 3x^2y + 3xy^2 - y^3 + 2y^3 + 3x^2y - 6xy^2 \\
&= (x-y)^3 + 3xy(x-y) - 3xy^2 + 2y^3 \\
&= (-y)^3 - 3(-y)(x-y)^2 + (x-y)^3,
\end{aligned}
$$

so $(-y, x-y)$ is the third solution of the equation.

We use these two transformations to solve the second part of the problem. Let (x,y) be a solution. Since 2891 is not divisible by 3, $x^3 + y^3$ is not divisible by 3, either. Thus either both of x and y give the same residue modulo 3 (different from 0), or exactly one of x and y is divisible by 3. Any of the two situations implies that one of the numbers $-x, y, x-y$ is divisible by 3, and by using the above transformations, we may assume that y is a multiple of 3. It follows that x^3 must be congruent to 2891 mod 9, which is impossible since 2891 has the residue 2, and the only cubic residues mod 9 are 0, 1, and 8.

(23rd IMO, 1982)

15. (a) Let (x,y,z) be a solution. Then

$$
\begin{aligned}
0 = x^2 + y^2 + 1 - xyz &= (x - yz)^2 + y^2 + 1 + xyz - y^2z^2 \\
&= (yz - x)^2 + y^2 + 1 - (yz - x)yz;
\end{aligned}
$$

hence $(yz - x, y, z)$ is also a solution provided that $yz - x > 0$. But this happens to be true, since $x(yz - x) = xyz - x^2 = y^2 + 1$. If $x > y$, then $x^2 > y^2 + 1 = x(yz - x)$. Hence $x > yz - x$, which shows that the newly obtained solution is smaller than the initial one (in the sense that $x + y > (yz - x) + y$). However this procedure can not be continued indefinitely, so we must hit a solution with $x = y$. However, this gives $x^2(z - 2) = 1$, hence $z = 3$ and $x = y = 1$.

(b) Clearly, $(1, 1)$ is a solution with $z = 3$, and all other solutions reduce to this one under the operation above. It follows that all solutions are obtained from $(x_1, y_1) = (1, 1)$ by the recursion

$$(x_{n+1}, y_{n+1}) = (y_n, 3y_n - x_n).$$

The sequences $\{x_n\}_n$ and $\{y_n\}_n$ satisfy the same recursion: $a_{n+1} = 3a_n - a_{n-1}, a_1 = 1, a_2 = 2$. This recursion characterizes the Fibonacci numbers of odd index. Therefore, $(x_n, y_n) = (F_{2n+1}, F_{2n-1}), n \geq 1$.

The solutions are $(1, 1)$, (F_{2n+1}, F_{2n-1}) and (F_{2n-1}, F_{2n+1}), for $n \geq 1$.

3.3   Repunits

1. (a) We can write

$$\underbrace{11\ldots 1}_{2n \text{ times}} = 1 + 5 + \ldots + 5^{2n-1} = \frac{5^{2n} - 1}{4} = \frac{5^n - 1}{2} \cdot \frac{5^n + 1}{2},$$

and $(5^n - 1)/2$ and $(5^n + 1)/2$ are consecutive integers.

(b) We have

$$\underbrace{11\ldots 1}_{n \text{ times}} = 1 + 9 + 9^2 + \cdots + 9^{n-1} = \frac{9^n - 1}{8} = \frac{1}{2} \cdot \frac{3^n - 1}{2} \cdot \frac{3^n + 1}{2},$$

and this is a triangular number, since $(3^n - 1)/2$ and $(3^n + 1)/2$ are consecutive integers.

2. Indeed,

$$\underbrace{11\ldots 1}_{2n \text{ times}} - \underbrace{22\ldots 2}_{n \text{ times}} = \underbrace{11\ldots 1}_{n \text{ times}}\underbrace{00\ldots 0}_{n \text{ times}} - \underbrace{11\ldots 1}_{n \text{ times}}$$

$$= \underbrace{11\ldots 1}_{n \text{ times}} \times (10^n - 1) = \underbrace{11\ldots 1}_{n \text{ times}} \times \underbrace{99\ldots 9}_{n \text{ times}} = \underbrace{33\ldots 3}_{n \text{ times}} \times \underbrace{33\ldots 3}_{n \text{ times}}.$$

3. Let A be the smallest repunit divisible by 19. If n is the number of digits of A, then $A = (10^n - 1)/9$. Since 9 and 19 are relatively prime, we are in fact looking after the smallest n such that $10^n - 1$ is divisible by 19. From Fermat's little theorem, it follows that $10^{18} - 1$ is divisible by 19. If there exists a smaller n such that

$10^n \equiv 1 \pmod{19}$, then n must divide 18. We have to check $2, 3, 6,$ and 9. The case $n = 2$ is easy. Also, $10^3 \equiv -7 \pmod{19}$, so $10^6 \equiv 11 \pmod{19}$ and $10^9 \equiv -1 \pmod{19}$. Hence none of the other numbers works. Thus the answer to the problem is $n = 18$.

4. If a repunit has $m \times n$ digits, with $m, n > 1$, then it can be factored as

$$\underbrace{11\ldots1}_{n} \times 1\underbrace{00\ldots0}_{n-1}1\underbrace{00\ldots0}_{n-1}1\ldots1\underbrace{00\ldots0}_{n-1}1,$$

where there are $m - 1$ groups of zeros. Hence such a repunit is not prime. Thus for a repunit to be prime, it must have a prime number of digits. The converse is not true, since for example, $111 = 3 \times 37$ and

$$\underbrace{11\ldots1}_{11 \text{ times}} = 21649 \times 513239.$$

5. It does:

$$\underbrace{111\ldots1}_{81 \text{ times}} = \underbrace{111\ldots1}_{9 \text{ times}} \times \underbrace{100\ldots0100\ldots010\ldots01}_{9 \text{ ones and } 64 \text{ zeros}}.$$

Both factors have the sum of the digits divisible by 9, so both are divisible by 9.
(Leningrad Mathematical Olympiad)

6. (a) Let $n = 5k + r$, where $r \in \{0, 1, 2, 3, 4\}$. Then

$$\underbrace{11\ldots1}_{n \text{ times}} = \underbrace{11\ldots1}_{5k \text{ times}}\underbrace{00\ldots0}_{r \text{ times}} + \underbrace{11\ldots1}_{r \text{ times}} = \underbrace{11\ldots1}_{5 \text{ times}} \times 100001\ldots00001 + \underbrace{11\ldots1}_{r \text{ times}}.$$

Since $11111 = 41 \times 271$, the latter is congruent to $\underbrace{11\ldots1}_{r \text{ times}}$ modulo 41. But $1, 11, 111,$ 1111 are not divisible by 41, so $\underbrace{11\ldots1}_{n \text{ times}}$ is divisible by 41 if and only if $r = 0$, that is, n is divisible by 5.

(b) Same idea, noting that $111111 = 91 \times 1221$.

7. Indeed, if a repunit $(10^n - 1)/9$ is a square, then it must be congruent to 1 mod 4. This implies that 10^n is congruent to 2 mod 4. But this happens only when $n = 1$.
The problem is still open if we replace the square by the mth power.

8. The important property that the numbers ending in 1, 3, 7, or 9 share is that they are all relatively prime to 10. Consider the numbers

$$1, 11, \ldots, \underbrace{11\ldots1}_{n+1 \text{ times}} .$$

By the pigeonhole principle, among these $n + 1$ numbers there are two giving the same residue when divided by n.

The difference of these two numbers is divisible by n and is of the form $a \cdot b$ with a a repunit and b a power of 10. Since n is relatively prime to 10, the repunit a must be divisible by n, and we are done.

9. *First solution.* We adapt Euclid's proof for the existence of infinitely many primes. The sequence is constructed inductively. Let $a_1 = 1$, and assume that we already have chosen the terms of the sequence up to a_n. By the previous problem, there is a repunit m divisible by the product $a_1 a_2 \cdots a_n$. The number $10m + 1$ is a repunit and is relatively prime to m; hence to any of the a_k's with $1 \le k \le n$. We let $a_{n+1} = 10m + 1$, and the proof is complete.

Second solution. We show that if n and m are relatively prime, then so are 10^n and 10^m, from which the conclusion follows. Indeed if for $m < n$, $10^n - 1$ and $10^m - 1$ have a common divisor d, then d also divides $10^n - 10^m = 10^m(10^{n-m} - 1)$, and thus divides $10^{n-m} - 1$. But $n - m$ and m are also relatively prime, which shows that an inductive argument produces the desired conclusion.

Remark. This shows that the number of primes dividing repunits is infinite, giving thus another proof to the fact that there are infinitely many prime numbers.

10. We prove by induction that the repunit with 3^n digits is divisible by 3^n. For $n = 1$, we have $111 = 3 \cdot 37$. Let us assume that the property is true for a certain n. Then

$$\underbrace{11\ldots1}_{3^{n+1}\text{ times}} = \underbrace{11\ldots1}_{3^n\text{ ones}} \times 1 \underbrace{00\ldots0}_{3^n-1\text{ zeros}} 1 \underbrace{00\ldots0}_{3^n-1\text{ zeros}} 1.$$

The first factor is divisible by 3^n by the induction hypothesis, and the second one is divisible by 3, since the sum of its digits is equal to 3.

11. The problem reduces to the computation of the greatest integer part of the number

$$B = \sqrt{\underbrace{11\ldots1}_{2n\text{ times}} \times 10^{2n}} = \sqrt{\underbrace{11\ldots1}_{2n\text{ times}}} \times 10^n.$$

Let k be the repunit with $2n$ digits. Then $9k < 10^{2n}$, or $9k^2 < 10^{2n}k$, which implies $9k^2 < B^2$; hence $3k < B$. Similarly, $9k + 6 > 10^{2n} + 1$; thus $9k^2 + 6k + 1 > 10^{2n}k$. It follows that $(3k+1)^2 > B^2$, or $3k + 1 > B$.

This shows that $3k < B < 3k + 1$, so $\lfloor B \rfloor = 3k$. The answer to the problem is $33\ldots3.33\ldots3$, with n digits before the decimal point and n digits after.

(P. Radovici-Mărculescu, *Probleme de teoria elementară a numerelor (Problems in elementary number theory)*, Editura Tehnică, Bucharest, 1986)

12. This is a generalization of the problem from the introduction to Section 3.3. Let $f(x)$ be such a polynomial. Note that from the hypothesis, it follows that there is a sequence $(a_n)_{n \ge 1}$ of positive integers such that $f\left(\frac{10^n - 1}{9}\right) = \frac{10^{a_n} - 1}{9}$. Let us analyze the sequence $(a_n)_{n \ge 1}$. Let $\deg(f) = d \ge 1$. Then there is a nonzero real number A such that $f(x) \approx Ax^d$ for large values of x. Therefore $f\left(\frac{10^n - 1}{9}\right) \approx \frac{A}{9^d} \cdot 10^{nd}$. We therefore have

$10^{a_n} \approx \frac{A}{9^{d-1}} \cdot 10^{nd}$. This shows that the sequence $(a_n - nd)_{n \geq 1}$ converges to a limit l such that $A = 9^{d-1} \cdot 10^l$. Because this sequence consists of integers, it becomes eventually equal to l. Thus from a certain point on, we have

$$f\left(\frac{10^n - 1}{9}\right) = \frac{10^{nd+l} - 1}{9}.$$

We deduce that the polynomial $f(x)$ coincides with the polynomial $\frac{(9x+1)^d \cdot 10^l - 1}{9}$ for infinitely many values of x. We conclude that the two polynomials are equal. It is not hard to see that any polynomial of the form

$$f(x) = \frac{(9x+1)^d \cdot 10^l - 1}{9}$$

with d, l integers, $d > 0$, $l \geq 0$ has the desired property, so these are all polynomials with the required property.

(W.L. Putnam Mathematical Competition, 2007)

3.4 Digits of Numbers

1. One computes easily $f(2) = f(3) = 2$, $f(4) = f(5) = f(6) = f(7) = 3$. As written, the defining recursion for f suggests a relationship with binary expansions. Rewriting, we have $f(10_2) = f(11_2) = 2$, $f(100_2) = f(101_2) = f(110_2) = f(111_2) = 3$. An inductive argument shows that f counts the number of digits in the binary expansion.

2. We use binary expansion and prove the identity by induction on n. If $n = 1_2$, the identity is clearly true. Assume that it holds for all numbers less than n and let us prove it for n. Let m be the number obtained from n by deleting the last digit. Then

$$\left\lfloor \frac{n + 10_2}{100_2} \right\rfloor + \left\lfloor \frac{n + 100_2}{1000_2} \right\rfloor + \left\lfloor \frac{n + 1000_2}{10000_2} \right\rfloor + \cdots$$
$$= \left\lfloor \frac{m + 1_2}{10_2} \right\rfloor + \left\lfloor \frac{m + 10_2}{100_2} \right\rfloor + \left\lfloor \frac{m + 100_2}{1000_2} \right\rfloor + \cdots = m,$$

where the last equality follows from the induction hypothesis. On the other hand

$$\left\lfloor \frac{n + 1_2}{10_2} \right\rfloor = m + a,$$

where a is the last digit of n. Thus the sum on the left side of the initial identity is equal to $m + m + a = 10_2 m + a = n$, and we are done.

(10th IMO, 1968)

3. Considering binary expansions, we compute $x_2 = 0.1_2$, $x_3 = 0.11_2$, $x_4 = 0.101_2$, $x_5 = 0.1011_2$, $x_6 = 0.10101_2$. An easy inductive argument shows that $x_{2k} = 0.101010\ldots101_2$, where there are $k - 1$ pairs 10 followed by a 1 at the end,

and $x_{2k+1} = 0.101010\ldots1011_2$, where there are $k-1$ pairs of 10 followed by a 11 at the end. This implies that $\lim_{n\to\infty} x_n = 0.101010\ldots10\ldots_2$, which, returning to decimal writing is equal to $\frac{2}{3}$.

4. We shall prove by induction on n that $a_n = n - t_n$, where t_n is the number of 1's in the binary representation of n. Suppose now that $a_n = n - t_n$ for all $n \le k-1$. If $k = 2l$, then $a_k = a_l + l = l - t_l + l = k - t_l = k - t_k$. If $k = 2l+1$, then $a_k = a_l + l = l - t_l + l = k - 1 - t_l = k - t_k$ and the statement follows. Because $\log_2(n+1) \ge t_n$, and $\lim_{n\to\infty} \frac{\log_2(n+1)}{n} = 0$, it follows that the sequence $\left\{ \frac{a_n}{n} \right\}_n$ converges to 1.

(Bulgarian Mathematical Olympiad, 2005)

5. Denote by $r(m)$ the length of the resting period before the mth catch. The problem says that $r(1) = 1$, $r(2m) = r(m)$, and $r(2m+1) = r(m)+1$. As seen in the introductory part to this section, $r(m)$ is equal to the number of 1's in the binary representation of m.

Denote also by $t(m)$ the moment of the mth catch and by $f(n)$ the number of flies caught after n minutes have passed. We notice that

$$t(m) = \sum_{i=1}^{m} r(i) \quad \text{and} \quad f(t(m)) = m$$

for every m. The following recurrence formulas hold:

$$t(2m+1) = 2t(m) + m + 1;$$
$$t(2m) = 2t(m) + m - r(m);$$
$$t(2^p m) = 2^p t(m) + p \cdot m \cdot 2^{p-1} - (2^p - 1)r(m).$$

The first follows from $\sum_{i=1}^{m} r(2i) = \sum_{i=1}^{m} r(i) = t(m)$ and $\sum_{i=0}^{m} r(2i+1) = 1 + \sum_{i=1}^{m}(r(i)+1) = t(m) + m + 1$. The second formula is justified by $t(2m) = t(2m+1) - r(2m+1) = 2t(m) + m - r(m)$. An easy induction on p proves the third formula.

(a) We have to find the first m such that $r(m+1) = 9$. The smallest number having 9 unit digits is $11\ldots1_2 = 2^9 - 1 = 511$, so the required m is 510.

(b) Using the above recurrence for $t(m)$, we obtain

$$t(98) = 2t(49) + 49 - r(49);$$
$$t(49) = 2t(24) + 25;$$
$$t(24) = 2^3 t(3) + 3 \cdot 3 \cdot 2^2 - (2^3 - 1)r(3);$$
$$r(1) = r(2) = 1, \quad r(3) = 2, \quad r(49) = r(110001_2) = 3.$$

Hence $t(3) = 4$, $t(24) = 54$, $t(49) = 133$, and $t(98) = 312$.

(c) Since $f(n) = m$ if and only if $n \in [t(m), t(m+1))$, we must find m_0 such that $t(m_0) \le 1999 < t(m_0+1)$. We start by computing more values of the function t:

$$t(2^p - 1) = t(2(2^{p-1} - 1) + 1) = 2t(2^{p-1} - 1) + 2^{p-1}.$$

Therefore, $t(2^p - 1) = p2^{p-1}$ and $t(2^p) = t(2^p - 1) + r(2^p) = p \cdot 2^{p-1} + 1$. Also,

$$t(\underbrace{11\ldots1}_{q}\underbrace{00\ldots0}_{p}) = t(2^p(2^q - 1))$$

$$= 2^p t(2^q - 1) + p(2^q - 1)2^{p-1} - (2^p - 1) \cdot r(2^q - 1)$$
$$= (p + q)2^{p+q-1} - p \cdot 2^{p-1} - q \cdot 2^p + q.$$

Now we can compute $f(2^8) = 8 \cdot 2^7 + 1 < 1999 < 9 \cdot 2^8 = t(2^9)$, so $2^8 < m_0 < 2^9$, which shows that the binary representation of m_0 has nine digits. Taking $q = 3$, $p = 6$, and $q = 4$, $p = 5$, we obtain $t(111000000_2) = 1923$ and $t(111100000_2) = 2100$. Therefore, the first binary digits of m_0 are 1110. Since $t(111010000_2) = 2004$, $t(111001111_2) = 2000$, and $t(111001110_2) = 1993$, it follows that $f(1999) = 111001110_2 = 462$.

(Short list 40th IMO, 1999; proposed by the United Kingdom)

6. An easy induction shows that, if $p_0, p_1, p_2, \ldots$ are the primes in increasing order and n has the base 2 representation $c_0 + 2c_1 + 4c_2 + \cdots$, then $x_n = p_0^{c_0} p_1^{c_1} p_2^{c_2} \cdots$. In particular $111111 = 3 \cdot 7 \cdot 11 \cdot 13 \cdot 37 = p_1 p_3 p_4 p_5 p_{11}$, so $x_n = 111111$ if and only if $n = 2^{11} + 2^5 + 2^4 + 2^3 + 2^1 = 2106$.

(Polish Mathematical Olympiad, 1996)

7. We would like to find an explicit description of the function f. The defining relations suggest that the binary expansion might be useful. With numbers written in base 2, we compute $f(1_2) = 1_2$, $f(10_2) = 1_2$, $f(11_2) = 11_2$, $f(100_2) = 1_2$, $f(101_2) = 101_2$, $f(110_2) = 11_2$, $f(1001_2) = 1001_2$, $f(1010_2) = 101_2$, $f(1011_2) = 1101_2, \ldots$. Eventually, one realizes that f inverts the order of digits, that is, $f(a_1 a_2 \ldots a_{k-1} a_k) = a_k a_{k-1} \ldots a_2 a_1$. Let us prove this by induction on n.

Assume that the property is true for all $m < n$ and let us prove it for n. If n is even, the property follows immediately from $f(2n) = f(n)$, since this tells us that the terminal zero is moved in front of the number. If $n = 4m + 1$ with $m = a_1 a_2 \ldots a_{k-1} a_k$, then $4m + 1 = a_1 a_2 \ldots a_{k-1} a_k 01$ and $2m + 1 = a_1 a_2 \ldots a_{k-1} a_k 1$. Thus

$$f(a_1 a_2 \ldots a_k 01) = 2f(2m + 1) - f(m)$$
$$= 1 a_k a_{k-1} \ldots a_2 a_1 0 - a_k a_{k-1} \ldots a_2 a_1$$
$$= 2^k + 2f(m) - f(m)$$
$$= 2^k + f(m) = 10 a_k a_{k-1} \ldots a_2 a_1.$$

Similarly, if $n = 4m + 3$ with $m = a_1 a_2 \ldots a_{k-1} a_k$, then we have $4m + 3 = a_1 a_2 \ldots a_{k-1} a_k 11$, and

$$f(a_1 a_2 \ldots a_k 11) = 3f(2m + 1) - 2f(m)$$
$$= f(2m + 1) + 2f(2m + 1) - 2f(m)$$
$$= 1 a_k a_{k-1} \ldots a_2 a_1 + 1 a_k a_{k-1} \ldots a_2 a_1 0 - a_k a_{k-1} \ldots a_2 a_1 0$$
$$= 1 a_k a_{k-1} \ldots a_2 a_1 + 2^{k+1}$$
$$= 11 a_k a_{k-1} \ldots a_2 a_1.$$

This proves our claim. Thus the fixed points of the function are the palindromic numbers, i.e., the numbers that remain unchanged when the order of their binary digits is reversed. For each k, there exist exactly $2^{\lfloor(k-1)/2\rfloor}$ palindromic numbers of length k. Indeed, the number is determined if we know its first $\lfloor(k+1)/2\rfloor$ digits, and the first digit is, of course, 1. The solutions to our equation are all palindromic numbers with at most 10 binary digits, together with those that have 11 digits and are less than $1988 = 2^{10} + 2^9 + 2^8 + 2^7 + 2^6 + 2^2$. There are $2(1 + 2 + \cdots + 16) = 62$ palindromic numbers in binary form with at most 10 digits. Also, a palindromic number with 11 digits is less than 1988 if and only if among its first five digits, at least one is equal to zero. There are $32 - 2 = 30$ such numbers, so the answer to the problem is 92.

(29th IMO, 1988; proposed by the United Kingdom)

8. From the statement, we deduce

$$a_n - a_{\lfloor n/2\rfloor} = \begin{cases} 1, & n \equiv 0,3 \pmod 4, \\ -1, & n \equiv 1,2 \pmod 4. \end{cases}$$

Since the binary representation of $\lfloor n/2\rfloor$ is obtained from that of n by deleting the last digit, we observe that a_n increases or decreases by 1 according as the last digit of n is the same as, or different from, the last digit of $\lfloor n/2\rfloor$. Let u_n denote the total number of pairs of consecutive zeros or consecutive ones in the binary expansion of n, and v_n the total number of pairs 10 or 01. Then

$$a_n = u_n - v_n.$$

(a) Of course, the number $n = 1111111111_2 = 1023$ has $v_n = 0$ and u_n maximal; hence has a_n maximal. Thus the maximum of a_n, with $n \leq 1996$, is 9. The number $m = 10101010101_2 = 1365$ has $u_m = 0$ and v_m maximal; hence for this number, the minimum of a_n is attained, which is equal to -10. These are the only examples where the extrema are attained.

(b) If $a_n = u_n - v_n = 0$, then n has an odd number of digits in the binary expansion. It can have $1,3,5,7,9,$ or 11 digits. If n has $2m + 1$ digits, then it must have exactly m pairs of consecutive different digits. Notice that specifying these pairs completely determines the number. Indeed, if we start from the left, the first pair changes the sequence of digits from 1 to 0, the second from 0 to 1, etc. Since for given m, the m pairs of consecutive different digits can be chosen in $\binom{2m}{m}$ ways, there are

$$\binom{0}{0} + \binom{2}{1} + \binom{4}{2} + \binom{6}{3} + \binom{8}{4} + \binom{10}{5} = 351$$

numbers n with at most 11 digits for which $a_n = 0$. From these, we must eliminate those between 1996 and 2047, which are

$$11111010010_2 = 2002, \ 11111010100_2 = 2004, \ 11111010110_2 = 2006$$
$$11111011010_2 = 2010, \ 11111101010_2 = 2026.$$

Thus the answer to the problem is 346.

(Short list 37th IMO, 1996)

9. A function with this property is called a Peano curve, after G. Peano, who in 1890 constructed a curve that passes through all points of a square. The example we have in mind was published by H. Lebesgue in 1928.

The example involves the Cantor set C, which can be constructed in the following way. One divides the interval $[0,1]$ into three equal intervals and removes the middle open interval. One divides each of the remaining closed intervals in three and removes again from each the middle open interval. The operation is repeated infinitely many times, and the set that is left after removing all those open intervals is the Cantor set.

An alternative description of C can be given by using the ternary expansion of numbers. As in the case of decimal expansions, the ternary expansion is sometimes ambiguous, for example, 1 can also be written as $0.2222\ldots_3$. We avoid this ambiguity by considering, whenever possible, the expansion with an infinite sequence of 2's in favor of the finite expansion ending with a 1. With this convention, the open interval $(\frac{1}{3},\frac{2}{3})$ consists exactly of those numbers that have their first ternary digit equal to 1, the open intervals $(\frac{2}{9},\frac{4}{9})$ and $(\frac{5}{9},\frac{5}{9})$ consist of those numbers that have their first ternary digit 0 or 2 and the second digit 1, etc. We conclude that in the process of removing open intervals, we remove the numbers that contain 1's in their ternary writing; hence C consists of those numbers that are written with only 0's and 2's.

For $x = 0.a_1a_2a_3a_4\ldots$, let $b_n = a_n/2$ and define

$$f(x) = (0.b_1b_3b_5\ldots, 0.b_2b_4b_6\ldots),$$

where the output of f is read in base 2. Note that the b_n's are 0 or 1, so the definition of f makes sense. Given a pair of numbers $(y_1,y_2) \in [0,1] \times [0,1]$, one can put their digits in an alternating sequence, then multiply the sequence by 2 and think of it as being the ternary expansion of a number. This shows that the function is onto (note that for 1 we use the binary expansion $0.11111\ldots$).

Let us prove that the function is continuous. The numbers in C have a unique ternary expansion that contains only 0's and 2's. If a sequence $\{x_n\}_n$ in C converges, then the difference of two terms of the sequence tends to 0 as their indices go to infinity. But the difference of two numbers in C whose digits differ at the nth position is at least $1/3^n$; hence the convergence implies the convergence of the digits. But then the convergence of the variable implies the convergence of the digits of the value of the function, so f is continuous.

Extend f over the open intervals that were removed as a linear function that joins the values of f at the endpoints of the interval to get a continuous function from the whole interval $[0,1]$ onto the square.

10. One computes $f(f(1)) = 3$; hence $f(1)$ cannot be 1 (otherwise, $f(f(1))$ would be 1 as well). Monotonicity implies $f(f(1)) > f(1)$, thus $1 < f(1) < 3$; hence $f(1) = 2$. Thus $f(2) = 3$, $f(3) = f(f(2)) = 6$, $f(6) = f(f(3)) = 9$; hence $f(4) = 7$ and $f(5) = 8$, $f(7) = f(f(4)) = 12$. We search for a pattern, but unfortunately, base 10 is not well suited for this. The presence of the number 3 in the definition of f suggests the use of base 3. Rewriting the above relations, we get $f(1_3) = 2_3$, $f(2_3) = 10_3$, $f(10_3) = 20_3$, $f(11_3) = 21_3$, $f(12_3) = 22_3$, $f(20_3) = 100_3$, $f(21_3) = 110_3$, etc. Thus f seems to be

unique, and its explicit definition is

$$f(1abc\ldots d_3) = 2abc\ldots d_3,$$
$$f(2abc\ldots d_3) = 1abc\ldots d0_3.$$

We will prove this formula by induction on the variable n of the function.

If n is of the form $f(m)$, then

$$f(2abc\ldots d_3) = f(f(1abc\ldots d_3)) = 1abc\ldots d0_3,$$

and

$$f(1abc\ldots d0_3) = f(f(2abc\ldots d_3)) = 2abc\ldots d0_3$$

shows that the property holds for n as well.

If n is not of the form $f(m)$, then it is of the form $1abc\ldots d$ with d equal to 1 or 2. We distinguish two cases. In the case where all digits of n except the first and the last are equal to 2, we have

$$f(122\ldots 20_3) = 222\ldots 20_3 < f(122\ldots 21_3)$$
$$< f(122\ldots 22_3) < f(200\ldots 00_3) = 1000\ldots 00_3.$$

A squeezing principle forces n to obey the formula. Otherwise, if n has more digits not equal to 2, then n starts again with a 1, and we have

$$f(1abc\ldots 0_3) = 2abc\ldots 0_3 < f(1abc\ldots 1_3)$$
$$< f(1abc\ldots 2_3) < f(1abc\ldots 0_3 + 10_3) = 2abc\ldots 0_3 + 10_3.$$

The squeezing argument shows that the formula holds again. This terminates the induction.

11. We will use a variation of Lebesgue's idea, similar to the one in the proof of Sierpiński's theorem. The function we construct is surjective when restricted to any interval, and the value of the function is read from the decimal expansion of the variable.

Let $x = 0.a_1a_2a_3\ldots$ be the decimal expansion of x. We read the value of the function from the digits in even positions in the following way. For numbers x with $a_{2n+1} = 0$ and $a_{2k+1} \neq 0$ for $k > n$, we define $f(x) = 0.a_{2n+2}a_{2n+4}\ldots a_{4n}\ldots$ if $a_{4n} \neq a_{2n}$ and $f(x) = 0.a_{2n+2}a_{2n+4}\ldots \hat{a}_{4n}\ldots$ where $\hat{a}_{4n} = 9 - a_{4n}$ if $a_{4n} = a_{2n}$. We also define $f(0) = 1$ and $f(x) = 0$ for all numbers to which the above algorithm cannot be applied.

Since the value of $f(x)$ depends only on the terminal structure of the digital expansion of x, in any interval we can find a number at which f attains a prescribed value. Moreover, the definition guarantees that x and $f(x)$ differ in some $2m$th position whenever the numerical algorithm is applied; hence in this case, $f(x)$ is different from x. Of course, for those x for which the algorithm is not applied, $f(x) = 0 \neq x$, except when $x = 0$, for which $f(0) = 1 \neq 0$; hence the function has no fixed point.

12. Yes, there is such a subset. If the problem is restricted to the nonnegative integers, it is clear that the set of integers whose representations in base 4 contains only the digits 0 and 1 satisfies the desired property. To accommodate the negative integers as well, we switch to "base -4." That is, we represent every integer in the form $\sum_{i=0}^{k} c_i(-4)^i$, with $c_i \in \{0,1,2,3\}$ for all i and $c_k \neq 0$. Let X be the set of numbers whose representations use only the digits 0 and 1. The set X will have the desired property, once we show that every integer has a unique representation in this fashion.

To show that base -4 representations are unique, let $\{c_i\}$ and $\{d_i\}$ be two distinct finite sequences of elements of $\{0,1,2,3\}$, and let j be the smallest integer such that $c_j \neq d_j$. Then

$$\sum_{i=0}^{k} c_i(-4)^i \not\equiv \sum_{i=0}^{k} d_i(-4)^i \pmod{4^j},$$

so in particular the two numbers represented by $\{c_i\}$ and $\{d_i\}$ are distinct. On the other hand, to show that n admits a base -4 representation, find an integer k such that $1 + 4^2 + \cdots + 4^{2k} \geq n$ and express $n + 4 + \cdots + 4^{2k-1}$ as $\sum_{i=0}^{2k} c_i 4^i$. Now set $d_{2i} = c_{2i}$ and $d_{2i-1} = 3 - c_{2i-1}$, and note that $n = \sum_{i=0}^{2k} d_i(-4)^i$. We are done.
   (USAMO, 1996; proposed by R. Stong)

13. The solution uses the following result.

Let $B = \{b_0, b_1, b_2, \ldots\}$, where b_n is the number obtained by writing n in base $p - 1$ and reading the result in base p. Then
(a) for every $a \notin B$, there exists $d > 0$ so that $a - kd \in B$ for $k = 1, 2, \ldots, p - 1$,
(b) B contains no p-term arithmetic progression.

The proof of this result proceeds as follows. First, note that $b \in B$ if and only if the representation of b in base p does not use the digit $p - 1$.

(a) Since $a \notin B$, when a is written in base p, at least one digit is $p - 1$. Let d be the positive integer whose representation in base p is obtained from that of a by replacing each $p - 1$ by 1 and each digit other than $p - 1$ by 0. Then none of the numbers $a - d, a - 2d, \ldots, a - (p-1)d$ has $p - 1$ as a digit when written in base p, and the result follows.

(b) Let $a, a + d, \ldots, a + (p-1)d$ be an arbitrary p-term arithmetic progression of nonnegative integers. Let δ be the rightmost nonzero digit when d is written in base p, and let α be the corresponding digit in the representation of a. Then $\alpha, \alpha + \delta, \ldots, \alpha + (p-1)\delta$ is a complete set of residues modulo p. It follows that at least one of the numbers $a, a + d, \ldots, a + (p-1)d$ has $p - 1$ as a digit when written in base p. Hence at least one term of the given arithmetic progression does not belong to B.

Let $\{a_n\}_{n \geq 0}$ be the sequence defined in the problem. To prove that $a_n = b_n$ for all $n \geq 0$, we use mathematical induction. Clearly, $a_0 = b_0 = 0$. Assume that $a_k = b_k$ for $0 \leq k \leq n - 1$, where $n \geq 1$. Then a_n is the smallest integer greater than b_{n-1} such that $\{b_0, b_1, \ldots, b_{n-1}, a_n\}$ contains no p-term arithmetic progression. By part (b) of the proposition, the choice of $a_n = b_n$ does not yield a p-term arithmetic progression with any of the preceding terms. It follows by induction that $a_n = b_n$ for all $n \geq 0$.
   (USAMO, 1995; proposed by T. Andreescu)

3.5   Residues

1. If $p = 2$, we have $2^2 + 3^2 = 13$ and $n = 1$. If $p > 2$, then p is odd, so $2^p + 3^p = (2 + 3)(2^{p-1} - 2^{p-2}3 + \cdots + 3^{p-1})$; hence 5 divides $2^p + 3^p$, and thus it divides w as well. Now, if $n > 1$, then 25 divides w^n; hence 5 divides

$$\frac{2^p + 3^p}{2 + 3} = 2^{p-1} - 2^{p-2}3 + \cdots + 3^{p-1}.$$

Since $(-1)^k 2^{p-k-1} 3^k$ is congruent to 2^{p-1} mod 5, for all k, the above sum is congruent to $p 2^{p-1}$ mod 5. But $p 2^{p-1}$ is divisible by 5 only if $p = 5$. However, $2^5 + 3^5 = 275$; hence $n = 1$ in this case as well.

(Irish Mathematical Olympiad, 1996)

2. We prove the stronger inequality by using residues mod 7. Let us first transform the given inequality as

$$\sqrt{7} - \frac{m}{n} > 0 \Leftrightarrow \sqrt{7} > \frac{m}{n} \Leftrightarrow 7n^2 - m^2 > 0.$$

The residue of m^2 mod 7 can be only 0, 1, 2, or 4, and since $7n^2 - m^2 > 0$, it follows that $7n^2 - m^2 \geq 7 - 4 = 3$. Hence $\sqrt{7}n \geq \sqrt{m^2 + 3}$. The inequality we need to prove is equivalent to

$$m + \frac{1}{m} \leq \sqrt{7}n.$$

For it to hold, it suffices to show that

$$m + \frac{1}{m} \leq \sqrt{m^2 + 3}.$$

This is obvious, since

$$m^2 + 3 \geq m^2 + 2 + \frac{1}{m^2} = \left(m + \frac{1}{m}\right)^2,$$

and we are done.

(Romanian Mathematical Olympiad, 1978; proposed by R. Gologan)

3. If $2\sqrt{28n^2 + 1} + 2$ is an integer, then $28n^2 + 1 = (2m + 1)^2$ for some nonnegative integer m. Then $7n^2 = m(m + 1)$, and since m and $m + 1$ are relatively prime, it follows that either $m = 7s^2$ and $m + 1 = t^2$ or $m = u^2$ and $m + 1 = 7v^2$. The second alternative is not possible, because $7v^2 - u^2 = 1$ does not have solutions, as can be seen reducing modulo 7 (see the previous problem). Thus $m + 1 = t^2$ and

$$2\sqrt{28n^2 + 1} + 2 = 2(2m + 1) + 2 = (2t)^2.$$

(J. Kűrshák Competition, 1969)

4. Suppose that the system has a nontrivial solution. Then, dividing by the common divisor of x, y, z, t, we can assume that these four numbers have no common divisor.

We add the two equations to get $7(x^2+y^2)=z^2+t^2$. The quadratic residues mod 7 are $0,1,2,4$. An easy check shows that the only way two residues can add up to 0 is if they are both equal to 0. Hence $z=7z_0$ and $t=7t_0$ for some integers z_0 and t_0. But then $x^2+y^2=7(z_0^2+t_0^2)$, which, by the same argument, implies that x and y are also divisible by 7. Thus each of x,y,z, and t is divisible by 7, a contradiction. Hence the system has no nontrivial solutions.

(W. Sierpiński, *250 problems in elementary number theory*, Państwowe Wydawnictwo Naukowe, Warszawa, 1970)

5. The sum of the digits of a number is congruent to the number modulo 9, so for a perfect square this must be congruent to 0, 1, 4, or 7. We show that any n that is a quadratic residue modulo 9 can occur as the digit sum of a perfect square. The cases $n=1$ and $n=4$ are trivial, so assume $n>4$.

If $n=9m$, then n is the sum of the digits of $(10^m-1)^2=10^{2m}-2\cdot10^m+1$, which looks like $9\ldots980\ldots01$. If $n=9m+1$, consider $(10^m-2)^2=10^{2m}-4\cdot10^m+4$, which looks like $9\ldots960\ldots04$. If $n=9m+4$, consider $(10^m-3)^2=10^{2m}-6\cdot10^m+9$, which looks like $9\ldots940\ldots09$. Finally, if $n=9m-2$, consider $(10^m-5)^2=10^{2m}-10^{m+1}+25$, which looks like $9\ldots900\ldots025$.

(Ibero–American Olympiad, 1995)

6. For y greater than 5, $y!$ is divisible by 9, so $y!+2001$ gives the residue 3 mod 9, which is not a quadratic residue. Hence the only candidates are $y=1,2,3,4,5$. Only $y=4$ passes, giving $x=45$.

7. Let us assume that there exist $m<n$ such that 2^m is obtained by permuting the digits of 2^n. Then 2^m and 2^n have the same number of digits, so $2^n<10\cdot2^m$. It follows that $n-m\le3$. On the other hand, 2^m and 2^n are congruent modulo 9, hence $2^n-2^m=2^m(2^{n-m}-1)$ is divisible by 9. But 2^m and 9 are coprime, while $2^{n-m}-1\le7$, which is impossible. It follows that the answer to the problem is negative.

(Iranian Mathematics Competition, 2001)

8. First we will show that the sum of the digits of B is quite small. From the inequality

$$4444^{4444}<10,000^{5000}$$

we deduce that 4444^{4444} has fewer than 20,000 digits; hence $A<9\cdot20,000=180,000$. Among the numbers less than 180,000, the one that has the largest sum of digits is 99,999 with the sum of digits equal to 45. Hence $B\le45$, so the sum of digits of B is at most the sum of digits of 39, which is 12. Thus we are looking at a number less than 12.

On the other hand, every number is congruent to the sum of its digits modulo 9. Therefore, the number we want to determine is congruent to 4444^{4444} modulo 9. But 4444 is congruent to $4+4+4+4=16$, hence to 7, modulo 9, and we have that $7^3\equiv1\pmod9$. It follows that

$$4444^{4444}\equiv7^{4444}\equiv7^{4443+1}\equiv7\pmod9,$$

so the solution to the problem is 7 (the only number less than 12 that is congruent to 7 mod 9).

(17th IMO, 1975)

9. We consider the equation modulo 11. Since $(x^5)^2 = x^{10} \equiv 0$ or 1 (mod 11), for all x, we have $x^5 \equiv -1, 0$ or 1 (mod 11). Thus the right side is either 6, 7, or 8 modulo 11. However, the square residues modulo 11 are 0, 1, 3, 4, 5, or 9 modulo 11, so the equation $y^2 = x^5 - 4$ has no integer solutions.

(Balkan Mathematical Olympiad, 1998)

10. The idea is to reduce modulo a number with very few residue classes that are cubes and fourth powers. One such number is 13, because the group of invertible elements in the field $\mathbf{Z}_{13}$ has order $12 = 3 \times 4$, and the perfect cubes and fourth powers form subgroups of orders 4 and 3, respectively.

Modulo 13, a cube gives the residues $0, 1, 5, 8, 12$, and a fourth power gives the residues $0, 1, 3, 9$. This shows that the sum of a cube and a fourth power can give only the residues $0, 1, 2, 3, 4, 5, 6, 8, 9, 10, 11, 12$ modulo 13. The residue 7 does not appear in this list.

By using Fermat's little theorem, we get $19^{19} \equiv 19^7 \equiv 6^7 \pmod{13}$. Since $6^7 = 2^7 \cdot 3^7 = 128 \cdot 2187$, an easy check shows that the residue of 6^7 mod 13 is 7, and we are done.

(P. Radovici-Marculescu, *Probleme de teoria elementară a numerelor (Problems in elementary number theory)*, Ed. Tehnică, Bucharest, 1986)

11. It is natural to try to reduce the given equation to a Pell-type equation. To complete the square, we multiply the equation by 4, then we rewrite it as

$$(2x + 3y)^2 - 17y^2 = 488.$$

Now we reduce this equation modulo 17. The quadratic residues modulo 17 are $0, 1, 4, 9, 16, 8, 2, 15, 13$, while the residue of 488 is 12; hence 488 cannot be the difference of a square and a multiple of 17. Therefore, the equation has no solutions.

(14th W.L. Putnam Mathematical Competition, 1954)

12. (a) It is not difficult to note that $1! + 2! + 3! = 3^2$. Let us show that this is the only possibility.

Suppose that there exists another n such that $1! + 2! + \cdots + n! = k^m$, for some $m > 1$. We can easily check that n must be greater than or equal to 9. For $n \geq 5$, the sum $1! + 2! + \cdots + n!$ ends in 3; hence m cannot be equal to 2 (squares can end only in $0, 1, 4, 5, 6$, or 9). This implies that $m \geq 3$. Also, for $n \geq 9$ the sum $1! + 2! + \cdots + n!$ is divisible by 3; hence k is divisible by 3. Thus k^m is divisible by 27, so the sum of the factorials is divisible by 27 as well. But since $a!$ is divisible by 27 for $a \geq 9$, the sum is congruent to $1! + 2! + \cdots + 8!$ modulo 27, and the latter is congruent to 9 modulo 27, a contradiction. This proves that $n = 3$ is the only answer.

(b) We can check that $(1!)^3 + (2!)^3 + (3!)^3 = 15^2$. Also $(1!)^3 + (2!)^3 + \cdots + (n!)^3$ is not a perfect power for $n = 2, 4, 5, 6$. For $n \geq 7$, the sum is congruent to $(1!)^3 + (2!)^3 + \cdots + (6!)^3$ modulo 49, and the latter is congruent to 7 modulo 49, which is not the residue of a perfect power.

(T. Andreescu)

13. We search for the smallest multiple of 29 that has the sum of digits as small as possible. No numbers with a single nonzero digit are divisible by 29. Of the numbers with two nonzero digits, the smallest sum of digits occurs when both nonzero digits are 1. The smallest such number must begin and end in 1, otherwise we could divide it by 10. Thus we are looking for the smallest number of the form $10^n + 1$ that is divisible by 29.

The powers of 10 modulo 29 are 1, 10, 13, 14, 24, 8, 22, 17, 25, 18, 6, 2, 20, 26, 28, 19, 16, 15, 5, 21, 7, 12, 4, 11, 23, 27, 9, 3, and we see that $10^{14} \equiv 28 \pmod{29}$. Therefore the desired number is $10^{14} + 1$, and $n = (10^{14} + 1)/29$.

(Lithuanian Mathematical Olympiad, 2000)

14. Set

$$m = 5^{5^{5^{5^5}}}.$$

Instead, it is better we first examine the residue of m modulo 2^5. Recall Euler's theorem, which states that if a and n are relatively prime, then $a^{\phi(n)} \equiv 1 \pmod n$.

We have

$$5^{\phi(2^5)} \equiv 1 \pmod{2^5}.$$

Here $\phi(2^5) = 2^4$, being equal to the number of odd numbers less than 2^5. Let us reduce the exponent $5^{5^{5^5}}$ modulo $2^4 = 16$. Repeating the same trick, we want to reduce its exponent modulo $\phi(16) = 8$. This amounts to computing 5^5 modulo 4, and this is easily seen to be 1. Going backwards, $5^{5^5} \equiv 5 \pmod 8$, $5^{5^{5^5}} \equiv 5 \pmod{16}$, and finally the given number is congruent to 5^5 modulo 2^5 is 5^5.

This shows that $m = 5^5 + 2^5 k$, where k is some integer. Since the left side is divisible by 5^5, it follows that k is divisible by 5^5 as well. Hence $m = 5^5 + 10^5 r$, for some r, and the residue of m modulo 10^5 is $5^5 = 3125$. The fifth-to-last digit of m is 0.

(P. Radovici-Marculescu, *Probleme de teoria elementară a numerelor (Problems in elementary number theory)*, Ed. Tehnică, Bucharest 1986)

15. The solution is similar to that of problem 2 in this section. It suffices to show that if $p/q < \sqrt{1998}$, then $p/q + 5/(pq) < \sqrt{1998}$. Put $n = 1998q^2 - p^2$. We will show that $n \notin \{1, \ldots, 10\}$, so that if $n > 0$, then in fact $n \geq 11$. Note that $1998 = 2 \times 27 \times 37$. Thus we have

$$n \equiv -p^2 \equiv 0, -1, -4, -7 \pmod 9$$

and so $n \neq 1, 3, 4, 6, 7, 10$. It then suffices to rule out $n = 2, 5, 8$, which we do by working modulo 37.

If $n = 2$, then

$$1 \equiv p^{36} \equiv (-2)^{18} = (-512)^2 \equiv 6^2 \equiv -1 \pmod{37},$$

a contradiction.

If $n = 5$, then

$$1 \equiv p^{36} \equiv (-5)^{18} = (-14)^6 \equiv 11^3 \equiv -1 \pmod{37},$$

again a contradiction.

If $n = 8$, the calculation for $n = 2$ implies that -8 also fails to be a quadratic residue modulo 37, a contradiction.

We conclude that $n \geq 11$, so that

$$1998 p^2 q^2 \geq p^2(p^2 + 11) > p^4 + 10p^2 + 25$$

for $p > 5$, and so $p/q + 5/(pq) < \sqrt{1998}$ in that case. For $p \leq 5$, we have $p/q + 5/(pq) \leq 10/q < \sqrt{1998}$, so the conclusion automatically holds in that case.

Remark. The value $n = 11$ is the smallest that cannot be ruled out by a congruence modulo 1998, since $787^2 + 11$ is divisible by 1998. However, using the continued fraction expansion of $\sqrt{1998}$, one can show that the smallest achievable value is $n = 26$, which occurs for $p = 134, q = 3$.

(Proposed by T. Andreescu and F. Pop for the IMO, 1998)

16. Since for $i \leq n - k$, $(i + k)$ and $(i + k) - k = i$ are both colored with the same color, it follows that the color of an element in M depends only on its residue class modulo k. For each i between 0 and $k - 1$, denote by A_i the set of elements of M that are congruent to i mod k. Also, for $i < k$, i and $k - i$ are colored by the same color, which implies that the residue classes A_i and A_{-i} are colored by the same color. Using also the fact that i and $n - i$ have the same color, we get inductively that all the residue classes in the sequence

$$A_i \to A_{n-i} \to A_{i-n} \to A_{2n-i} \to A_{i-2n} \to A_{3n-i} \to A_{i-3n} \to \cdots$$

are colored by the same color. In particular, for a fixed i, the residue classes of the form A_{rn-i} are all colored by the same color. Since k and n are relatively prime, these are all the residue classes mod k; hence M is monochromatic.

(26th IMO, 1985; proposed by Australia)

3.6   Diophantine Equations with the Unknowns as Exponents

1.   For $3^x - 2^y = 1$, we have the solutions $x = 1, y = 1$, and $x = 2, y = 3$. If $y \geq 2$, then x must be even. Setting $x = 2z$, $z > 1$ and $3^z = 2m + 1$, $m > 1$, yields $2^y = (2m + 1)^2 - 1 = 4m(4m + 1)$, a contradiction.

For $3^x - 2^y = -1$, working modulo 8 we see that $3^x + 1$ is equivalent to either 2 or 4 modulo 8. Hence the only possibilities are $y = 1$ or 2. Checking, we obtain $x = 1$, $y = 2$ as the only solution.

2.   Let us assume first that $y \geq 3$. Reducing mod 8, we deduce that 3^x must give the residue 7. However, 3^x can be congruent only to 3 or 1 mod 8, depending on the

parity of x. We are left with the cases $y = 1$ and $y = 2$, which are immediate. The only solution is $x = 2$, $y = 1$.

3. *First solution:* The case $n = 1$ clearly does not yield a solution, so let us assume $n > 1$. Without loss of generality, we may assume $x \leq y \leq z \leq t$. Dividing by n^x we get the equivalent equation

$$1 + n^{y-x} + n^{z-x} = n^{t-x}.$$

Reducing the equation mod n, we conclude that $y = x$. Thus the equation can be reduced to $2 + n^a = n^b$, where $a = z - x$ and $b = t - x$. If $a = 0$, then $n = 3, b = 1$, and we obtain the solution $(n,x,y,z,t) = (3,x,x,x,x+1)$, parameterized by $x \in \mathbf{N}$. If $a > 0$, then again by reducing mod n, we conclude that $n = 2$, which produces the solution $(n,x,y,z,t) = (2,x,x,x+1,x+2)$, $x \in \mathbf{N}$.

Second solution: Rewrite the equation as $n^{x-t} + n^{y-t} + n^{z-t} = 1$. Since each term on the left is positive, they are less than 1. Hence $x,y,z < t$. Thus the left-hand side is at most $3/n$, and we see $n = 2$ or 3. For $n = 3$, we must have equality so $n = 3$, $x = y = z = t - 1$. For $n = 2$, only one of the three terms can be $1/n$, and we find $x = y = t - 2$ and $z = t - 1$ and permutations.

(W. Sierpiński, *250 problems in elementary number theory*, Państwowe Wydawnictwo Naukowe, Warszawa, 1970, second solution by R. Stong)

4. (a) It is easy to see that $x = y = 0$ is a solution to the given equation. For positive integers x and y, rewrite the equation as

$$3^x = y^3 + 1.$$

The right side can be factored as $(y+1)(y^2 - y + 1)$. Since this product is a power of 3, each factor must be a power of three. Thus there exist two positive integers α and β, such that

$$y + 1 = 3^\alpha \quad \text{and} \quad y^2 - y + 1 = 3^\beta.$$

Squaring the first equality, and subtracting the second one from it gives $3y = 3^\beta(3^{2\alpha-\beta} - 1)$. From the initial equation, we deduce that y is not divisible by 3; hence $\beta = 1$. The equality $y^2 - y + 1 = 3^\beta$ then implies $y = 2$, and the initial equation gives $x = 2$. In conclusion, there are two solutions to the given equation: $x = y = 0$ and $x = y = 2$.

(Romanian Mathematical Contest, 1983; proposed by T. Andreescu)

(b) If (x,y) is a solution, then

$$p^x = y^p + 1 = (y+1)(y^{p-1} - y^{p-2} + \cdots + y^2 - y + 1),$$

and so $y + 1 = p^n$ for some n. If $n = 0$, then $x = y = 0$, and p may be arbitrary. If $n \geq 1$,

$$p^x = (p^n - 1)^p + 1$$

$$= p^{np} - p \cdot p^{n(p-1)} + \binom{p}{2} p^{n(p-2)} + \cdots - \binom{p}{p-2} p^{2n} + p \cdot p^n.$$

Since p is prime, all of the binomial coefficients are divisible by p. Hence all terms are divisible by p^{n+1}, and all but the last by p^{n+2}. Therefore, the highest power of p dividing the right side is p^{n+1}, and so $x = n+1$. We then have

$$0 = p^{np} - p \cdot p^{n(p-1)} + \binom{p}{2} p^{n(p-2)} + \cdots - \binom{p}{p-2} p^{2n}.$$

For $p = 3$ this reads $0 = 3^{3n} - 3 \cdot 3^{2n}$, which occurs only for $n = 1$, yielding $x = y = 2$. For $p \geq 5$, the coefficient $\binom{p}{p-2}$ is not divisible by p^2, so every term but the last on the right side is divisible by p^{2n+2}, whereas the last term is not. Since the terms sum to 0, this is impossible.

Hence the only solutions are $x = y = 0$ for all p and $x = y = 2$ for $p = 3$.
(Czech–Slovak Match, 1995)

Remark: A conjecture of Catalan proved by P. Mihăilescu states that the only consecutive powers of natural numbers are 8 and 9, i.e., that the unique solutions $x, y, m, n > 1$ of the equation $x^m - y^n = 1$ are $x = 3, y = 2, m = 2, n = 3$.

5. First we will examine the last digit of each term. The last digits of 1^n, 5^n, 6^n, 10^n, and 11^n are always 1, 5, 6, 0, and 1, respectively. Since the sum of the last digits of the terms on the right side is 12, it follows that 9^n must end in a 1; hence n is even. One can easily check that $n = 2$ and $n = 4$ are solutions.

For $n \geq 6$, we have

$$11^n + 6^n + 5^n > 11^n = (10+1)^n$$
$$= 10^n + n \cdot 10^{n-1} + \cdots + 1 \geq 10^n + 9^n + 1^n,$$

since

$$n \cdot 10^{n-1} \geq 6 \cdot 10^4 \cdot 10^{n-5} \geq 9^5 \cdot 10^{n-5} \geq 9^n.$$

This shows that the only solutions are $n = 2$ and $n = 4$.
(*Kőzépiskolai Matematikai Lapok (Mathematics Gazette for High Schools, Budapest)*)

6. If a solution exists, x is greater than 2. Reducing mod 4, we see that m must be odd. We obtain

$$2^x = z^m + 1 = (z+1)(z^{m-1} - z^{m-2} + \cdots + 1).$$

Since the left side is a power of 2, it follows that each of the factors on the right is a power of 2 as well; hence $z = 2^y - 1$ for some y. But from the hypothesis, $z^m = 2^x - 1$. These two equalities yield $2^x - 1 = (2^y - 1)^m$. Recall that m is odd, so when we expand using the binomial formula, we get a number of terms divisible by at least 2^{2y} plus the last two terms, which are $m2^y - 1$. Reducing modulo 2^y, and using the fact that $x > y$, it follows that $m2^y$ is congruent to zero, which is impossible, since m is odd. This proves that the equation has no solutions.

7. Remembering the existence of the Pythagorean triple $(5, 12, 13)$, one can guess that $x = y = 2$ and $z = 13$ is a solution. We will show that this is the only solution.

Reducing the equation modulo 5, we obtain $2^y \equiv z^2 \pmod{5}$. Since a perfect square is $0, 1$, or 4 modulo 5, it follows that y must be even. In particular, we have $y > 1$. Looking at the equation mod 3, we see that x is also even, so we may assume $x = 2k$, with k an integer. The equation can be rewritten as

$$12^y = (z - 5^k)(z + 5^k).$$

It follows that $z - 5^k = 2^\alpha 3^\beta$ and $z + 5^k = 2^\gamma 3^\delta$ for some nonnegative integers $\alpha, \beta, \gamma, \delta$. Subtracting the first equality from the second gives

$$2 \cdot 5^k = 2^\gamma 3^\delta - 2^\alpha 3^\beta.$$

Since the left side is not divisible by 3 and is divisible by 2 but not by 4, the same must be true for the right side. Thus the right side is either $2^{2y-1} - 2 \cdot 3^y, 2 \cdot 3^y - 2^{2y-1}$, or $2^{2y-1} 3^y - 2$.

Let us show that the first situation cannot occur. Indeed, if the right side has this form, then dividing by 2 we obtain $5^k = 4^{y-1} - 3^y$. Looking at this equality mod 4, we conclude that y must be odd. On the other hand, looking at it mod 5 gives y even, a contradiction.

Let us consider the case $5^k = 3^y - 4^{y-1}$. For the right side to be positive, y should be less than or equal to 2. This leads to the solution $x = y = 2, z = 13$.

Finally, the case $5^k = 2^{2y-2} 3^y - 1$ is excluded modulo 8. Hence we have the unique solution $x = y = 2, z = 13$.

8. Since $z = 1, 2$, or 3 does not yield a solution, we assume $z \geq 4$. Factoring, we obtain $(z + 1)(z - 1) = 2^x 3^y$. Since the numbers $z - 1$ and $z + 1$ differ by 2, at most one of them is divisible by 3. Also, since the product is even, the factors are both even, and exactly one of them is divisible by 4. Thus either $z + 1 = 2 \cdot 3^y$ and $z - 1 = 2^{x-1}$ or $z + 1 = 2^{x-1}$ and $z - 1 = 2 \cdot 3^y$.

By subtracting the two equations, the first case yields $2 = 2 \cdot 3^y - 2^{x-1}$; hence $3^y - 2^{x-2} = 1$, which appeared in Problem 1. This has the solution $x = 3, y = 1$, for which $z = 5$.

The second case leads to the equation $2^{x-2} - 3^y = 1$, which also appeared in Problem 1. We obtain the solution $x = 4, y = 1$, and $z = 7$.

(Romanian IMO Team Selection Test, 1984)

9. Either x or y is nonzero, and looking at the equality modulo 5 or modulo 7, we conclude that z must be even (in the first case it must be of the form $4k + 2$, in the second of the form $6k + 4$). Set $z = 2z_1$ and rewrite the equation as $5^x 7^y = (3^{z_1} - 2)(3^{z_1} + 2)$. The two factors are divisible only by powers of 5 and 7, and since their difference is 4, they must be relatively prime. Hence either $3^{z_1} + 2 = 5^x$ and $3^{z_1} - 2 = 7^y$ or $3^{z_1} + 2 = 7^y$ and $3^{z_1} - 2 = 5^x$.

In the first case, assuming $y \geq 1$, by subtracting the two equalities we get $5^x - 7^y = 4$. Looking at residues mod 7, we conclude that x is of the form $6k + 2$; hence even. But then, with $x = 2x_1$, we have $7^y = (5^{x_1} - 2)(5^{x_1} + 2)$. This is impossible, since the difference between the two factors is 4, and so they cannot both be powers of 7. It follows that $y = 0$, and consequently $x = 1$, $z = 2$.

In the second case, again by subtracting the equalities we find $7^y - 5^x = 4$. Looking modulo 5, we conclude that y must be even, and the same argument as above works *mutatis mutandis* to show that there are no solutions in this case.

(Bulgarian Mathematical Olympiad)

10. We start by noting that $(1,1,0)$, $(2,3,0)$, and $(3,3,1)$ are solutions. Let us show that there are no other solutions.

Clearly, any other solution must have $x \geq 2$, so by looking at the equation mod 9, we see that $y = 6k + 3$, where k is some integer.

As a consequence of problem 1, there are no other solutions with $z = 0$. Let us look for solutions with $z \geq 1$. Reducing the equation mod 19 gives $3^x \equiv 2^y \pmod{19}$. By Fermat's little theorem, $2^{18} \equiv 1 \pmod{19}$, so the residues of 2^y repeat with period 18. From the above, it follows that the only possible residues of 2^{6k+3} are those of 2^3, 2^9, and 2^{15}. The first one is 8, the second one is 18, and the third is 12. For the equation to be satisfied, these must also be the residues of 3^x.

Of course, we could check all residue classes mod 19. To simplify computations, note that 3^3 gives the same residue as 2^3, so the same is true for 3^9 and 3^{15}. If there is another number 3^a that gives the same residue as 3^b where $b = 3, 9$, or 15, then 3^{a-b} is congruent to 1 mod 19. If c is the greatest common divisor of $a - b$ and 18, then 3^c is congruent to 1. But c can be only $2, 3, 6$, or 9, and none of these numbers has the desired property. Hence x gives the residue 3, 9, or 15 modulo 18.

All this work was necessary in order to show that x and y are multiples of 3, so that we are able to factor the left side. Now let $x = 3m$ and $y = 3n$ with m, n nonnegative integers. We have

$$(3^m - 2^n)(3^{2m} + 3^m 2^n + 2^{2n}) = 19^z.$$

It follows that $3^m - 2^n = 19^\gamma$, $3^{2m} + 3^m 2^n + 2^{2n} = 19^\delta$. We square the first equality and subtract it from the second to get $3^{m+1} 2^n = 19^\delta - 19^{2\gamma}$. Since the left-hand side is not divisible by 19, this equality can hold only for $\gamma = 0$. Thus $3^m - 2^n = 1$. We saw in Problem 1 that the only solutions to this equation are $m = n = 1$ and $m = 2, n = 3$. Only the first leads to a solution of the initial equation, namely $x = y = 3$ and $z = 1$, which was already listed at the beginning.

(Proposed by R. Gelca for the USAMO, 1998)

11. Let us first observe that 2 and 3 are not quadratic residues modulo 5, 2 is a quadratic residue modulo 7, 3 and 6 are not quadratic residues modulo 7, 7 is a quadratic residue modulo 3, and 5 is not a quadratic residue modulo 3. With this in mind, we proceed as follows:

(a) Assume that all x, y, z, w are nonzero. As $2^x 3^y \equiv 1 \pmod{5}$, we deduce $x + y$ is even (because 1 is a square residue modulo 5 whereas 2, 3 are not). As $2^x 3^y \equiv 1 \pmod{7}$, we see that y is even, hence so is x. Next as $5^z 7^w \equiv -1 \pmod{3}$, we find that z is odd. If w were even, then $5^z 7^w + 1$ would be 2 modulo 8 and x would not be even. Thus w is odd, and we get the equation $2^{2x} 3^{2y} - 1 = 5^{2z+1} 7^{2w+1}$, thus $(2^x 3^y - 1)(2^x 3^y + 1) = 5^{2z+1} 7^{2w+1}$. From here we deduce $\{2^x 3^y - 1, 2^x 3^y + 1\} = \{5^{2z+1}, 7^{2w+1}\}$.

For $z = 0$, we obtain immediately the solution $2^2 3^2 - 5 \cdot 7 = 1$. For $z > 0$, we note that $7^2 \equiv -1 \pmod{5^2}$ so 7^{2w+1} is ± 7 modulo 5^2, so we cannot have $5^{2z+1} - 7^{2w+1} = \pm 2$. This case is exhausted.

(b) Assume now that only w is zero. We obtain $2^x 3^y - 5^z = 1$. We conclude again that z is odd, so $2^x 3^y = 5^z + 1$, and we deduce that $x = 1$. Thus $2 \cdot 3^y = 5^z + 1$. For $z = 1$, we get $y = 1$. For $z > 1$, we get that $y > z$. However, it is easy to prove by induction on k that $5^z + 1$ is divisible by 3^k if and only if z is odd and is divisible by 3^{k-1} (this is done proving by induction that the exact power of 3 dividing $5^{3^{k-1}} + 1$ is 3^k). Thus z must be divisible by $3^{y-1} \geq 3^z > z$, a contradiction.

(c) Assume that only z is zero. We obtain the equation $2^x 3^y = 7^w + 1$, which is impossible modulo 3.

(d) Assume that only y is zero. We get $2^x - 5^z 7^w = 1$, and again x is even so we have the equation $2^{2x_1} - 1 = 5^z 7^w$, so $(2^{x_1} - 1)(2^{x_1} + 1) = 5^z 7^w$. Clearly, $x_1 > 1$, then modulo 8 we get that w is odd. Then we get $\{5^z, 7^w\} = \{2^{x_1} - 1, 2^{x_1} + 1\}$, but as in (a) we deduce that for $z > 1$, we cannot have $|5^z - 7^w| = 2$ for w odd, and $z = 1$ yields $w = 1$, which does not give a solution.

(e) If only x is zero, we obtain $3^y - 5^z 7^w = 1$. Again we deduce that y is even, thus we get $3^{2y_1} - 1 = 5^z 7^w$, that is, $\{3^{y_1} - 1, 3^{y_1} + 1\} = \{5^z, 7^w\}$. Hence $3^{y_1} + 1 = 7^w$, $3^{y_1} - 1 = 5^z$. The equation $3^{y_1} - 1 = 5^z$ is solved exactly as the equation $2 \cdot 3^{y_1} = 5^z + 1$ from (b) and has no solutions.

(f) If two of x, y, z, w are zero, we obtain one of the four equations $2^x - 5^z = 1$, $2^x - 7^w = 1, 3^y - 5^z = 1, 3^y - 7^w = 1$. The equation $3^y - 5^z = 1$ was already solved. If $2^x - 5^z = 1$, we get x even, so $2^{2x} - 1 = 5^z$, or $(2^x - 1)(2^x + 1) = 5^z$, which is impossible. If $2^x - 7^w = 1$, then either w is even and $7^w + 1$ is 2 mod 4, so $x = 1$ and $w = 0$, which is impossible, or w is odd and $7^w + 1$ is 8 modulo 16, so $x = 3, w = 1$. Finally $3^y - 7^w = 1$ has no solutions, as $7^w + 1$ is 2 modulo 3.

(g) The final case: three of x, y, z, w equal to zero yields the only possibility: $2^1 \cdot 3^0 - 5^0 7^0 = 1$. The problem is solved.

(Chinese Mathematical Olympiad, 2005)

12. Let us assume that there exist positive integers x, y, z such that $x^x + y^y = z^z$. Then $z^z > x^x$ and $z^z > y^y$, so $z > x$ and $z > y$. Because x, y, z are integers, this implies that $z \geq x + 1$ and $z \geq y + 1$. It follows that

$$z^z \geq (x+1)^{x+1} = (x+1)(x+1)^x \geq 2(x+1)^x \geq 2x^x.$$

Similarly, one proves that $z^z > 2y^y$. Adding the two inequalities and dividing by 2, we obtain $z^z > x^x + y^y$, a contradiction. Hence the equation does not admit positive integer solutions.

(Russian Mathematical Olympiad, 1977–1978)

13. Examining the sign of both sides, we deduce that $y^x < 19$. The case $y = 1$ is ruled out by

$$x^{x^{x^x}} = (19 - 1) \cdot 1 - 74 < 0.$$

Hence $y \geq 2$. But then $x \leq 4$.

For $x = 1$, the equation becomes $1 = (19 - y)y - 74$, or $y^2 - 19y + 75 = 0$, which has no integer solutions. If $x > 2$, by requiring that $19 - y^x$ be positive, we obtain the cases $(x,y) = (2,2), (2,3), (3,2), (4,2)$. Performing the computations, we see that only $x = 2$, $y = 3$ yields a solution.

(Leningrad Mathematical Olympiad)

14. We will use the fact that for each $k > 0$ and each positive integer n, the inequality $(1 + k/n)^n < e^k$ holds. Let (x,y) be a solution. Since obviously $x \neq y$, we distinguish two cases.

Case I: $x > y$. Set $x = y + k$ for some $k \geq 1$ and write the left side as

$$x^y - y^x = (y+k)^y - y^{y+k} = y^y \left[\left(1 + \frac{k}{y}\right)^y - y^k \right].$$

Assuming that $y \geq 3$, we obtain $(1 + k/y)^y < e^k < 3^k \leq y^k$. It follows that $x^y - y^x$ is negative, a contradiction. Thus we are left with the possibilities $y = 1$ and $y = 2$, the first one leading to the solution $x = 2$, $y = 1$. If $y = 2$, we have to solve the equation $x^2 - 2^x = 1$. It can be rewritten as $(x-1)(x+1) = 2^x$, and we conclude that $x - 1 = 2$ and $x + 1 = 4$. Thus $(x,y) = (3,2)$.

Case II: $x < y$. Then $y = x + k$, where $k \geq 1$, so

$$x^y - y^x = x^{x+k} - (x+k)^x = x^x \left[x^k - \left(1 + \frac{k}{x}\right)^x \right].$$

If $x \geq 3$, then $x^x \geq 3^3$, $x^k \geq 3^k$. Combined with $(1 + k/x)^x < e^k$, these imply

$$x^y - y^x > 3^3 \left[3^k - e^k \right] \geq 3^3(3 - e) > 1$$

(the last inequality follows from, say, $e < 2.8$). Thus the only possibilities are $x = 1$ and $x = 2$. They can be handled as in the previous case but yield no solutions.

(Communicated by S. Savchev)

3.7 Numerical Functions

1. The idea is to use the given identity in order to write a recursive relation for f. This can be done by plugging $n = 1$ in the relation. We deduce that $f(m+1) = f(m) + m + 1$. An easy induction yields $f(m) = m(m+1)/2$.

(*Matematika v Škole (Mathematics in school)*, 1982)

2. If we let $g : \mathbf{N_0} \to \mathbf{N_0}$, $g(n) = n + (-1)^n$, then g satisfies the equation. Moreover, g is bijective. We will show that for any solution f, we must have $f = g$.

In fact, we will prove a more general property, namely that if f and g are two functions defined on the nonnegative integers such that $f(n) \geq g(n)$ for all n, and f is surjective and g bijective, then $f = g$. The proof is based on the well ordering of the set of positive integers, namely on the fact that any set of positive integers has a smallest element.

Assume $f \neq g$, and let n_0 be such that $f(n_0) > g(n_0)$. If we let $M = g(n_0)$, then the set $A = \{k, \ g(k) \leq M\}$ has exactly $M + 1$ elements, since g is bijective. On the other hand, since $f \geq g$ and n_0 does not belong to A, the set $B = \{k, \ f(k) \leq M\}$ is included in A but has at least one less element, namely n_0. Hence the values of f do not exhaust all numbers less than $M + 1$, which contradicts the surjectivity of f. Therefore, $f(n) = g(n) = n + (-1)^n$ is the only solution.

(Romanian contest, 1986; proposed by M. Burtea)

3. If $f(a) = f(b)$, then $a + 1 - f(f(a) + f(1)) = f(f(b) + f(1)) = b + 1$ or $a = b$. Hence f is injective. For all $m \geq 1$, $f(f(m) + f(1)) = m + 1$. Hence the inputs $f(1) + f(m)$ that are all at least 2 give every possible output bigger than 1. Thus by process of elimination, $f(1) = 1$. Now we show by induction on k that $f(k) = k$. Since $f(m) = m$ for $m < k$, by injectivity $f(m) \geq k$ for $m \geq k$. Hence the inputs $f(1) + f(m)$ with $m \geq k$ are all at least $k + 1$ and give all outputs $f(f(m) + f(1)) = m + 1$ greater than k. Thus again by the process of elimination, $f(k) = k$.

(G. Andrei, I. Cucurezeanu, C. Caragea and G. Bordea, *Exerciţii şi probleme de algebră (Exercises and problems in algebra)*, Universul, Constanţa, 1990)

4. Arguing by contradiction, let us assume that such a function f exists. Then the function $n \to g(5n + 2)$ would be increasing, while the function $n \to h(3n + 1)$ would be decreasing. Thus the function $n \to g(5n + 2) - h(3n + 1)$ would be increasing. But

$$g(5n + 2) - h(3n + 1) = f(15n + 7) - 5n - 2 - f(15n + 7) + 3n + 1$$
$$= -2n - 1,$$

which is strictly decreasing. This contradiction implies that a function f with the required property does not exist, and we are done.

(C. Mortici, *Probleme Pregătitoare pentru Concursurile de Matematică (Training Problems for Mathematics Contests)*, GIL, 1999)

5. One possibility is that f is identically equal to 0 and g is arbitrary. Another possibility is that g is identically equal to zero and $f(n) = 2^n f(0)$.

Let us find the remaining pairs of functions. Note that the identity implies $f(n + 1) \geq f(n)$ for all n; hence f is nondecreasing. If for a certain n, $g(n) \geq 1$, then $f(n + 1) \leq f(n + g(n))$; hence $f(n) = 0$. A backwards induction shows that $f(n - 1) = f(n - 2) = \cdots = f(0) = 0$.

Hence in order for f not to be identically zero, there must exist m such that $g(k) = 0$ for all $k \geq m$. Assume m minimal, that is, $g(m - 1) \neq 0$. Then, on the one hand, $f(k) = 0$ for $k \leq m - 1$, and on the other hand, $f(k) = 2^{k-m} f(m)$ for $k > m$, so for $k \geq m$ the function is strictly increasing. We have $f(n) + f(n + g(n)) = f(n + 1)$, and by setting $n = m - 1$, we obtain $f(m - 1 + g(m - 1)) = f(m)$, so $g(m - 1) = 1$. Taking $n = k < m - 1$, we get $f(k) + f(k + g(k)) = f(k + 1)$, hence $f(k + g(k)) = 0$ and $g(k) < m - k$. Thus all other solutions (f, g) satisfy $f(0) = f(1) = \cdots = f(m - 1) = 0$, $f(k) = 2^{(k-m)}a$ for $k \geq m$ and a arbitrary, and $g(k) < m - k$, for $k \leq m$, $g(m - 1) = 1$, and $g(k) = 0$ for $k \geq m$.

(Mathematical Olympiad Summer Program, 1996)

6. Clearly, $f = 0$ is a solution; hereafter we assume that f is not identically 0. Setting $m = n = 0$, we obtain $f(0) = 0$. Now setting only $n = 0$, we find that $f(f(m)) = f(m)$ for all $m \geq 0$.

Let T be the range of f. By previous observation, T consists precisely of those n for which $f(n) = n$. If $m, n \in T$, the functional equation implies $f(m + n) = m + n$, so $m + n \in T$. Conversely, if $m, n \in T$, and $m \geq n$, then $f(m) + f(n - m) = f(n) = n$, yielding $f(n - m) = n - m$, so $n - m \in T$. It is a standard fact that T consists of all multiples of a certain natural number a. For completeness, we will prove this fact below.

Let a be the smallest nonzero element in T; such an element exists because f is not identically zero. Then every multiple of a belongs to T. Conversely, any $m \in T$ can be expressed as $qa + r$ with $0 \leq r < a$, and since m and qa belong to T, so does r. However, a is the smallest nonzero element of T and $r < a$, so we must have $r = 0$. Therefore, T consists of precisely the multiples of a.

Since $f(n)$ is a multiple of a for all n, we can write $f(r) = n_r a$ for $r = 0, 1, \ldots, a - 1$, where $n_0 = 0$. Then for every m expressed as $qa + r$ with $0 \leq r < a$,

$$f(m) = f(qa + r) = f(r + f(qa)) = f(f(r)) + f(qa) = (n_r + q)a.$$

On the other hand, any function of the form $f(qa + r) = (n_r + q)a$ satisfies the given equation, for if $m = qa + r$ and $n = sa + t$ with $0 \leq r, t < a$, then

$$f(m + f(n)) = f(qa + r + sa + n_t a)$$
$$= qa + sa + n_t a + n_r a$$
$$= f(qa + n_r a) + f(sa + t) = f(f(m)) + f(n).$$

Hence the desired functions are $f = 0$ and all the functions of the form given above.
(37th IMO, 1996; proposed by Romania)

7. Fix n and consider the sequence $\{x_n\}_n$ defined by $x_0 = n$, $x_k = f^{(k)}(n)$, $k \geq 1$, where $f^{(k)}$ denotes f composed with itself k times. This sequence satisfies the linear recursive relation

$$6x_{n+1} = 5x_n - x_{n-1}, \quad n \geq 1.$$

To find the general term formula, we associate to the recursive relation its characteristic equation

$$6\lambda^2 - 5\lambda + 1 = 0.$$

The roots of this equation are $\lambda_1 = \frac{1}{2}$ and $\lambda_2 = \frac{1}{3}$. It follows that the general term of the sequence is of the form $x_n = c_1 \left(\frac{1}{2}\right)^n + c_2 \left(\frac{1}{3}\right)^n$, where c_1 and c_2 are determined by the initial condition $c_1 + c_2 = x_0 = n$ and $\frac{1}{2}c_1 + \frac{1}{3}c_2 = x_1 = f(n)$. But

$$\lim_{n \to \infty} x_n = \lim_{n \to \infty} c_1 \left(\frac{1}{2}\right)^n + c_2 \left(\frac{1}{3}\right)^n = 0,$$

which is impossible since the terms of the sequence are the positive integers $n, f(n)$, $f(f(n)), \ldots$. We conclude that such a function does not exist.

8. This problem might look easy to people familiar with the axiomatic description of the set of positive integers. The solution uses again the property that every set of natural numbers has a smallest element and is similar to the one given above to Problem 3.

Let us look at the set

$$\{f(f(1)), f(2), f(f(2)), f(3), f(f(3)), \ldots, f(n), f(f(n)), \ldots\}.$$

Note that these are exactly the numbers that appear in the inequality $f(f(n)) < f(n+1)$. This set has a smallest element, which cannot be of the form $f(n+1)$ because then it would be larger than $f(f(n))$. Thus it is of the form $f(f(n))$. The same argument shows that for this n, $f(n) = 1$. If n itself were greater than 1, we would get $1 = f(n) > f(f(n-1))$, which is impossible. Hence $f(1) = 1$ and $f(n) > 1$ for $n > 1$.

Considering the restriction $f : \{n \geq 2\} \to \{n \geq 2\}$, the same argument applies *mutatis mutandis* to show that $f(2) = 2$ and $f(n) > 2$ for $n > 2$. By induction, one shows that $f(k) = k$, and $f(n) > k$ for $n > k$, thus the unique solution to the problem is the identity function.

(19th IMO, 1977)

9. The equality from the statement reminds us of the well-known identity

$$\frac{1}{1 \cdot 2} + \frac{1}{2 \cdot 3} + \cdots + \frac{1}{n(n+1)} = \frac{n}{n+1},$$

which shows that the function $f : \mathbf{N} \to \mathbf{N}$, $f(n) = n$ is a solution. Let us prove that this is the only function with the required property.

The ratio $f(f(n))/f(n+1)$ reminds us of the previous problem. In fact, we will reduce the current problem to the previous one.

Plugging in $n = 1$ into the given relation yields $f(f(1))f(1) = 1$; hence $f(1) = 1$. Replacing the given equality for n into the one for $n+1$, we obtain

$$\frac{f(f(n))}{f(n+1)} + \frac{1}{f(n+1)f(n+2)} = \frac{f(f(n+1))}{f(n+2)}.$$

This is equivalent to

$$f(f(n))f(n+2) + 1 = f(f(n+1))f(n+1).$$

Note that $f(n+1) = 1$ implies that $f(f(n+1)) = 1$; hence $f(f(n))f(n+2) = 0$, which is impossible. Therefore, $f(n) > 1$ for $n > 1$.

We use induction to show that $f(f(n)) < f(n+1)$. The inequality is true for $n = 1$, since $f(2) > 1 = f(f(1))$. Also, if $f(n+1) > f(f(n))$, then $f(n+1) \geq f(f(n)) + 1$. Hence

$$f(f(n))f(n+2) + 1 \geq f(f(n+1))f(f(n)) + f(f(n+1)).$$

Since $n+1 > 1$, we have $f(n+1) > 1$, thus $f(f(n+1)) > 1$, which implies that $f(n+2) > f(f(n+1))$.

Therefore the function satisfies $f(n+1) > f(f(n))$ for all $n \in \mathbf{N}$. In view of Problem 6, the only function with this property is the identity function, and we are done.

(R. Gelca)

10. The idea is to look at the orbits of the function. First, observe that in the equality $f(f(n)) = n + 1987$, the right side is a one-to-one function. Consequently, f is one-to-one.

Define $g(n) = f(f(n)) = n + 1987$. Considering the orbit of k through g (i.e., the set $\{ k, g(k), g(g(k)), \ldots \}$) for all $k < 1988$, we see that the union of these orbits exhausts $\mathbf{N}$.

It follows that for each k there exist an m and n, with $n < 1988$, such that $f(k) = g^{(m)}(n)$. From the injectivity of f, it follows that either $m = 0$ or $m = 1$. Thus either $f(k) = n$ or $f(k) = n + 1987$. In the latter case, injectivity implies $f(n) = k$. This shows that all integers from 1 to 1987 can be arranged in pairs of the form $(n, f(n))$. This is impossible, since there are 1987 such numbers, and 1987 is odd. Hence there exists no such function f.

(28th IMO, 1987)

11. Substituting successively $m = 0$ and $n = 0$ in (a) and subtracting the two relations yields

$$f(m)^2 - f(n)^2 = 2(f(m^2) - f(n^2)),$$

which together with (b) implies that f is increasing, i.e., if $m \geq n$, then $f(m) \geq f(n)$. Plugging $m = n = 0$ into (b) yields $f(0) = 0$ or 1.

Case 1. $f(0) = 1$. Then $2f(m^2) = f(m)^2 + 1$, so $f(1) = 1$. Plugging $m = n = 1$ in (a), we get $f(2) = 1$. Also,

$$f\left(2^{2^n}\right) = \frac{1}{2}\left(f\left(2^{2^{n-1}}\right)^2 + 1\right).$$

This implies that $f(2^{2^k}) = 1$ for all nonnegative integers k. By the monotonicity of f, we conclude that $f(n) = 1$ for all nonnegative integers n.

Case 2. $f(0) = 0$. Then $2f(m^2) = f(m)^2$, or $f(m^2)/2 = (f(m)/2)^2$. Since $f(2) = f(1)^2$, we obtain

$$\frac{f(2^{2^n})}{2} = \left(\frac{f(2^{2^{n-1}})}{2}\right)^2 = \left(\frac{f(2^{2^{n-2}})}{2}\right)^{2^2} = \cdots$$

$$= \left(\frac{f(2)}{2}\right)^{2^n} = \frac{f(1)^{2^{n+1}}}{2^{2^n}}.$$

On the other hand, (a) implies that $2f(1) = f(1)^2$, so either $f(1) = 0$ or $f(1) = 2$. If $f(1) = 0$, the above chain of equalities implies that $f(2^{2^n}) = 0$ for $n \geq 0$. Monotonicity implies that f is identically equal to zero.

If $f(1) = 2$, then $f(2^{2^n}) = 2 \cdot 2^{2^n}$. Since $f(m^2)/2 = (f(m)/2)^2$, $f(m)$ is always even. We have

$$f(m+1)^2 = 2f((m+1)^2) \geq 2f(m^2+1) = f(m)^2 + f(1)^2 > f(m)^2,$$

which implies that $f(m+1) > f(m)$. Consequently, $f(m+1) - f(m) - 2 \geq 0$. But

$$\sum_{m=0}^{2^{2^n}-1} (f(m+1) - f(m) - 2) = f(2^{2^n}) - f(0) - 2 \cdot 2^{2^n} = 0.$$

Varying n, we conclude that $f(m+1) = f(m) + 2$ for all $m \geq 0$. Thus $f(n) = 2n$ for all $n \in \mathbf{N}_0$.

In conclusion, $f(n)$ identically equal to zero, $f(n)$ identically equal to 1, or $f(n) = 2n$ for all n, are the only possible solutions.
(Korean Mathematical Olympiad, 1996)

12. One can easily see that $f(n) = n$ satisfies the given property. Let us show that this is the only function. The proof is based on factorizations of positive integers.

We start by computing the value of $f(3)$. Since the function is increasing, $f(3)f(5) = f(15) < f(18) = f(2)f(9)$, hence $f(3)f(5) < 2f(9)$ and $f(9) < f(10) = f(2)f(5) = 2f(5)$. Combining the two inequalities, we get $f(3)f(5) < 4f(5)$; hence $f(3) < 4$. We also have that $f(3) > f(2) = 2$; thus $f(3)$ can be equal only to 3.

Since 2 and 3 are relatively prime, we deduce that $f(6) = 6$, and from monotonicity it follows that $f(4) = 4$ and $f(5) = 5$. We will prove by induction that $f(n) = n$ for all $n \in \mathbf{N}$. For $n = 1, 2, 3, 4, 5, 6$, the property is true, as shown above. Now suppose there is an n with $f(n-1) = n-1$ and $f(n) = n$. Since $n-1$ and n are relatively prime, we conclude $f(n(n-1)) = n(n-1)$. Since f is strictly increasing, it follows that $f(m) = m$ for $m \leq n(n-1)$. Define the sequence (a_k) by $a_1 = 3$ and $a_{k+1} = a_k(a_k - 1)$. Then it follows by induction on k that $f(m) = m$ for $m \leq a_k$. Since this sequence tends to infinity (for example, because $a_{k+1} \geq 2a_k > 2^{k+1}$), it follows that $f(m) = m$ for all m.
(24th W.L. Putnam Mathematical Competition, 1963)

13. The solution, as in the case of the previous problem, uses the factorization of positive integers. Suppose that a function f having the required property has been found. We use f to define a function $g : 3\mathbf{N}_0 + 1 \to 4\mathbf{N}_0 + 1$ by

$$g(x) = 4f\left(\frac{x-1}{3}\right) + 1.$$

This is certainly well-defined, and one can check immediately that g is a bijection from $3\mathbf{N}_0 + 1$ onto $4\mathbf{N}_0 + 1$, with the inverse function given by

$$g^{-1}(y) = 3f^{-1}\left(\frac{y-1}{4}\right) + 1.$$

For $m, n \in \mathbf{N_0}$, by using the definition of f and g, we obtain

$$
\begin{aligned}
g((3m+1)(3n+1)) &= g(3(3mn+m+n)+1) = 4f(3mn+m+n)+1 \\
&= 4(4f(m)f(n)+f(m)+f(n))+1 \\
&= (4f(m)+1)(4f(n)+1) = g(3m+1)g(3n+1).
\end{aligned}
$$

Thus g is multiplicative, in the sense that $g(xy) = g(x)g(y)$ for all $x, y \in 3\mathbf{N_0}+1$.

Conversely, given any multiplicative bijection from $3\mathbf{N_0}+1$ onto $4\mathbf{N_0}+1$, we can construct a function f having the required property by letting $f(x) = (g(3x+1)-1)/4$.

It remains only to exhibit such a bijection. Let P_1 and P_2 denote the sets of primes of the form $3n+1$ and $3n+2$, respectively, and let Q_1 and Q_2 denote the sets of primes of the form $4n+1$ and $4n+3$, respectively. Since each of these sets is infinite, there exists a bijection h from $P_1 \cup P_2$ to $Q_1 \cup Q_2$ that maps P_1 bijectively onto Q_1 and P_2 onto Q_2. Define g as follows: $g(1) = 1$, and for $n > 1$, $n \in 3\mathbf{N_0}+1$, let the prime factorization of n be $n = \prod p_i$ (with possible repetitions among the p_i's), then define $g(n) = \prod h(p_i)$.

Note that g is well-defined, because if $n \in 3\mathbf{N_0}+1$, then there must be an even number of P_2-type primes that divide n. Each of these primes gets mapped by h to a prime in Q_2, and since there are an even number of such primes, their product lies in $4\mathbf{N_0}+1$. The multiplicativity of g follows easily.

(Short list 36th IMO, 1995; proposed by Romania)

14. Such a function does exist. Let $P(n) = n^2 - 19n + 99$, and note that $P(n) = P(19-n)$ and that $P(n) \geq 9$ for all $n \in \mathbf{N}$. We first set $f(9) = f(10) = 9$ and $f(8) = f(11) = 11$. (One could alternatively set $f(9) = f(10) = 11$ and $f(8) = f(11) = 9$.)

Write $P^{(k)}(n)$ for the kth composite of P. That is, $P^{(0)}(n) = n$ and $P^{(k+1)}(n) = P(P^{(k)}(n))$. For $n \geq 12$, let $g(n)$ be the smallest integer k such that n is not in the image of $P^{(k)}$. Such a k exists because aside from 9 and 11, every integer in the image of $P^{(k)}(n)$ for $k > 0$ is greater than or equal to $P^{(k)}(12)$, and an easy induction shows that $P^{(k)}(n) \geq n+k$ for $n \geq 12$.

Let $12 = s_1 \leq s_2 \leq \cdots$ be the integers greater than or equal to 12 not in the image of P, in increasing order. Then for every integer $n \geq 12$, there exists a unique integer $h(n)$ such that $n = P^{(g(n))}(s_{h(n)})$. For $n \geq 12$, set

$$
f(n) = \begin{cases} P^{(g(n))}(s_{h(n)+1}) & h(n) \text{ odd} \\ P^{(g(n)+1)}(s_{h(n)-1}) & h(n) \text{ even.} \end{cases}
$$

For $n \leq 7$, put $f(n) = f(19-n)$. To show that $f(f(n)) = P(n)$, we need only consider $n \geq 12$, and we may examine two cases. If $h(n)$ is odd, then $g(f(n)) = g(n)$ and $h(f(n)) = h(n)+1$ is even, so

$$
f(f(n)) = f(P^{(g(n))}(s_{h(n)+1})) = P^{(g(n)+1)}(s_{h(n)}) = P(n).
$$

Similarly, if $h(n)$ is even, then $g(f(n)) = g(n+1)$ and $h(f(n)) = h(n)-1$ is odd, so

$$
f(f(n)) = f(P^{(g(n)+1)}(s_{h(n)-1})) = P^{(g(n)+1)}(s_{h(n)}) = P(n).
$$

(Proposed by T. Andreescu for the IMO, 1999)

15. Setting $m = n = 1$ in the relation from the statement, we find that $f^2(1) + f(1)$ is a (positive) divisor of $(1^2 + 1)^2$. The equation $t^2 + t = 4$ has no integer roots, moreover, the number $f^2(1) + f(1)$ is greater than one. The only possibility is $f^2(1) + f(1) = 2$, so $f(1) = 1$.

Now set just $m = 1$ in the functional equation to obtain that $f(n) + 1$ divides $(n+1)^2$ for all positive integers n. Setting $n = 1$ implies that $(f(m))^2 + 1$ divides $(m^2 + 1)^2$ for all positive integers m.

To prove that f is the identity function, it suffices to find infinitely many k such that $f(k) = k$. Indeed, suppose that this is true, and fix an arbitrary positive integer n. For each k such that $f(k) = k$, the number $k^2 + f(n) = (f(k))^2 + f(n)$ divides $(k^2 + n)^2$ by hypothesis. On the other hand, $(k^2 + n)^2$ can be written as

$$(k^2 + n)^2 = [(k^2 + f(n)) + (n - f(n))]^2 = A(k^2 + f(n)) + (n - f(n))^2$$

for some integer A. It follows that $(n - f(n))^2$ is divisible by $k^2 + f(n)$ for all k with the property that $f(k) = k$. Since there are infinitely many such k by assumption, we conclude that $(n - f(n))^2$ is divisible by arbitrarily large numbers, hence $(n - f(n))^2 = 0$. And so $f(n) = n$ for all $n \in \mathbf{N}$, as claimed.

We will prove that $f(p - 1) = p - 1$ for each prime number p. Indeed, as seen above, $f(p - 1) + 1$ divides p^2, so $f(p - 1) + 1$ equals either p or p^2. In the latter situation, since $(f(p - 1))^2 + 1$ divides $(p^2 + 1)^2$, it follows that $(p^2 - 1)^2 + 1 = p^4 - 2p^2 + 2$ is a divisor of $((p - 1)^2 + 1)^2 = p^4 - 4p^3 + 8p^2 - 8p + 4$. However for $p \geq 2$, the second of these numbers is smaller than the first. Consequently, $f(p - 1) + 1 = p$, proving the claim. The conclusion follows.

(Short list, 45th IMO 2004)

16. This problem comes as a warning that sometimes solving a functional equation for integer-valued functions could require techniques of real analysis. We begin by observing that $f(0) = f(1) = 0$. For $n \geq 2$, let us define $g(n) = f(n) + 1$. Then $g(2) = 8$ and

$$g(mn) = f(mn) + 1 = f(m) + f(n) + f(mn) + 1 = (f(m) + 1)(f(n) + 1)$$
$$= g(m)g(n)$$

for all $m, n \geq 2$.

Fix an integer $n > 2$ and consider a sequence p_k/q_k, $k \geq 1$, of rational numbers that are greater than $\log_2 n$ and converge to $\log_2 n$ (here p_k and q_k are integers). Then from $n < 2^{p_k/q_k}$, we deduce that $n^{q_k} < 2^{p_k}$, so by the monotonicity of g,

$$g(n^{q_k}) \leq g(2^{p_k}).$$

The multiplicativity of g implies that

$$g(n) \leq g(2)^{p_k/q_k} = 2^{3p_k/q_k} = \left(2^{p_k/q_k}\right)^3.$$

Passing to the limit with $k \to \infty$, we find that $g(n) \leq n^3$. A similar argument shows that $g(n) \geq n^3$. Hence $g(n) = n^3$, and so $f(n) = n^3 - 1$ for all n is the unique solution to the functional equation.

3.8 Invariants

1, The answer is negative. As in the example from the introduction to Section 3.8, we choose the invariant of the path to be the color of its last square. A path that goes through all squares and returns to where it started passes through 5×5 squares. Since it has odd length, it must end on a square whose color is opposite to that from the initial square; hence it cannot end where it started.

2. The invariant we use is the difference between the number of black squares and the number of white squares. Since each domino contains a white square and a black square, only boards with this invariant equal to zero can be covered by dominoes. Our chessboard has the invariant equal to ± 2, hence cannot be covered with dominoes.

3. Color the board by two colors in the usual manner. The invariant that gives the obstruction is the parity of the number of black squares. Each of the pieces covers 3 squares of one color and 1 square of the other color, hence it covers an odd number of black squares. The 25 pieces cover an odd number of black squares, while the entire board has an even number of black squares, and we are done.
(Leningrad Mathematical Olympiad)

4. The invariant is the parity of the number of pluses in the 2×2 lower left square. In the first configuration this parity is even, whereas in the second it is odd. Hence the first configuration cannot be transformed into the second by applying the specified operations.
(Russian Mathematical Olympiad, 1983–1984)

5. The invariant used is the residue of $\sum_{i=1}^{1997}(R_i + C_i)$ modulo 4. This residue is invariant under changing the sign of one of the numbers written on the board, since the sign change of the element in the ith row and jth column transforms $R_i + C_j$ into $-(R_i + C_j)$ keeping the rest unchanged. Since $R_i + C_j$ is equal to 2, 0, or -2, the whole sum changes by a multiple of 4.

Thus the invariant does not depend on the particular choice of $+1$'s and -1's. Making all numbers equal to $+1$, the invariant is equal to 2, so it is not zero. Since this property remains true for an arbitrary configuration, the sum can never be equal to zero.
(Colorado Mathematical Olympiad, 1997)

6. Notice that the difference between the joint number of stones at the first and third vertices and the joint number of stones at the second and fourth vertices changes by a multiple of 3 under the given operation. Thus it is natural to choose as an invariant the residue of this difference mod 3. Since the configuration $(1,1,1,1)$ has the invariant equal to 0 and the configuration $(1989, 1988, 1990, 1989)$ has the invariant equal to 2, one cannot transform the first into the second.
(Hungarian Mathematical Olympiad, 1989)

7. Place the n bulbs in the plane so that their complex coordinates are the nth roots of unity. At a move, one changes a bulb of coordinate w, together with all bulbs having coordinates $we^{2\pi di/n}$ for some divisor d of n. Since $e^{2\pi di/n}$ are the n/dth roots of unity, we see that at each move, the sum of the coordinates of the bulbs that are on does not

change. Choose the invariant to be the sum of the coordinates of the bulbs that are on. The invariant of the initial configuration is equal to a root of unity, thus is nonzero, whereas the configuration with all bulbs on has the invariant equal to 0, so the two configurations cannot be transformed one into the other.

(Proposed by J. Propp for the USAMO)

8. Let $C_1, C_2, \ldots, C_{2n+1}$ be the cards, identified with the vertices of a regular polygon. If we denote by $\text{dist}(C_i, C_j)$ the number of cards between C_i and C_j counted clockwise, we see that at each stage, $\text{dist}(C_i, C_{i+1})$ does not depend on i (here C_{2n+2} is identified with C_1). Thus each configuration depends only on C_1 and C_2; hence there are at most $2n(2n+1)$ distinct configurations. By applying the two moves successively, we see that we can obtain all $2n(2n+1)$ distinct configurations, and the problem is solved.

9. The myth of Sisyphus suggests the answer: zero. The invariant we consider for this problem is $a(u-1)/2 + b(b-1)/2 + c(c-1)/2 + s$, where a, b, c are the numbers of stones in the three piles, and s is the net income of Sisyphus. To see that this number is invariant, consider the move of a stone from the pile with a stones to the one with b stones. The income of Sisyphus increases by $a - b$, and we have

$$(a-1)(a-2)/2 + (b+1)b/2 - 1 + c(c-1)/2 + s + a - b$$
$$= a(a-1)/2 - a + b(b-1)/2 + b + c(c-1)/2 + s + a - b$$
$$= a(a-1)/2 + b(b-1)/2 + c(c-1)/2 + s.$$

Since at the beginning Sisyphus had no money, he will not have any money at the end, either.

(Russian Mathematical Olympiad, 1995)

10. We will show that the necessary and sufficient condition is that $\gcd(x, y) = 2^s$ for some nonnegative integer s. Indeed, since $\gcd(p, q) = \gcd(p, q - p)$, we see that the number of odd common divisors is invariant under the two transformations. Since initially this number is 1, it remains the same, and so the greatest common divisor of x and y can be only a power of 2.

As for sufficiency, suppose $\gcd(x, y) = 2^s$. Of those pairs (p, q) from which (x, y) can be reached, choose the one to minimize $p + q$. Neither p nor q can be even, else one of $(p/2, q)$ or $(p, q/2)$ contradicts the minimality. If $p > q$, then (p, q) is reachable from $((p+q)/2, q)$, again a contradiction; similarly, $p < q$ is impossible. Hence $p = q$, but $\gcd(p, q)$ is a power of 2, and neither p nor q is even. We conclude that $p = q = 1$, and so (x, y) is indeed reachable.

(German Mathematical Olympiad, 1996)

11. We can reformulate the problem as follows:

The sextuple $(x_1, x_2, x_3, x_4, x_5, x_6)$ is transformed into $(x_2, x_3, x_4, x_5, x_6, x_7)$, where x_7 is the last digit of $x_1 + x_2 + x_3 + x_4 + x_5 + x_6$. Can one obtain $(0, 1, 0, 1, 0, 1)$ from $(1, 0, 1, 0, 1, 0)$ by applying this transformation finitely many times?

We will prove that the answer is negative by defining an invariant of the sextuple that does not change under the described move. Let $s(x_1, x_2, x_3, x_4, x_5, x_6)$ be the last digit of the number $2x_1 + 4x_2 + 6x_3 + 8x_4 + 10x_5 + 12x_6$. Since

$$s(x_2, x_3, x_4, x_5, x_6, x_7) - s(x_1, x_2, x_3, x_4, x_5, x_6)$$
$$= 2x_2 + 4x_3 + \cdots + 10x_6 + 12(x_1 + x_2 + \cdots + x_6)$$
$$- 2x_1 - 4x_2 - \cdots - 12x_6$$
$$\equiv 10(x_1 + x_2 + \cdots + x_6) \equiv 0 \pmod{10},$$

it follows that $s(x_1, x_2, x_3, x_4, x_5, x_6)$ is invariant under the move. Since $s(1, 0, 1, 0, 1, 0) = 18$ and $s(0, 1, 0, 1, 0, 1) = 24$, the two cannot be transformed one into the other.

(*Kvant (Quantum)*)

12. Color the board with the elements of the Klein four group as shown in Figure 3.8.1. The product of all elements on the board is equal to the unit e.

a	b	a	b	a	b	a	b	a	b	a
c	e	c	e	c	e	c	e	c	e	c
b	a	b	a	b	a	b	a	b	a	b
e	c	e	c	e	c	e	c	e	c	e

Figure 3.8.1

When an L-shaped piece is placed on the board, the product of the elements that it covers is either a or b. Thus if a covering is possible, then an identity of the form $a^x b^y = e$ holds with $x + y = 11$. One of x and y must be even, and the other must be odd. Since $a^2 = b^2 = e$, the equality $a^x b^y = e$ becomes $a = e$ or $b = e$, which are both impossible. This shows that no covering exists.

13. (a) If $n > 1994$, it follows from the pigeonhole principle that there is one girl holding at least two cards. Hence the game cannot end. Suppose $n = 1994$. Label the girls $G_1, G_2, \ldots, G_{1994}$ and let G_1 hold all the cards initially. Define the current value of a card to be i if it is being held by G_i, $1 \le i \le 1994$. Initially, the sum of the total current values of the cards is 1994. If a girl other than G_1 or G_{1994} passes cards, the sum does not change. If G_1 or G_{1994} passes cards, S increases or decreases by 1994. Hence it is a good idea to choose as an invariant the residue S of the sum of the values of all cards modulo 1994. If the game is to end, each girl must be holding exactly one card, and $S \equiv 997 \cdot 1995 \pmod{1994}$. But this cannot happen, since the final S is not 0 modulo 1994, whereas the initial one is. Hence the game cannot terminate.

(b) Whenever a card is passed from one girl to another for the first time, both sign their names on it. Thereafter, if one of them is passing cards and holding this one, she must pass it back to the other. Thus a signed card is stuck between two neighboring girls. If $n < 1994$, there are two neighboring girls who never exchanged cards. For the game to go on forever, at least one girl must pass cards infinitely often. Hence there exists a girl who does so, while a neighbor of hers passes cards only a finite number of times. When the neighbor eventually stops passing cards, she will continue to accumulate cards indefinitely. This is clearly impossible.

(Short list 35th IMO, 1994; proposed by Sweden)

14. If either m or n is odd, we can remove all markers as follows. By symmetry, we may assume that $m = 2k - 1$. Denote the markers by their locations (i, j), $1 \leq i \leq m$, $1 \leq j \leq n$. We may assume that $(1, 1)$ has its black side up. We remove the markers $(i, 1)$ in the order $i = 1, 2, \ldots, m$. If $n = 1$, the task is accomplished. Suppose $n \geq 2$. Then each $(i, 2)$ has its black side up. We remove the markers $(2i - 1, 2)$ in the order $i = 1, 2, \ldots, k$. Now each $(2i, 2)$ has flipped twice, so that they can be removed, independent of one another. If $n = 2$, the task is accomplished. Otherwise each $(i, 3)$ had its black side up, and the same procedure can be repeated (see Figure 3.8.2).

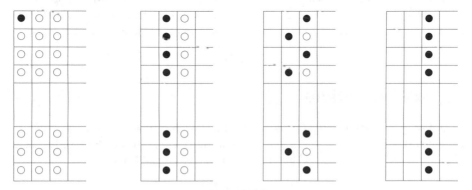

Figure 3.8.2

Consider now the case where m and n are both even. We will find an invariant obstructing the removal of all markers. We construct a graph with mn vertices representing the markers and connect two vertices by an edge if and only if the markers they represent occupy adjacent squares. Assign -1 to each edge and to each vertex representing a marker with its white side up, and 1 to each vertex representing a marker with its black side up. Let P be the product of all these numbers, on the edges as well as on the vertices. We claim that P remains unchanged as the game progresses.

For now, take the claim for granted. Note that as markers are removed, we delete the vertices representing them, along with the edges incident to these vertices. Initially, the number of -1's on the vertices is $mn - 1$, and the number of -1's on edges is $m(n - 1) + n(m - 1)$. Since the total $3mn - m - n - 1$ is odd, $P = -1$. If we have succeeded in removing all markers, the last move must involve the removal of the last marker, which must have its black side up. At that point, we have an isolated vertex with a 1, so $P = 1$. This is a contradiction.

Now let us justify the claim. Note that we can remove a marker only if its black side is up. If the vertex representing it is isolated, P is unchanged. Suppose this vertex is adjacent to another. The deletion of the connecting edge causes a change in sign in P. However, since the marker represented by the other vertex is flipped over, the change in sign is negated. Thus the claim is justified.

(Short list 39th IMO, 1998; proposed by Iran)

15. The expression $a^2 + b^2 + c^2 - 2(ab + bc + ca)$ is invariant under the operation, and it takes different values for our two triples. Another invariant is $a + b + c \pmod 2$.

16. Every vertex belongs to two sides of different colors. If these colors are, clockwise, red and blue or blue and yellow or yellow and red, we assign 1 to that vertex, otherwise we assign 2 to that vertex. Now assume that we change the color of a side $[AB]$ from color 1 to color 2. We infer the other side containing A must be neither of color 1 nor of color 2, thus it has the remaining color 3. So does that other side containing B. Then before the move, the sides containing A had colors 1 and 2, whereas the sides containing B had colors 2 and 1 (clockwise), so A and B were assigned different numbers. Analogously, we deduce that A and B will have different numbers assigned after the move. The numbers assigned to the other vertices do not change, so that, total number of 1's and 2's assigned is invariant. We are left to note that initially, the numbers assigned were $1, 2, 1, 2 \ldots 1, 2, 1, 1, 1$ whereas in the final state they are $1, 2, 1, 2, \ldots, 1, 2, 2, 2, 2$, so one configuration cannot be changed into the other.
(Communicated by I. Boreico)

3.9 Pell Equations

1. Using the sum formula for the terms of an arithmetic progression, we transform the equality into $2k(k+1) = n(n+1)$. This is equivalent to the negative Pell equation $(2n+1)^2 - 2(2k+1)^2 = -1$. The solution can be developed from

$$(2n+1) + (2k+1)\sqrt{2} = (1+\sqrt{2})^{2t+1}, \quad t = 1, 2, \ldots.$$

From this we derive

$$n = \frac{(1+\sqrt{2})^{2t+1} + (1-\sqrt{2})^{2t+1} - 2}{4}$$

and

$$k = \frac{(1+\sqrt{2})^{2t+1} - (1-\sqrt{2})^{2t+1} - 2\sqrt{2}}{4\sqrt{2}}.$$

(*College Mathematics Journal*, 1993; proposed by Zh. Zaiming, solution by W. Blumberg)

2. Let $m(m+1)/3 = n^2$ for some positive integer n. Then

$$(2m+1)^2 - 3(2n)^2 = 1.$$

We see that $2m+1$ and $2n$ must be solutions to Pell equation $X^2 - 3Y^2 = 1$, hence $(2m+1) + 2n\sqrt{3} = (2+\sqrt{3})^k$. Parity forces k to be even, say $k = 2j$. Then $(2m+1) + 2n\sqrt{3} = (7+2\sqrt{3})^j$. Taking conjugates gives $(2m+1) - 2n\sqrt{3} = (7-2\sqrt{3})^j$, and we obtain the answer to the problem

$$m = [(7+2\sqrt{3})^j + (7-2\sqrt{3})^j - 2]/4, \quad j = 1, 2, 3, \ldots.$$

(*Gazeta Matematică (Mathematics Gazette, Bucharest)*, proposed by D. Andrica)

3. The problem amounts to solving the equation $n(n+1) = 2m^2$, which after multiplication by 4 can be transformed into the Pell equation

$$(2n+1)^2 - 2(2m)^2 = 1.$$

The solutions (x_k, y_k) to the equation $x^2 - 2y^2 = 1$ are generated by any of the equalities $x_k + y_k\sqrt{2} = (3+2\sqrt{2})^{k+1}$ and $x_k - y_k\sqrt{2} = (3-2\sqrt{2})^{k+1}$. Note that x_k is always odd and y_k is always even. It follows that the nth triangular number is a perfect square if and only if

$$n = \frac{(3+2\sqrt{2})^k + (3-2\sqrt{2})^k}{2}$$

for some nonnegative integer k.

4. Two examples of such pairs are $(8,9)$ and $(288,289)$. One can see that in both of them the second number is a square, while the first is twice a square. It is thus natural to consider the Pell equation $x^2 - 2y^2 = 1$. This equation has infinitely many solutions (x_n, y_n). Reducing modulo 4 shows that x_n is odd and y_n is even. The pairs of consecutive numbers $(2y_n^2, x_n^2)$ have the required property, since the power of 2 in the first term is at least 3, and all other primes appear at even powers, whereas the second term is a square.

Note that the first term of the pairs we constructed is of the form $8k$, with k a triangular number, and since $8k+1$ is automatically a perfect square, the problem follows immediately from the previous problem if we choose k to be a perfect square as well.

(W.L. Putnam Mathematical Competition)

5. We reduce the problem to a Pell equation. Since $(x+1)^3 - x^3 = 3x^2 + 3x + 1$, the equation becomes $3x^2 + 3x + 1 = y^2$. Multiplying by 4 and completing the square, we obtain

$$(2y)^2 - 3(2x+1)^2 = 1,$$

which is a Pell equation in $u = 2y$ and $v = 2x + 1$. The solutions to the equation $u^2 - 3v^2 = 1$ are given by $u_n + v_n\sqrt{3} = (2+\sqrt{3})^n$, but the parity condition implies that only every other solution leads to a solution to the initial equation. We obtain the solutions

$$x = \frac{1}{4\sqrt{3}}[(2+\sqrt{3})^{2k+1} - (2-\sqrt{3})^{2k+1} - 2],$$

$$y = \frac{1}{4}[(2+\sqrt{3})^{2k+1} + (2-\sqrt{3})^{2k+1}], k \geq 1.$$

(W. Sierpiński, *250 problems in elementary number theory*, Państwowe Wydawnictwo Naukowe, Warszawa, 1970)

6. The recurrence relations show that

$$u_n + v_n\sqrt{2} = (3+2\sqrt{2})^n;$$

hence u_n and v_n are the solutions to the Pell equation $u^2 - 2v^2 = 1$. Since $y_n^2 = u_n^2 + 4v_n^2 + 4u_n v_n$ and $2x_n^2 = 2u_n^2 + 2v_n^2 + 4u_n v_n$, it follows that $y_n^2 - 2x_n^2 = 2v_n^2 - u_n^2 = -1$; hence $y_n^2 = 2x_n^2 - 1$ is the largest perfect square less than $2x_n^2$. This implies $y_n = \lfloor x_n \sqrt{2} \rfloor$, and we are done.

(D. Andrica)

7. Let $x_1^2 - 5y_1^2 = a$ and $x_2^2 - 5y_2^2 = b$ for some integers x_1, x_2, y_1, y_2. Then

$$
\begin{aligned}
ab &= (x_1^2 - 5y_1^2)(x_2^2 - 5y_2^2) \\
&= x_1^2 x_2^2 + 25 y_1^2 y_2^2 - 5y_1^2 x_2^2 - 5x_1^2 y_2^2 \\
&= (x_1 x_2)^2 + (5y_1 y_2)^2 + 10 x_1 x_2 y_1 y_2 - 5(y_1 x_2)^2 - 5(x_1 y_2)^2 - 10 x_1 x_2 y_1 y_2 \\
&= (x_1 x_1 + 5y_1 y_2)^2 - 5(y_1 x_2 + x_1 y_2)^2.
\end{aligned}
$$

The integer numbers $x = x_1 x_2 + 5y_1 y_2$ and $y = y_1 x_2 + x_1 y_2$ satisfy the general Pell equation $x^2 - 5y^2 = ab$.

(Leningrad Mathematical Olympiad)

8. Let $2n + 1 = x^2$ and $3n + 1 = y^2$. Multiply the first equation by 3 and the second by 2 and subtract them to obtain the negative Pell equation

$$
3x^2 - 2y^2 = 1.
$$

The smallest solution to this equation is $x = y = 1$, and we have the general solution it follows that

$$
x_m = \frac{1}{\sqrt{3}} \left((\sqrt{3} + \sqrt{2})^{2m+1} + (\sqrt{3} - \sqrt{2})^{2m+1} \right),
$$

$$
y_m = \frac{1}{\sqrt{2}} \left((\sqrt{3} + \sqrt{2})^{2m+1} - (\sqrt{3} - \sqrt{2})^{2m+1} \right).
$$

Consequently,

$$
\begin{aligned}
n &= y_m^2 - x_m^2 \\
&= \left(\frac{1}{\sqrt{2}} - \frac{1}{\sqrt{3}} \right) (\sqrt{3} + \sqrt{2})^{2m+1} - \left(\frac{1}{\sqrt{2}} + \frac{1}{\sqrt{3}} \right)(\sqrt{3} - \sqrt{2})^{2m+1} \\
&= \frac{(\sqrt{3} + \sqrt{2})^{2m} - \sqrt{3} - \sqrt{2})^{2m}}{\sqrt{6}}.
\end{aligned}
$$

An inductive argument based on the recurrence relations for x_m and y_m can prove that all such n are divisible by 40. This can be proved also using residues. The quadratic residues modulo 8 are 0, 1, and 4. Hence $2n$ is congruent to -1, 0, or 3. Of course, only the second situation is possible, since $2n$ is even, so n is congruent to 0 or 4. But $3n + 1$ must also be a quadratic residue, so n is congruent to 0 modulo 8. The quadratic residues modulo 5 are 0, 1, and 4. So $2n$ must be congruent to -1, 0, or 3; thus n is congruent to 2, 0, or 4. This implies that $3n + 1$ is congruent to 2, 1, or 3. Since this is again a perfect square, the only admissible residue is 1, which implies that n is divisible by 5. Therefore, n is divisible by $5 \cdot 8 = 40$.

(*American Mathematical Monthly*, 1976; proposed by R.S. Luthar)

9. For the proof we will rely on Pell's equation. Let $3n+1 = p^2$ and $4n+1 = q^2$. Multiplying the first equation by 4 and the second by 3 and subtracting them, we obtain $(2p)^2 - 3q^2 = 1$. Thus $2p$ and q are solutions to the Pell equation

$$x^2 - 3y^2 = 1.$$

The minimal solution is $x = 2$, $y = 1$, corresponding to the case $n = 0$. The general solution (x_m, y_m) is given by

$$x_m + y_m\sqrt{3} = (2+\sqrt{3})^m.$$

However, we are interested only in those solutions for which x_n is even, and these are the ones for which the index is *even*. Hence the numbers p_n and q_n we are looking for are defined by

$$2p_n + q_n\sqrt{3} = ((2+\sqrt{3})^2)^n = (7+4\sqrt{3})^n,$$

so they satisfy the recurrence

$$p_{k+1} = 7p_k + 6q_k, \qquad q_{k+1} = 7q_k + 8p_k.$$

This implies $p_{k+1}^2 \equiv q_k^2 \pmod 7$ and $q_{k+1}^2 \equiv p_k^2 \pmod 7$, and inductively we obtain $q_k^2 \equiv 1 \pmod 7$; hence n is a multiple of 7. The fact that n is divisible by 8 can be proved in a similar way.

10. Since the sum of squares of the first n positive integers is equal to $n(n+1)$ $(2n+1)/6$, the problem reduces to finding the smallest n for which there is an integer m such that

$$\frac{(n+1)(2n+1)}{6} = m^2.$$

Multiplying by 48 and completing the square, this becomes

$$(4n+3)^2 - 3(4m)^2 = 1.$$

Thus we must find the smallest solution to the Pell equation $x^2 - 3y^2 = 1$ with $x > 7$ for which x gives the residue 3 when divided by 4 and y is a multiple of 4. The smallest solution to the equation is $(2,1)$, and all others are obtained by the recursion $x_{k+1} = 2x_k + 3y_k$, $y_{k+1} = x_k + 2y_k$. The next solutions are $(7,4)$, $(26,15)$, $(97,56)$, $(362,209)$, $(1351,780)$. The last pair is the first one to satisfy the above requirements and yields the answer to the problem $n = 337$.
    (USAMO, 1986)

11. By induction, one easily shows that

$$A^n = \begin{pmatrix} a_n & b_n \\ 2b_n & a_n \end{pmatrix}$$

for all n. Since $\det A^n = 1$, a_n and b_n are solutions of the Pell equation

$$x^2 - 2y^2 = 1.$$

It follows that a_n is odd and $(a_n - 1)((a_n + 1)/2) = b_n^2$; thus $a_n - 1$ divides b_n^2. This shows that the greatest common divisor of $a_n - 1$ and b_n, which is also the greatest common divisor of the four entries of the matrix $A^n - I$, is at least $\sqrt{a_n - 1}$. Since $a_n \to \infty$ as $n \to \infty$, the conclusion follows.

(54th W.L. Putnam Mathematical Competition, 1994).

12. The recurrence relation can be rewritten as

$$(a_{n+1} - 2a_n)^2 - 3a_n^2 = -2.$$

The problem reduces to showing that if (b_n, a_n) is the general solution to the general Pell equation $x^2 - 3y^2 = -2$, then $b_n = a_{n+1} - 2a_n$. The solutions to this equation are given by

$$b_n + \sqrt{3}a_n = (2 + \sqrt{3})^n(1 + \sqrt{3}).$$

Adding to the equality $b_{n+1} + \sqrt{3}a_{n+1} = (2 + \sqrt{3})^{n+1}(1 + \sqrt{3})$, the equality $b_n + \sqrt{3}a_n = (2 + \sqrt{3})^n(1 + \sqrt{3})$ multiplied by -2 yields

$$(b_{n+1} - 2b_n) + \sqrt{3}(a_{n+1} - 2a_n) = \sqrt{3}(2 + \sqrt{3})^n(1 + \sqrt{3})$$
$$= \sqrt{3}(b_n + \sqrt{3}a_n).$$

Equating the coefficients of $\sqrt{3}$, we obtain $b_n = a_{n+1} - 2a_n$, which solves the problem.

(*Revista Matematică din Timişoara (Timişoara's Mathematics Gazette)*, 1979; proposed by T. Andreescu)

13. Observe that if we choose n to be a solution to the negative Pell equation

$$n^2 - 5m^2 = -1$$

that is large enough, then it satisfies the condition from the statement. Indeed, if n and m are solutions and $m > 5$, then $2m$ is smaller than n, so $5, m$, and $2m$ are among the factors of $n!$. Hence $5m^2$ divides $n!$. Consequently, $n^2 + 1$ divides $n!$. The conclusion follows from the fact that the given Pell equation has infinitely many solutions (since it has the minimal solution $(2, 1)$).

(*Kvant(Quantum)*)

14. Completing the square, we have

$$x^2 + y^2 + (z + xy)^2 - x^2y^2 - 1 = 0.$$

After factorization, this becomes

$$(x^2 - 1)(y^2 - 1) = (z + xy)^2.$$

For the right side to be a square, we should have $x^2 - 1 = p^2q$ and $y^2 - 1 = r^2q$, for some integers p, q, r, with q not divisible by any square. Fixing q, we obtain the Pell equations $x^2 - qp^2 = 1$ and $y^2 - qr^2 = 1$, which have infinitely many solutions. Each of these solutions leads to a solution of the initial equation, with $z = -xy$.

15. Yes, there are infinitely many such rows. For example,

$$\binom{203}{68} = 2\binom{203}{67} \text{ and } \binom{203}{85} = 2\binom{203}{83}.$$

There are infinitely many rows having two adjacent elements in ratio 1:2, for $2\binom{n}{k} = \binom{n}{k+1}$ reduces to $2(k+1) = n-k$, or $n = 3k+2$. Thus, as long as $n \equiv 2 \pmod 3$, there will be two adjacent elements in a 1:2 ratio.

Next, we search for pairs a, b, with $b = 2a$ that are not adjacent. The next easiest case to try is

$$2\binom{n}{r} = \binom{n}{r+2},$$

which reduces to

$$2(r+2)(r+1) = (n-r)(n-r-1).$$

If we substitute $u = n - r$ and $v = r + 2$, the equation becomes

$$2v^2 - 2v = u^2 - u.$$

Multiplying both sides by 4, completing the square, and making the substitution $x = 2v - 1$, $y = 2u - 1$, we obtain the negative Pell equation

$$2x^2 - y^2 = 1.$$

This equation has infinitely many solutions, and they are given by

$$x_m\sqrt{2} + y_m = (\sqrt{2}+1)^{2m-1}, m = 1, 2, 3\ldots.$$

There exist infinitely many m with $n = (x_m + y_m)/2 - 1$ congruent to 2 modulo 3, since one can show by induction that the residues modulo 6 of the pairs (x_m, y_m) have the repeating pattern

$$(1,1), \quad (-1,1), \quad (-1,-1), \quad (1,-1), \quad (1,1).$$

Let us show that infinitely many pairs $c = \binom{n}{r+2}$, $d = \binom{n}{r}$ obtained this way are disjoint from the pairs $a = \binom{n}{k+1}$, $b = \binom{n}{k}$ obtained by the procedure described at the beginning. Indeed, the relations $k = r$, $k = r+1$, $k = r+2$, and $k+1 = r$ produce linear relations between the solutions (x_m, y_m) of the Pell equation, which lead to four quadratic equations in x_m. These equations have finitely many solutions, and hence there are infinitely many solutions to the Pell equations for which a, b, c, and d are all distinct.

(Proposed by P. Zeitz for the USAMO, 1997)

16.   Let the sides be $a = y + z$, $b = x + z$, $c = x + y$, the semiperimeter $p = \frac{a+b+c}{2} = x + y + z$, the inradius r, and the area A. We have $A = rp$,

$A^2 = p(p-a)(p-b)(p-c) = pxyz$. Thus $\frac{p}{r} = \frac{p}{\frac{A}{p}} = \frac{p^2}{A} = \sqrt{\frac{p^3}{xyz}}$. So our relation

is equivalent to $n^2 = \frac{(x+y+z)^3}{xyz}$.

Note that $2x = b+c-a, 2y = a+c-b, 2z = a+b-c$ are integers. The existence of a triangle with integer sides whose semiperimeter divided by its inradius is n is equivalent to the existence of $2x, 2y, 2z \in \mathbf{N}$ such that $\frac{(x+y+z)^3}{xyz} = n^2$. As the relation is homogeneous, it is enough to find three positive rational numbers x, y, z satisfying this relation (and then we multiply them by a suitable integer to make them positive integers).

We look for solutions with $z = k(x+y)$. We should have

$$\frac{(k+1)^3 (x+y)^3}{xyk(x+y)} = \frac{(k+1)^3}{k} \frac{(x+y)^2}{xy} = n^2.$$

For simplicity, we can try $n = m(k+1)$, in which case the equation becomes

$$\frac{(x+y)^2}{xy} = \frac{m^2 k}{k+1}$$

thus

$$\frac{(x-y)^2}{(x+y)^2} = \frac{\frac{m^2 k}{k+1} - 4}{\frac{m^2 k}{k+1}} = \frac{(m^2-4)k-4}{m^2 k}.$$

It suffices for this number to be a perfect square, as then by setting $x+y = 1, x-y = \sqrt{\frac{(m^2-4)k-4}{m^2 k}}$, we obtain a rational pair of solutions. Thus $((m^2-4)k-4)k$ should be a perfect square. The smallest possible m is 3, for which $(5k-4)k$ should be a perfect square. This happens if $k = u^2, 5k-4 = v^2$, which is possible only when $v^2 - 5u^2 = -4$. This is a generalized Pell equation that has a particular solution $v = 1, u = 1$, thus it has infinitely many solutions (u, v). For each such solution, the number $n = 3(u^2+1)$ satisfies the required properties.

(Short list, 42nd IMO, 2001)

3.10   Prime Numbers and Binomial Coefficients

1. Because $\binom{1}{0} = \binom{1}{1} = \binom{0}{0} = 1$ and $\binom{0}{1} = 0$, by Lucas's theorem $\binom{n}{k}$, $0 \le k \le n$ is odd if and only if the set of the positions of 1's in the binary representation of k is a subset of the set of the positions of 1's in the binary representation of n. Thus the number of odd binomial coefficients is 2 raised to the number of nonzero digits from the binary expansion of n.

(16th W.L. Putnam Mathematical Competition, 1956)

2. Let us first count the number of coefficients that are not divisible by p. If $n = n_1 n_2 \ldots n_m$ and $k = k_1 k_2 \ldots k_m$ are the representations of n and k in base p (with some of the first k_i's possibly equal to zero), then by Lucas's theorem, $\binom{n}{k}$ is congruent

to zero modulo p if and only if at least one of the binomial coefficients $\binom{n_i}{k_i}$ is equal to zero, hence if and only if $n_i < k_i$ for some i. Hence $\binom{n}{k}$ is not divisible by p if and only if $k_i \le n_i$ for all i.

Since there are $(n_i + 1)$ nonnegative integers less than n_i, there are $(n_1 + 1)$ $(n_2 + 1) \cdots (n_m + 1)$ m-tuples of digits $(k_1, k_2, \ldots k_m)$, such that $k_i \le n_i$, and hence so many binomial coefficients are not multiples of p. It follows that $(n + 1) - (n_1 + 1)$ $(n_2 + 1) \cdots (n_m + 1)$ binomial coefficients are divisible by p.

3. All these numbers are even, since

$$\binom{2^n}{k} = \frac{2^n}{k} \binom{2^n - 1}{k - 1}$$

and $2^n / k$ is different from 1 for all $k = 1, 2, \ldots, 2^n - 1$.

From the same relation, it follows that $\binom{2^n}{k}$ is a multiple of 4 for all k different from 2^{n-1}. For $k = 2^{n-1}$, we have

$$\binom{2^n}{2^{n-1}} = 2 \binom{2^n - 1}{2^{n-1} - 1}.$$

But from Lucas's theorem, it follows that $\binom{2^n - 1}{2^{n-1} - 1}$ is odd, since $2^n - 1$ contains only 1's in its binary representation and $\binom{1}{k} = 1$ if $k = 0$ or 1. This solves the problem.

(Romanian Mathematical Olympiad, 1988; proposed by I. Tomescu)

4. *First solution:* Write

$$\binom{p^n}{p} = p^{n-1} \binom{p^n - 1}{p - 1}.$$

Now the conclusion follows by applying Lucas's theorem.

Second solution: Expanding, we obtain

$$\binom{p^n}{p} = \frac{p^n \cdot (p^n - 1)(p^n - 2) \cdots (p^n - (p - 1))}{p \cdot (p - 1)(p - 2) \cdots 1}$$
$$= p^{n-1} \cdot \frac{(p^n - 1)(p^n - 2) \cdots (p^n - (p - 1))}{(p - 1)(p - 2) \cdots 1}.$$

The denominator of the fraction is not divisible by p; hence the fraction alone is an integer. Moreover, modulo p, the numerator is congruent to the denominator, so the fraction is congruent to $(-1)^{p-1} = 1$ modulo p. It follows that $\binom{p^n}{p} \equiv p^{n-1} \pmod{p^n}$, and we are done.

5. Let $n = \sum_{j=1}^{t} n_j 3^j$, and let $A = \{j \mid n_j = 1\}$, $B = \{j \mid n_j = 2\}$. Denote by r and s the number of elements in A, respectively B. We will show that the difference between the number of coefficients that give the residue 1 when divided by 3 and the number of coefficients that give the residue 2 when divided by 3 is 2^r.

Let $k = \sum_{j=1}^{t} k_j 3^j$. Then by Lucas's theorem, $\binom{n}{k} \equiv 1 \pmod 3$ if and only if $\binom{n_t}{k_t} \binom{n_{t-1}}{k_{t-1}} \cdots \binom{n_0}{k_0} \equiv 1 \pmod 3$. For the latter to hold, we must have $\binom{n_j}{k_j} \equiv 1$ or 2, and the number of j such that $\binom{n_j}{k_j} \equiv 2$ must be even. For this to happen, k_j has to be 0 whenever $n_j = 0$, $k_j = 0$ or 1 for $j \in A$ and $k_j = 0$, 1 or 2 for $j \in B$, with an even number of j equal to 1 when j is in B.

Once we have chosen $2m$ of the k_j's to be equal to 1, for those j that are in B, there are $2^r 2^{s-2m}$ possibilities for the remaining k_j's to be chosen (equal to 0 or 1 for indices in A, or to 0 and 2 for indices in B). Thus there are

$$2^r \left(2^s \binom{s}{0} + 2^{s-2} \binom{s}{2} + 2^{s-4} \binom{s}{4} + \cdots \right)$$

binomial coefficients that give the residue 1.

To count coefficients that give the residue 2, note that there should be an odd number of $\binom{n_j}{k_j}$ that are congruent to 2; hence $k_j = 1$ for an odd number of j's in B. Using a similar counting argument as above, we obtain that the number of coefficients giving the residue 2 is

$$2^r \left(2^{s-1} \binom{s}{1} + 2^{s-3} \binom{s}{3} + 2^{s-5} \binom{s}{5} + \cdots \right).$$

If we subtract this number from the number of coefficients congruent to 1, we find that

$$2^r \left(2^s \binom{s}{0} - 2^{s-1} \binom{s}{1} + 2^{s-2} \binom{s}{2} - 2^{s-3} \binom{s}{3} + 2^{s-4} \binom{s}{4} + \cdots \right)$$
$$= 2^r (2-1)^s = 2^r 1^s = 2^r.$$

Since the difference is positive, there are more coefficients that give the residue 1 than coefficients that give the residue 2.

(British Mathematical Olympiad, 1984)

6. *First solution:* Because the congruence is mod p^2, we can no longer apply Lucas's theorem. The proof is by induction on m. First we prove the congruence for $m = 0$. We compute

$$\binom{np}{p} = \frac{np(np-1)\cdots(np-n+1)}{n!}$$
$$= p \cdot \frac{(np-1)(np-2)\cdots(np-n+1)}{(n-1)!}.$$

Since $(np-1)(np-2)\cdots(np-n+1) \equiv (-1)^{n+1} \pmod p$, it follows from the above computation that

$$\binom{np}{p} \equiv p \frac{(-1)^{n+1}(n-1)!}{(n-1)!} \pmod{p^2}.$$

Note that it is essential that p be relatively prime to all numbers less than n, so that it can be brought in front of the fraction.

Assume that the relation is true for $m-1$ and let us prove it for m. Using the identities

$$\binom{n}{k} = \frac{n}{n-k}\binom{n-1}{k} \quad \text{and} \quad \binom{n}{k} = \frac{n-k+1}{k}\binom{n}{k-1}$$

it follows that the binomial coefficient $\binom{np+m}{mp+n}$ is congruent to

$$\frac{((n-m)p-(n-m)+1)\cdots(n-m)p-(n-m)+p-1)}{(mp+n)(mp+n-1)\cdots(mp+n-p+1)}$$
$$\times \frac{np+m}{(n-m+1)p-(n-m)} \times \binom{np+m-1}{(m-1)p+n} \pmod{p^2}.$$

The numerator and the denominator contain only factors relatively prime to p, except for $(n-m)p$ in the numerator and mp in the denominator. Cancel this p. The induction hypothesis implies that the binomial coefficient is divisible by p, so let us divide the expression by p and denote the remaining expression by A. It remains to show that A is congruent to $(-1)^{m+n+1}$ modulo p. Since A is an integer, it is congruent modulo p to the number obtained by deleting the denominator and multiplying the new expression by the inverse modulo p of the denominator. Since the inverse modulo p depends only on the residue class, it follows that A is congruent modulo p to

$$(p-1)![(p-1)!]^{-1}(n-m)m[m(m-n)]^{-1} \times \frac{1}{p}\binom{np+m-1}{(m-1)p+n},$$

and this is further congruent to -1. From the induction hypothesis, it follows that

$$\binom{np+m}{mp+n} \equiv -\binom{np+m-1}{(m-1)p+n} \equiv (-1)^{m+n+1}p \pmod{p^2},$$

and we are done.

Second solution: First look at factors of $(np+m)!$. There are n factors that are multiples of p, whose product is $n!p^n$. For $k = 0,\ldots,n-1$, the factors $(kp+1),\ldots,(kp+p-1)$ multiply to $(p-1)!$ mod p, and the factors $np+1,\ldots,np+m$ multiply to $m!$ mod p. Hence

$$(np+m)! \equiv m!n![(p-1)!]^n p^n \pmod{p^{n+1}}.$$

Similarly

$$(mp+n)! \equiv m!n![(p-1)!]^m p^m \pmod{p^{m+1}}$$

and

$$((n-m)p+(m-n))! \equiv (n-m-1)!(p+m-n)![(p-1)!]^{n-m-1}p^{n-m-1} \pmod{p^{n-m}}.$$

Hence dividing the first by the other two gives

$$\binom{np+m}{mp+n} \equiv p\binom{p-1}{n-m-1} \pmod{p^2}.$$

Now since
$$\binom{p-1}{n-m-1} = \frac{(p-1)(p-2)\cdots(p-(n-m-1))}{1\cdot 2\cdots (n-m-1)},$$

we see that the binomial coefficient is $(-1)^{n+m-1}$ mod p and we are done.

(*Gazeta Matematică (Mathematics Gazette, Bucharest)*, proposed by M.O. Drimbe, second solution by R. Stong)

7. The sum $\sum_{j=0}^{p} \binom{p}{j}\binom{p+j}{j}$ is the coefficient of x^p in the expansion of

$$\sum_{j=0}^{p} \binom{p}{j}(x+1)^{p+j}.$$

Note that this expression can be rewritten as

$$\left(\sum_{j=0}^{p} \binom{p}{j}(x+1)^j\right)(x+1)^p = ((x+1)+1)^p(x+1)^p = (x+2)^p(x+1)^p.$$

The coefficient of x^p in this is

$$\sum_{k=0}^{p} \binom{p}{k}\binom{p}{p-k}2^k.$$

We were able to transform $\sum_{j=0}^{p}\binom{p}{j}\binom{p+j}{j}$ into this latter sum. Because $\binom{p}{k}$ is divisible by p for all $1 < k < p$ in this sum all but the first and the last term are divisible by p^2. Therefore

$$\sum_{j=0}^{p}\binom{p}{j}\binom{p+j}{j} \equiv \binom{p}{0}\binom{p}{p}2^0 + \binom{p}{p}\binom{p}{0}2^p \equiv 1+2^p \pmod{p^2}.$$

(52nd W.L. Putnam Mathematical Competition, 1991)

8. The proof is based on the following identity:

$$\binom{2p}{p} = \binom{p}{0}^2 + \binom{p}{1}^2 + \cdots + \binom{p}{p}^2.$$

This identity can be proved as follows. First rewrite it as

$$\binom{2p}{p} = \sum_{k=0}^{p} \binom{p}{k}\binom{p}{p-k},$$

and then use the fact that the left side is the coefficient of x^p in the expansion of $(1+x)^{2p}$, whereas the right side is the coefficient of x^p in the expansion of the product $(1+x)^p(1+x)^p$, and of course the two are the same.

Since $\binom{p}{k}$ is divisible by p for all $k = 1, 2, \ldots, p-1$, each of the terms of the sum is divisible by p^2, except for the first and the last, which are equal to 1; hence the conclusion.

9. Recall that every integer that is relatively prime to p has a multiplicative inverse modulo p. Denote the inverse of x modulo p by x^{-1}. We start by improving the conclusion of the previous problem. We have

$$\binom{2p}{p} - 2 = \sum_{k=1}^{p-1} \binom{p}{k}^2 = \sum_{k=1}^{p-1} \left(\frac{p}{k}\binom{p-1}{k-1}\right)^2.$$

Note that $(1/k)\binom{p-1}{k-1}$ is an integer, since it is equal to $(1/p)\binom{p}{k}$ and p divides $\binom{p}{k}$. Hence the latter sum is congruent modulo p^3 to $p^2 \sum_{k=1}^{p-1} (k^{-1}\binom{p-1}{k-1})^2$. We will show that the sum is divisible by p. Modulo p we have

$$\sum_{k=1}^{p-1} \left(k^{-1}\binom{p-1}{k-1}\right)^2$$

$$\equiv \sum_{k=1}^{p-1} (k^{-1})^2 ((p-1)(p-2)\cdots(p-k+1)[(k-1)(k-2)\cdots 1]^{-1})^2$$

$$\equiv \sum_{k=1}^{p-1} (k^{-1})^2 ((-1)(1)^{-1}(-2)(2)^{-1}\cdots(-(k-1))(k-1)^{-1})^2$$

$$\equiv \sum_{k=1}^{p-1} (k^{-1})^2 (-1)^{2k-2}.$$

But the inverses of the numbers $1, 2, \ldots, p-1$ modulo p are the same numbers in some order. Hence the sum is congruent to $\sum_{k=1}^{p-1} k^2$, which is equal to $(p-1)p(2p-1)/6$. This is divisible by p, since $p \neq 3$. Thus $\binom{2p}{p} - 2$ is divisible by p^3. Since

$$\binom{2p-1}{p-1} - 1 = \frac{1}{2}\left(\binom{2p}{p} - 2\right),$$

it follows that $\binom{2p-1}{p-1} - 1$ is divisible by p^3, as desired.

10. If $k = 2^m$, then $(1+x)^k \equiv 1 + x^k \pmod{2}$. Note that if the degree of R is less than k, then $w(R + x^k S) = w(R) + w(S)$.

The solution is by induction on i_n. For $i_n = 0$ or 1, the result follows easily. Let $i_n > 1$ and let $k = 2^m \le i_n < 2^{m+1}$. Set $Q = Q_{i_1} + Q_{i_2} + \cdots + Q_{i_n}$. We distinguish two cases:

(a) $i_1 < k$. Let r be such that $i_r < k \le i_{r+1}$. Write $Q = R + (1+x)^k S$, where $R = Q_{i_1} + Q_{i_2} + \cdots + Q_{i_r}$. Note that the degrees of R and S are less than k. We have

$$w(Q) = w(R + S + x^k S) = w(R+S) + w(S) \ge w(R) \ge w(Q_{i_1}),$$

which follows by the induction hypothesis for R and by noting that the "triangle inequality" $w(R+S) + w(S) \ge w(R)$ holds.

(b) $i_1 \geq k$. Write $Q_{i_1} = (1+x)^k R$ and $Q = (1+x)^k S$. Then the degrees of R and S are both less than k, so $w(S + x^k S) = 2w(S)$ and $w(R + x^k R) = 2w(R)$. We have

$$w(Q) = w(S + x^k S) = 2w(S) \geq 2w(R) = w(R + x^k R) = w(Q_{i_1}),$$

where we have used again the induction hypothesis applied to S. This ends the proof.

(26th IMO, 1985; proposed by The Netherlands)

11. Denote $a_{n,k} = \sum_{i \in \mathbb{Z}} \binom{n}{k+(p-1)i}(-1)^i$. We see that $a_{n,k+p-1} = -a_{n,k}$(*) and $a_{n,k+2(p-1)} = a_{n,k}$. Let $A_n(x) = \sum_{k \in \mathbb{Z}} a_{n,k} x^k$. Obviously we obtain $a_{n,k} = a_{n-1,k-1} + a_{n-1,k}$. Hence we get $A_{n+1}(x) = (1+x)A_n(x)$. Thus $A_{n+p+1}(x) = (1+x)^{p+1}A_n(x)$. Working modulo p, we have $(1+x)^p \equiv 1+x^p$, from where $A_{n+p+1}(x) \equiv (1+x^p)(1+x)A_n(x) \equiv (1+x+x^p+x^{p+1})A_n(x)$. This means that $a_{n+p+1,k} \equiv a_{n,k} + a_{n,k-1} + a_{n,k-p} + a_{n,k-p-1} \equiv a_{n,k} - a_{n,k-2}$ (here we used (*)).

Now let us prove that $a_{n+p+1,j} \equiv 0$ for any odd j is equivalent to $a_{n,j} \equiv 0$ for any odd j (**). Indeed if $a_{n,j} \equiv 0$ for any odd j, then $a_{n+p+1,j} \equiv a_{n,j} - a_{n,j-2} \equiv 0$. Conversely, if $a_{n+p+1,j} \equiv 0$ for any odd j, then $a_{n,j} - a_{n,j-2} \equiv 0$ so $a_{n,j} \equiv a_{n,j-2}$. Further, we obtain $a_{n,j} \equiv a_{n,j+p-1}$. But from (*) we have $a_{n,j} \equiv -a_{n,j+p-1}$ and hence $a_{n,j} \equiv 0$. The property (**) is proved. Now (**) says we should investigate the validity of the problem conclusion just for $0 \leq n \leq p$, and this is easy.

Appendix A

Definitions and Notation

A.1 Glossary of Terms

AM–GM inequality
   If $a_1, a_2, \ldots, a_n$ are n nonnegative numbers, then

$$\frac{1}{n} \sum_{i=1}^{n} a_i \geq (a_1 a_2 \cdots a_n)^{\frac{1}{n}}$$

with equality if and only if $a_1 = a_2 = \cdots = a_n$.

Binomial coefficient

$$\binom{n}{k} = \frac{n!}{k!(n-k)!},$$

the coefficient of x^k in the expansion of $(x+1)^n$.

Binomial theorem

$$(x+y)^n = \sum_{k=0}^{n} \binom{n}{k} x^{n-k} y^k.$$

Cauchy–Schwarz inequality
   For any real numbers $a_1, a_2, \ldots, a_n$, and $b_1, b_2, \ldots, b_n$,

$$\sum_{i=1}^{n} a_i^2 \cdot \sum_{i=1}^{n} b_i^2 \geq \left(\sum_{i=1}^{n} a_i b_i \right)^2,$$

with equality if and only if a_i and b_i are proportional, $i = 1, 2, \ldots, n$.

Centroid of a triangle
   Point of intersection of the medians.

Centroid of a tetrahedron
   Point of the intersection of the segments connecting the midpoints of the opposite edges, which is the same as the point of intersection of the segments connecting each vertex with the centroid of the opposite face.

Circumcenter
   Center of the circumscribed circle or sphere.

Circumcircle
   Circumscribed circle.

Congruence
   $a \equiv b \pmod{p}$, $a - b$ is divisible by p.

Concave up (down) function
   A function $f(x)$ is concave up (down) on $[a,b]$ if $f(x)$ lies under (above) the line connecting $(a_1, f(a_1))$ and $(b_1, f(b_1))$ for all $a \le a_1 < x < b_1 \le b$.

Cyclic polygon
   Polygon that can be inscribed in a circle.

Egyptian fraction
   Fraction with numerator equal to 1.

Eigenvalue of a matrix
   λ is an eigenvalue of an $n \times n$ matrix A if $\det(\lambda I_n - A) = 0$.

Euler's theorem
   If m is relatively prime to a, then $a^{\phi(m)} \equiv a \pmod{m}$, where $\phi(m)$ is the number of positive integers less than a and relatively prime to a.

Exterior angle bisector
   The line through the vertex of the angle that is perpendicular to the angle bisector.

Fermat's little theorem
   If p is prime, then $a^p \equiv a \pmod{p}$, for all integers a.

Fibonacci sequence
   Sequence defined recursively by $F_1 = F_2 = 1$, $F_{n+1} = F_n + F_{n-1}$.

Hero's formula
   The area of a triangle with sides a, b, c is equal to $\sqrt{s(s-a)(s-b)(s-c)}$, where $s = (a+b+c)/2$.

Homothety of center O and ratio r
   Geometric transformation that maps each point M to a point M' on the half-line (ray) OM such that $OM' = rOM$.

Incenter
   Center of inscribed circle.

Incircle
   Inscribed circle.

Inversion of center O and ratio r
   Geometric transformation that maps each point M different from O to a point M' on the half-line (ray) OM such that $OM \cdot OM' = r^2$.

Jensen's inequality
   If f is concave up on an interval $[a,b]$ and $\lambda_1, \lambda_2, \ldots, \lambda_n$ are nonnegative numbers with sum equal to 1, then

$$\lambda_1 f(x_1) + \lambda_2 f(x_2) + \cdots + \lambda_n f(x_n) \ge f(\lambda_1 x_1 + \lambda_2 x_2 + \cdots + \lambda_n x_n)$$

for any $x_1, x_2, \ldots, x_n$ in the interval $[a,b]$. If the function is concave down, the inequality is reversed.

Law of sines
   In a triangle ABC with circumradius equal to R, one has

$$\frac{BC}{\sin A} = \frac{CA}{\sin B} = \frac{AB}{\sin C} = 2R.$$

De Moivre's formula
   For any angle α and for any integer n,

$$(\cos\alpha + i\sin\alpha)^n = \cos n\alpha + i\sin n\alpha.$$

Orthocenter of a triangle
   Point of intersection of altitudes.

Periodic function
   $f(x)$ is periodic with period $T > 0$ if $f(x+T) = f(x)$ for all x.

Pigeonhole principle
   If n objects are distributed among $k < n$ boxes, some box contains at least two objects.

Polynomial in x of degree n
   Function of the form $f(x) = \sum_{k=0}^{n} a_k x^k$.

Ptolemy's theorem
   In a quadrilateral $ABCD$, $AC \cdot BD \le AB \cdot CD + AD \cdot BC$, where equality holds if and only if the quadrilateral is cyclic.

Root of an equation
   Solution to the equation.

Root of unity
   Solution to the equation $z^n - 1 = 0$.

Triangular number
   A number of the form $n(n+1)/2$, where n is some positive integer.

Trigonometric identities

$$\sin^2 x + \cos^2 x = 1,$$
$$\tan x = \frac{\sin x}{\cos x},$$
$$\cot x = \frac{1}{\tan x}$$

   Addition and subtraction formulas:

$$\sin(a \pm b) = \sin a \cos b \pm \cos a \sin b,$$
$$\cos(a \pm b) = \cos a \cos b \mp \sin a \sin b,$$
$$\tan(a \pm b) = \frac{\tan a \pm \tan b}{1 \mp \tan a \tan b};$$

Double-angle formulas:

$$\sin 2a = 2\sin a \cos a,$$
$$\cos 2a = 2\cos^2 a - 1 = 1 - 2\sin^2 a,$$
$$\tan 2a = \frac{2\tan a}{1 - \tan^2 a};$$

Triple-angle formulas:

$$\sin 3a = 3\sin a - 4\sin^3 a,$$
$$\cos 3a = 4\cos^3 a - 3\cos a,$$
$$\tan 3a = \frac{3\tan a - \tan^3 a}{1 - 3\tan^2 a};$$

Half angle formulas:

$$\sin a = \frac{2\tan \frac{a}{2}}{1 + \tan^2 \frac{a}{2}},$$
$$\cos a = \frac{1 - \tan^2 \frac{a}{2}}{1 + \tan^2 \frac{a}{2}},$$
$$\tan a = \frac{2\tan \frac{a}{2}}{1 - \tan^2 \frac{a}{2}};$$

Sum-to-product formulas:

$$\sin a + \sin b = 2\sin \frac{a+b}{2} \cos \frac{a-b}{2},$$
$$\cos a + \cos b = 2\cos \frac{a+b}{2} \cos \frac{a-b}{2},$$
$$\tan a + \tan b = \frac{\sin(a+b)}{\cos a \cos b};$$

Difference-to-product formulas:

$$\sin a - \sin b = 2\sin \frac{a-b}{2} \cos \frac{a+b}{2},$$
$$\cos a - \cos b = -2\sin \frac{a-b}{2} \sin \frac{a+b}{2},$$
$$\tan a - \tan b = \frac{\sin(a-b)}{\cos a \cos b};$$

Product-to-sum formulas:

$$2\sin a \cos b = \sin(a+b) + \sin(a-b),$$
$$2\cos a \cos b = \cos(a+b) + \cos(a-b),$$
$$2\sin a \sin b = -\cos(a+b) + \cos(a-b).$$

A.2   Glossary of Notation

$a \mid b$	a divides b
$\lvert x \rvert$	the absolute value of x
$\lfloor x \rfloor$	the greatest integer not exceeding x
$\{x\}$	the fractional part of x, equal to $x - \lfloor x \rfloor$
$a \equiv b \pmod{c}$	a is congruent to b modulo c, i.e., $a - b$ is divisible by c
$n!$	n factorial, equal to $1 \cdot 2 \cdots n$
$\binom{n}{k}$	the binomial coefficient n choose k
$[a,b]$	the closed interval, i.e., all x such that $a \le x \le b$
(a,b)	the open interval, i.e., all x such that $a < x < b$
$abc \cdots d_m$	the number written in base m with the digits $a, b, c, \cdots d$
$\sum_{i=1}^{n} a_i$	the sum $a_1 + a_2 + \cdots + a_n$
$\prod_{i=1}^{n} a_i$	the product $a_1 \cdot a_2 \cdots a_n$
$A - B$	for two sets A and B the elements of A not in B
$A \cup B$	the union of the sets A and B
$\mathbf{N}$	the set of positive integers $1, 2, 3, \ldots$
$\mathbf{N}_0$	the set of nonnegative integers $0, 1, 2, \ldots$
$\mathbf{Z}$	the set of integers
$\mathbf{Q}$	the set of rational numbers
$\mathbf{R}$	the set of real numbers
$\mathbf{C}$	the set of complex numbers
$\angle ABC$	the angle ABC
$\overparen{AB}$	the arc of a circle with extremities A and B

(ABC)	the plane determined by the points A, B, and C
$\mathscr{O}_n$	the $n \times n$ zero matrix
$\mathscr{I}_n$	the $n \times n$ identity matrix
$\det A$	the determinant of the matrix A

Printed in the United States